Control of Distributed Parameter and Stochastic Systems

IFIP - The International Federation for Information Processing

IFIP was founded in 1960 under the auspices of UNESCO, following the First World Computer Congress held in Paris the previous year. An umbrella organization for societies working in information processing, IFIP's aim is two-fold: to support information processing within its member countries and to encourage technology transfer to developing nations. As its mission statement clearly states,

IFIP's mission is to be the leading, truly international, apolitical organization which encourages and assists in the development, exploitation and application of information technology for the benefit of all people.

IFIP is a non-profitmaking organization, run almost solely by 2500 volunteers. It operates through a number of technical committees, which organize events and publications. IFIP's events range from an international congress to local seminars, but the most important are:

· The IFIP World Computer Congress, held every second year;
· open conferences;
· working conferences.

The flagship event is the IFIP World Computer Congress, at which both invited and contributed papers are presented. Contributed papers are rigorously refereed and the rejection rate is high.

As with the Congress, participation in the open conferences is open to all and papers may be invited or submitted. Again, submitted papers are stringently refereed.

The working conferences are structured differently. They are usually run by a working group and attendance is small and by invitation only. Their purpose is to create an atmosphere conducive to innovation and development. Refereeing is less rigorous and papers are subjected to extensive group discussion.

Publications arising from IFIP events vary. The papers presented at the IFIP World Computer Congress and at open conferences are published as conference proceedings, while the results of the working conferences are often published as collections of selected and edited papers.

Any national society whose primary activity is in information may apply to become a full member of IFIP, although full membership is restricted to one society per country. Full members are entitled to vote at the annual General Assembly, National societies preferring a less committed involvement may apply for associate or corresponding membership. Associate members enjoy the same benefits as full members, but without voting rights. Corresponding members are not represented in IFIP bodies. Affiliated membership is open to non-national societies, and individual and honorary membership schemes are also offered.

Control of Distributed Parameter and Stochastic Systems

Proceedings of the IFIP WG 7.2 International Conference, June 19-22, 1998 Hangzhou, China

Edited by

Shuping Chen
Zhejiang University

Xunjing Li
Fudan University

Jiongmin Yong
Fudan University

Xun Yu Zhou
The Chinese University of Hong Kong

SPRINGER SCIENCE+BUSINESS MEDIA, LLC

Library of Congress Cataloging-in-Publication Data

A C.I.P. Catalogue record for this book is available
from the Library of Congress.

ISBN 978-1-4757-4868-0 ISBN 978-0-387-35359-3 (eBook)
DOI 10.1007/978-0-387-35359-3

Printed on acid-free paper.

Table of Contents

Part II. Stochastic Systems

Preface

The *Conference on Control of Distributed Parameter and Stochastic Systems* was held in Hangzhou, China, June 19–22, 1998. It was devoted to the current issues on the controls of distributed parameter and stochastic systems. The objective of the conference was to provide a forum where leading researchers from around the world can converge to disseminate new ideas and discusses the latest as well as the future trends in the areas. The following subjects were covered: Adaptive Control, Controllability, Filtering, Identification, Manufacturing Systems, Mathematical Finance/Insurance, Numerical Approximation, Optimal Control, Stabilization, and Stochastic Analysis.

There were two parallel sessions, distributed parameter systems and stochastic systems, in the conference. Totally 52 invited speakers delivered their 40-minute lectures. Participants came from the following 9 countries: Australia, Canada, China, France, Germany, Japan, Korea, Spain, and USA.

In the distributed parameter session, the following were the main topics: (1) optimal control for PDE systems (by J. Burns, E. Casas, H. Gao, J. Lagnese, S. M. Lenhart, X. Li, T. I. Seidman, F. Tröltzsch, G. Wang, X. Xiang, and J. Ye). (2) controllability/stabilizability of PDE systems (by M. Delfour, S. Hansen, M. A. Horn, I. Lasiecka, K. Liu, Z. Liu, T. Nambu, R. Triggiani, and E. Zuazua). (3) numerical study/approximation (by R. H. Fabiano, F. Fahroo, S. Kang, C. Wang, and J. Zhou).

In the stochastic session, the following two topics were very strongly represented: (1) backward stochastic differential equations (by M. Kohlmann, J. Ma, S. Peng, R. Situ, S. Tang, and J. Yong), and (2) mathematical finance/insurance (by T. Bielecki, N. El Karoui, U. Haussmann, S. Stojanovic, M. Taksar, and X. Y. Zhou). Other topics included: linear-quadratic controls (S. Chen, J. B. Moore, Q. Zhang, and X. Y. Zhou), risk-sensitive controls (M. James and H. Nagai), ergodic controls (Y. Fujita, H. Morimoto and H. Nagai), adaptive controls (L. Guo and T. Duncan), Zakai equations (H. Kunita), singular perturbation (G. Yin), and robust stabilization (H. Yang).

The conference was financially sponsored by the Education Ministry of China, the National Natural Science Foundation of China, Zhejiang University and Fudan University. We also acknowledge the International Federation for Information and Processing (IFIP).

This volume contains 39 written version of the talk presented at the conference.

SHUPING CHEN
XUNJING LI
JIONGMIN YONG
XUN YU ZHOU

Participants of the Conference

Tomasz R. Bielecki
Mathematics Department, Northeastern Illinois University, Chicago, USA
email: trbielec@gamut.neiu.edu

John Burns
ICAM, Virginia Tech, Blacksburg, USA
email: burns@sun.icam.vt.edu

Eduardo Casas
Departamento de Matematica Aplicada y C.C., E.T.S.I. Industriales y de Telecomunicacion, Universidad de Cantabria, Av. Los castros s/n 39005-Santander Spain
email: casas@macc.unican.es

Shuping Chen
Department of Applied Mathematics, Zhejiang University, Hangzhou 310027 China

Michel Delfour
Centre de recherches mathematiques, Universite de Montreal, CP 6128, succ Centre-ville Montreal (Qc), CANADA H3C 3J7
email: delfour@crm.umontreal.ca

Tyrone Duncan
Department of Mathematics, University of Kansas, Lawrence, Kansas 66045, USA
email: duncan@math.ukans.edu

Nicole El Karoui
Ecole Polytechnique, CMAPX, 91128 Palaiseau Cedex, France

Richard H. Fabiano
Department of Mathematics, University of North Carolina at Greensboro Greensboro, NC 27412, USA
email: rhfabian@euler.uncg.edu

Fariba Fahroo
Naval Postgraduate school, Math Department, Code MA/Ff, Monterey, CA 93943, USA
email: fahroo@math.nps.navy.mil

Y. Fujita
Department of Mathematics, Faculty of Science, Toyama University, Toyama 930, Japan

Hang Gao
Department of Mathematics, Jilin University, Changchun 130023, China

Lei Guo
Institute of System Science, Chinese Academy of Science, Beijing, China
email: lguo@iss03.iss.ac.cn

Scott Hansen
Department of Mathematics, Iowa State University, Ames, IA 50011, USA
email: shansen@iastate.edu

Ulrich G. Haussmann
Department of Mathematics, University of British Columbia, Vancouver, British Columbia, V6T 1Z2, Canada
email: uhaus@math.ubc.ca

Mary Ann Horn
Department of Mathematics, Vanderbilt University, Nashville, TN 37240, USA
email: horn@math.vanderbilt.edu

M. James
Department of Engineering, Faculty of Engineering and Information Technology, Australian National University, Canberra 0200, AUSTRALIA
email: matthew.james@anu.edu.au

S. Kang
College of Natural Science, Chosun University, Kwangju 501-759, Korea
email: skang@chonnam.chonnam.ac.kr

M. Kohlmann
Fakultat fur Wirtschaftswissenschaften und Statistik Universitaet Konstanz, Postfach 560 ¡D126¿, 75434 Konstanz, Germany
email: michael.kohlmann@uni-konstanz.de

H. Kunita
Graduate School of Mathematics, Kyushu University-36, Fukuoka 812-81, Japan
email: kunita@math.kyushu-u.ac.jp

John Lagnese
Department of Mathematics, Georgetown University, Washington, DC 20057, USA
email: lagnese@gumath1.math.georgetown.edu

Irena Lasiecka
Department of Mathematics, University of Virginia, Charlottesville, VA 22903-2442, USA
email: il2v@virginia.edu

Suzanne M. Lenhart
Department of Mathematics, University of Tennessee, Knoxville, TN 37996, USA
email: lenhart@math.utk.edu

Xunjing Li
Department of Mathematics, Fudan University, Shanghai 200433, China
email: xjli@ms.fudan.edu.cn

Kangsheng Liu
Department of Applied Mathematics, Zhejiang University, Hangzhou 310027, China

Zhuangyi Liu
Department of Mathematics and Statistics, University of Minnesota, Duluth, MN 55812, USA
email: zliu@d.umn.edu

Jin Ma
Department of Mathematics, Purdue University, West Lafayette, IN 47097, USA
email: majin@math.purdue.edu

J. B. Moore
Department of Systems Engineering, Research School of Information Sciences and Engineering, Australian National University, Canberra, ACT 0200, Australia
email: john@keating.anu.edu.au

B. Mordukhovich
Department of Mathematics, Wayne State University, Detroit, MI 48202, USA
email: boris@math.wayne.edu

H. Morimoto
Department of Mathematical Sciences, Ehime University, 2-5 Bunkyo, Matsuyama 790, Japan
email: morimoto@SCI.sci.ehime-u.ac.jp

H. Nagai
Department of Mathematical Science, Graduate School of Engineering Science, Osaka University, Toyonaka 560, Osaka, Japan
email: nagai@sigmath.es.osaka-u.ac.jp

Takao Nambu
Department of Applied Mathematics, Faculty of Engineering, Kobe University, Nada, Kobe 657, Japan
email: nambu@godel.seg.kobe-u.ac.jp

Bozenna Pasik-Duncan
Department of Mathematics, University of Kansas, Lawrence, Kansas 66045, USA
email: bozenna@kuhub.cc.ukans.edu

Shige Peng
Department of Mathematics, Shandong University, Jinan 250100, China

T. I. Seidman
Department of Mathematics and Statistics, University of Maryland Baltimore County, Baltimore, MD 21250, USA

Rong Situ
Department of Mathematics, Zhongshan University, Guangzhou 510275, China

Srdjan Stojanovic
Department of Mathematical Sciences, University of Cincinnati, Cincinnati, OH 45221-0025, USA

email: Srdjan.Stojanovic@Math.UC.Edu

M. Taksar
Department of Applied Mathematics, SUNY at Stony Brook, Stony Brook, NY 11794, USA
email: taksar@ams.sunysb.edu

Shanjian Tang
Department of Mathematics, Fudan University, Shanghai 200433, China

Roberto Triggiani
Department of Mathematics, University of Virginia, Charlottesville, VA 22903-2442, USA
email: rt7u@virginia.edu

Dr. Fredi Tr"oltzsch
Technische Universit"at Chemnitz-Zwickau, Fakult"at f"ur Mathematik, D-09107 Chemnitz, Germany
email: f.troeltzsch@mathematik.tu-chemnitz.de

C. Wang
Department of Mathematics, University of Southern California, Los Angeles, CA 90089-1113, USA
email: cwang@mtha.usc.edu

G. Wang
Department of Mathematics, Huazhong Normal University, Wuhan, China

Xiaoling Xiang
Department of Mathematics, Guizhou University, Guiyang, China

Masahiro Yamamoto
Department of Mathematical Sciences, The University of Tokyo, 3-8-1 Komaba, Meguro, Tokyo 153, Japan
email: myama@ms.u-tokyo.ac.jp

Hailiang Yang
Department of Mathematics, University of Hong Kong, Hong Kong, China

Jane Ye
Department of Mathematics and Statistics, University of Victoria, Victoria, B.C., P.O. Box 3045, V8W 3P4, Canada
email: janeye@uvic.ca

George Yin
Department of Mathematics, Wayne State University, Detroit, MI 48202, USA
email: gyin@math.wayne.edu

Jiongmin Yong
Department of Mathematics, Fudan University, Shanghai 200433, China
email: jyong@ms.fudan.edu.cn

Qing Zhang
Department of Mathematics, University of Georgia, Athens, GA 30602, USA
email: qingz@control.math.uga.edu

Xun Yu Zhou
Department of Systems Engineering and Engineering Management, The Chinese University of Hong Kong, Shatin, N.T., Hong Kong, China
email: xyzhou@se.cuhk.edu.hk

Jianxin Zhou
Department of Mathematics, Texas A & M University, 123 Milner Hall, College Station, TX 77843, USA
email: Jianxin.Zhou@math.tamu.edu

Enrique Zuazua
Departmento de Mathematica Aplicada, Universidad Complutense, 28040 Madrid, Spain
email: zuazua@sunma4.mat.ucm.es

Part I. Distributed Parameter Systems

EXACT–APPROXIMATE BOUNDARY CONTROLLABILITY OF THERMOELASTIC SYSTEMS UNDER FREE BOUNDARY CONDITIONS

George Avalos

Department of Mathematics and Statistics
Texas Tech University
Lubbock, Texas 79409 USA
email: avalos@math.ttu.edu

Irena Lasiecka

Department of Applied Mathematics
University of Virginia
Charlottesville, Virginia 22903 USA
email: il2v@amsun.apma.virginia.edu

Abstract: Controllability properties of a partial differential equation (PDE) model describing a thermoelastic plate are studied. The PDE is comprised of a Kirchoff plate equation coupled to a heat equation on a bounded domain, with the coupling taking place on the interior and boundary of the domain. The coupling in this PDE is parameterized by $\alpha > 0$. Control is exerted through the (two) free boundary conditions of the plate equation, and through the Robin boundary condition of the temperature. These controls have the physical interpretation, respectively, of inserted forces and moments, and prescribed temperature, all of which act on the edges of the plate. The main result here is that under such boundary control, and with initial data in the basic space of wellposedness, one can simultaneously control the displacement of the plate *exactly*, and the temperature *approximately*. Moreover, the thermal control may be taken to be arbitrarily smooth in time and space, and the thermal control region may be any nonempty subset of the boundary. This controllability holds for arbitrary values of the coupling parameter α.

1 INTRODUCTION

Statement of the Problem

Let Ω be a bounded open subset of $\mathbb{R}^2$ with its sufficiently smooth boundary denoted as Γ. The boundary will be decomposed as $\Gamma = \overline{\Gamma_0} \cup \overline{\Gamma_1}$, with both Γ_0 and Γ_1 being open and nonempty, and further satisfying $\overline{\Gamma_0} \cap \overline{\Gamma_1} = \varnothing$. In addition, let Γ_2 be any open and nonempty subset of Γ. With this geometry, we shall consider here the following thermoelastic system on finite time $(0,T)$:

$$\left\{\begin{array}{l} \left\{\begin{array}{l} \omega_{tt} - \gamma\Delta\omega_{tt} + \Delta^2\omega + \alpha\Delta\theta = 0 \\ \beta\theta_t - \eta\Delta\theta + \sigma\theta - \alpha\Delta\omega_t = 0 \end{array}\right. \quad \text{on } (0,T)\times\Omega; \\ \\ \omega = \dfrac{\partial\omega}{\partial\nu} = 0 \quad \text{on } (0,T)\times\Gamma_0; \\ \\ \left\{\begin{array}{l} \Delta\omega + (1-\mu)B_1\omega + \alpha\theta = u_1 \\ \dfrac{\partial\Delta\omega}{\partial\nu} + (1-\mu)\dfrac{\partial B_2\omega}{\partial\tau} - \gamma\dfrac{\partial\omega_{tt}}{\partial\nu} + \alpha\dfrac{\partial\theta}{\partial\nu} = u_2 \end{array}\right. \quad \text{on } (0,T)\times\Gamma_1; \\ \\ \dfrac{\partial\theta}{\partial\nu} + \lambda\theta = \left\{\begin{array}{ll} u_3 & \text{on } (0,T)\times\Gamma_2 \\ 0 & \text{on } (0,T)\times\Gamma\backslash\Gamma_2 \end{array}\right. \qquad \lambda \geq 0; \\ \\ \omega(t=0) = \omega_0,\ \omega_t(t=0) = \omega_1,\ \theta(t=0) = \theta_0 \text{ on } \Omega. \end{array}\right. \tag{1.1}$$

Here, α, β, η and σ are positive parameters. The positive constant γ is proportional to the thickness of the plate and assumed to be small with $0 < \gamma \leq M$. The boundary operators B_i are given by $B_1\omega \equiv 2\nu_1\nu_2\dfrac{\partial^2\omega}{\partial x\partial y} - \nu_1^2\dfrac{\partial^2\omega}{\partial y^2} - \nu_2^2\dfrac{\partial^2\omega}{\partial x^2}$, and $B_2\omega \equiv (\nu_1^2 - \nu_2^2)\dfrac{\partial^2\omega}{\partial x\partial y} + \nu_1\nu_2\left(\dfrac{\partial^2\omega}{\partial y^2} - \dfrac{\partial^2\omega}{\partial x^2}\right)$. The constant $\mu \in (0, \frac{1}{2})$ is the familiar Poisson's ratio, and $\nu = [\nu_1, \nu_2]$ denotes the outward unit normal to the boundary. Here we shall also make the following geometric assumption on the (uncontrolled) portion of the boundary Γ_0:

$$\exists\ \{x_0, y_0\} \in \mathbb{R}^2 \text{ such that } \overline{h}(x,y)\cdot\nu \leq 0 \ \text{ on } \Gamma_0, \tag{1.2}$$

where $\overline{h}(x,y) \equiv [x - x_0, y - y_0]$.

The PDE model (1.1), with boundary functions $u_1 = u_2 = 0$, and $u_3 = 0$, mathematically describes an uncontrolled Kirchoff plate subjected to a thermal damping, with the displacement of the plate represented by the function $\omega(t,x,y)$, and the temperature given by the function $\theta(t,x,y)$ (see [8]). The given control variables $u_1(t,x)$ and $u_2(t,x)$ are defined on the portion of the boundary $(0,T)\times\Gamma_1$; the control $u_3(t,x)$ is defined on $(0,T)\times\Gamma_2$.

With the denotation

$$H^k_{\Gamma_0}(\Omega) \equiv \left\{\varpi \in H^k(\Omega) : \left.\frac{\partial^j\varpi}{\partial\nu^j}\right|_{\Gamma_0} = 0 \text{ for } j = 0, ..., k-1\right\},$$

we will throughout take the initial data $[\omega_0, \omega_1, \theta_0]$ to be in $H^2_{\Gamma_0}(\Omega) \times H^1_{\Gamma_0}(\Omega) \times L^2(\Omega)$. In this paper, we will study controllability properties of solutions of (1.1) under the influence of boundary control functions in preassigned spaces. In particular we intend to address, on the finite time interval $[0, T]$, the following problem of *exact–approximate controllability* with respect to the basic energy space $H^2_{\Gamma_0}(\Omega) \times H^1_{\Gamma_0}(\Omega) \times L^2(\Omega)$ (this term being coined in [5]): For given data $[\omega_0, \omega_1, \theta_0]$ (initial) and $[\omega_0^T, \omega_1^T, \theta_0^T]$ (terminal) in $H^2_{\Gamma_0}(\Omega) \times H^1_{\Gamma_0}(\Omega) \times L^2(\Omega)$ and arbitrary $\epsilon > 0$, we seek a suitable control triple $[u_1, u_2, u_3] \in L^2(0, T; L^2(\Gamma_1) \times H^{-1}(\Gamma_1)) \times C^r(\Sigma_{2,T})$ (where $r > 0$ and $\Sigma_{i,T} = (0, T) \times \Gamma_i$, $i = 0, 1, 2$) such that the corresponding solution $[\omega, \omega_t, \theta]$ of (1.1) satisfies the steering property at terminal time T

$$[\omega(T), \omega_t(T)] = [\omega_0^T, \omega_1^T] \; ; \text{ and } \; \left\|\theta(T) - \theta_0^T\right\|_{L^2(\Omega)} \leq \epsilon. \tag{1.3}$$

In regards to the literature on this particular problem, the most relevant work is that of J. Lagnese in [9]. Therein, Lagnese shows that if the coupling parameter α is small enough and the boundary Γ is "star–shaped", then the boundary–controlled system (1.1) is (partially) exactly controllable with respect to the displacement ω. Also in [14], a boundary–controlled thermoelastic wave equation is studied, with a coupling parameter α likewise present therein, and a result of partial exact controllability (again for the displacement) for this PDE is cited. This controllability result is quoted to be valid for all sizes of α; however in [15], the author has acknowledged a flaw in the controllability proof, the correction of which necessitates a smallness criterion on α, akin to the situation in [9]. The chief contribution of the present paper is to remove restrictions on the size of the coupling parameter (see **Theorem 3** below), at the expense of adding the arbitrarily smooth boundary control u_3 in the thermal component. For a 1–D version of (1.1), S. Hansen and B. Zhang in [6], via a moment problem approach, show the system's exact null controllability with boundary control in either the plate or thermal component. Other controllability results for the thermoelastic system which do not assume any "smallness" condition on the coupling parameters deal with *distributed/internal* controls. Such include that in [5], in which interior control is placed in the Kirchoff plate component subject to *clamped* boundary conditions; with such control, one obtains exact controllability for the displacement ω, and approximate controllability for the temperature θ. Alternatively in [3], interior control is placed in the heat equation of (1.1) so as to obtain exact controllability for both components ω and θ. In addition, the work in [13] deals with obtaining a result of null controllability for both components of a coupled wave and heat equation, in the case that interior control is inserted in the wave component only.

So again, the main contribution and novelty of this paper is that we consider *boundary controls* acting via the higher order *free* boundary conditions, and we do not assume any size restriction on the coupling parameter α. Moreover, we do not impose any geometric "star–shaped" conditions on the controlled portion of the geometry.

It should be noted that the particular type of boundary conditions imposed on the mechanical variables greatly affects the analysis of the problem, even in the case of internal control. Indeed, in the case of all boundary conditions, *save for the free case*, it is known that the thermoelastic plate semigroup can be decomposed into a damped Kirchoff plate semigroup and a compact perturbation (see [5] and [12]). Since controllability estimates are invariant with respect to compact perturbations (at least in the case of approximately controllable systems, which we are dealing with here), the aforesaid decomposition, valid for the case of lower order boundary conditions, reduces the problem of exact controllability for the mechanical variable to that of uncoupled Kirchoff plates. Thus, the case of lower order boundary conditions allows a reduction of the coupled problem into one which has been much studied in the past. This strategy, while successfully employed in the case of clamped or hinged boundary conditions (see [5]), is not applicable here. Indeed, in our present case of free boundary conditions, there is no decomposition with a compact part, as in the lower order case (see [12]); moreover, the controllability operator corresponding to the given boundary controls is *not bounded* on the natural energy space. This latter complication is due to the fact that the Lopatinski conditions are not satisfied for the Kirchoff model under free boundary conditions.

The strategy adopted in this paper consists of the following steps: Initially, a suitable transformation of variables is made and applied to the equation (1.1); subsequently, a multiplier method is invoked with respect to the transformed equation. The mulitiplers employed here are the differential multipliers used in the study of exact controllability for the Kirchoff plate model, together with the nonlocal (ΨDO) multipliers used in the study of thermoelastic plates in [1] and [2]. This multiplier method allows the attainment of preliminary estimates for the energy of the system. However, these estimates are "polluted" by certain boundary terms which are not majorized by the energy. To cope with these, we use the sharp trace estimates established in [11] for Kirchoff plates. The use of this PDE result introduces lower order terms into the energy estimate, which are eventually eliminated with the help of a new unique continuation result in [7]. It is *only* at the level of invoking this uniqueness result that the thermal control u_3 on Γ_2 must be introduced.

We post our main result here on controllability.

Theorem 1 *Let the assumption (1.2) stand. There is then a $T^* > 0$ so that for $T > T^*$ the following controllability property holds true: For given initial data $[\omega_0, \omega_1, \theta_0]$ and terminal data $[\omega_0^T, \omega_1^T, \theta_0^T]$ in the space $H^2_{\Gamma_0}(\Omega) \times H^1_{\Gamma_0}(\Omega) \times L^2(\Omega)$, and arbitrary $\epsilon > 0$, one can find control functions $[u_1^*, u_2^*, u_3^*] \in L^2(0,T; L^2(\Gamma_1) \times H^{-1}(\Gamma_1)) \times C^r(\Sigma_{2,T})$ (where given $r > 0$) such that the corresponding solution $[\omega^*, \omega_t^*, \theta^*]$ to (1.1) satisfies (1.3) at terminal time T.*

Remark 2 *The presence of the control u_3 in* ***Theorem 1*** *is owing solely to the need to invoke the aforementioned uniqueness result of Isakov in the proof below; it plays no part whatsoever in obtaining the preliminary (lower order term-tainted) estimate on the energy. Consequently, we have the freedom to* *prescribe the thermal control region to be as small as we wish, and the control* u_3 *to be as smooth in time and space as desired.*

2 PROOF OF THEOREM 1

A preponderant portion of the proof of **Theorem 1** is wrapped up in showing the following result of exact controllability for the displacement only:

Theorem 3 *With the coupling parameter* α *in (1.1) being arbitrary and the assumption (1.2) in place, there is then a* $T^* > 0$ *so that for* $T > T^*$*, the following property holds true: For all initial data* $[\omega_0, \omega_1, \theta_0] \in H^2_{\Gamma_0}(\Omega) \times H^1_{\Gamma_0}(\Omega) \times L^2(\Omega)$ *and terminal data* $[\omega_0^T, \omega_1^T] \in H^2_{\Gamma_0}(\Omega) \times H^1_{\Gamma_0}(\Omega)$*, there exists* $[u_1, u_2, u_3] \in L^2(0,T; L^2(\Gamma_1) \times H^{-1}(\Gamma_1)) \times H^s(\Sigma_{2,T})$*, where arbitrary* $s \geq 0$*, such that the corresponding solution* $[\omega, \omega_t, \theta]$ *to (1.1) satisfies* $[\omega(T), \omega_t(T)] = [\omega_0^T, \omega_1^T]$. .

Indeed, if **Theorem 3** is shown to be true, then using the miminal norm steering control (see Appendix B of [10]), one can, in a straightforward fashion, construct a control $[u_1^*, u_2^*, u_3^*]$ such that the corresponding trajectory $[\omega^*, \omega_t^*, \theta^*]$ has the desired reachability property (1.3). (See [4] for the precise details). Accordingly, the sequel is devoted to showing the validity of **Theorem 3**.

The theme of the proof of **Theorem 3** is a classical one. Denoting the control space $\mathcal{U}_s \equiv L^2(\Gamma_1) \times H^{-1}(\Gamma_1) \times H^s(\Gamma_2)$, one defines the operator $\mathcal{L}_T : D(\mathcal{L}_T) \subset \mathcal{U} \to H^2_{\Gamma_0}(\Omega) \times H^1_{\Gamma_0}(\Omega)$ to be that which takes the terminal control to the terminal state; i.e., $\mathcal{L}_T [u_1, u_2, u_3] = [\omega(T), \omega_t(T)]$, where $[\omega(T), \omega_t(T)]$ is the plate component of the solution to (1.1) at time $t = T$. As it is defined, $\mathcal{L}_T$ is a closed, unbounded operator, with its domain being densely defined. By a principle of functional analysis then (see e.g., [16]), to prove **Theorem 3**, which is essentially a statement of the surjectivity of $\mathcal{L}_T$, it is enough to establish the PDE inequality

$$\int_0^T \left\| \left| \nabla \widehat{\phi}_t \right| \right\|^2_{L^2(\Gamma_1)} dt + \|\psi\|^2_{[H^s(\Sigma_{2,T})]'} \geq C_T \, \|[\phi_0, \phi_1]\|^2_{H^2_{\Gamma_0}(\Omega) \times H^1_{\Gamma_0}(\Omega)} ; \qquad (2.1)$$

where $[\phi, \phi_t, \psi]$ is the solution to the following *backwards* system, corresponding to terminal data $[\phi_0, \phi_1] \in D(\mathcal{L}_T^*)$:

$$
\begin{cases}
\begin{cases} \phi_{tt} - \gamma\Delta\phi_{tt} + \Delta^2\phi + \alpha\Delta\psi = 0 \\ \beta\psi_t + \eta\Delta\psi - \sigma\psi - \alpha\Delta\phi_t = 0 \end{cases} & \text{on } (0,T)\times\Omega; \\
\phi = \dfrac{\partial\phi}{\partial\nu} = 0 \text{ on } \Sigma_{0,T}; & \\
\begin{cases} \Delta\phi + (1-\mu)B_1\phi + \alpha\psi = 0 \\ \dfrac{\partial\Delta\phi}{\partial\nu} + (1-\mu)\dfrac{\partial B_2\phi}{\partial\tau} - \gamma\dfrac{\partial\phi_{tt}}{\partial\nu} + \alpha\dfrac{\partial\psi}{\partial\nu} = 0 \end{cases} & \text{on } \Sigma_{1,T}; \\
\dfrac{\partial\psi}{\partial\nu} + \lambda\psi = 0 \qquad \text{on } (0,T)\times\Gamma,\ \lambda \geq 0; & \\
[\phi(T), \phi_t(T), \psi(T)] = [\phi_0, \phi_1, 0] &
\end{cases}
\tag{2.2}
$$

The proof of **Theorem 3** is then based upon showing the inequality (2.1), at least for $T > 0$ large enough, where $\left[\left.\dfrac{\partial\phi_t}{\partial\nu}\right|_{\Gamma_1}, \phi_t|_{\Gamma_1}, \psi|_{\Gamma_2} \right]$ are traces of the solution $[\phi, \phi_t, \psi]$ to the backwards system (2.2). Because of space constraints, we give here only a broad sketch of the proof of **Theorem 3**; the full particulars are provided in [4].

Step 1. We start by making the substitution

$$
\widehat{\phi}(t) = e^{-\xi t}\phi(t); \text{ and } \widehat{\psi}(t) = e^{-\xi t}\psi(t), \tag{2.3}
$$

where parameter $\xi \equiv \frac{\alpha^2}{2\gamma\eta}$. This particular choice of parameter allows the PDE (2.2) to be transformed into the following plate equation whose forcing function is comprised in part of the high order term $\widehat{\psi}_t$:

$$
\begin{cases}
\widehat{\phi}_{tt} - \gamma\Delta\widehat{\phi}_{tt} + \Delta^2\widehat{\phi} = c_0\widehat{\psi} + c_1\widehat{\psi}_t + c_2\widehat{\phi} + c_3\widehat{\phi}_t + c_4\Delta\widehat{\phi} \quad \text{on } (0,T)\times\Omega; \\
\widehat{\phi} = \dfrac{\partial\widehat{\phi}}{\partial\nu} = 0 \text{ on } \Sigma_{0,T}; \\
\begin{cases} \Delta\widehat{\phi} + (1-\mu)B_1\widehat{\phi} = -\alpha\widehat{\psi} \\ \dfrac{\partial\Delta\widehat{\phi}}{\partial\nu} + (1-\mu)\dfrac{\partial B_2\widehat{\phi}}{\partial\tau} = \gamma\dfrac{\partial}{\partial\nu}\left(\xi^2\widehat{\phi} + 2\xi\widehat{\phi}_t + \widehat{\phi}_{tt}\right) - \alpha\dfrac{\partial\widehat{\psi}}{\partial\nu} \end{cases} \quad \text{on } \Sigma_{1,T}; \\
\left[\widehat{\phi}(T), \widehat{\phi}_t(T), \widehat{\psi}(T)\right] = \left[e^{-\xi T}\phi_0, -\xi e^{-\xi T}\phi_0 + e^{-\xi T}\phi_1, 0\right],
\end{cases}
\tag{2.4}
$$

where the c_i are constants depending on the physical parameters.

Step 2. At this point we invoke a multiplier method with respect to the uncoupled equation in (2.4), using two multipliers which are, respectively, standard (as in [9]) and nonstandard (as in [1] and [2]). To wit, we multiply this equation by $\widehat{\phi}_t$, and subsequently integrate in time and space so as to arrive at

$$\frac{1}{2}\left[\left\|\widehat{\phi}(t)\right\|^2_{H^2_{\Gamma_0}(\Omega)}+\left\|\widehat{\phi}_t(t)\right\|^2_{H^1_{\Gamma_0}(\Omega)}\right]_{t=s}^{t=\tau}=-\alpha\int_s^\tau\left(\widehat{\psi},\frac{\partial\widehat{\phi}_t}{\partial\nu}\right)_{L^2(\Gamma_1)}dt$$
$$-\int_s^\tau\left(\gamma\xi^2\frac{\partial\widehat{\phi}}{\partial\nu}+2\gamma\xi\frac{\partial\widehat{\phi}_t}{\partial\nu}+\alpha\lambda\gamma_0\widehat{\psi},\widehat{\phi}_t\right)_{L^2(\Gamma_1)}dt$$
$$+\int_s^\tau\left(c_0\widehat{\psi}+c_2\widehat{\phi}+c_3\widehat{\phi}_t+c_4\Delta\widehat{\phi}+c_1\widehat{\psi}_t,\widehat{\phi}_t\right)_{L^2(\Gamma_1)}dt. \tag{2.5}$$

Moreover, letting A_D denote the Laplacian operator with Dirichlet boundary conditions, and $A_D^{-1}\in\mathcal{L}(L^2(\Omega))$ its corresponding (smoothing) inverse, we then multiply the PDE in (2.4) by $-\frac{c_1}{\gamma}A_D^{-1}\widehat{\psi}$, and thereafter integrate in time and space. Adding the resultant expression to (2.5), and subsequently majorizing the sum, we obtain the following:

Lemma 4 *The solution $\left[\widehat{\phi},\widehat{\phi}_t,\widehat{\psi}\right]$ to (2.4) satisfies the following relation for all s and $\tau\in[0,T]$:*

$$\left[\left\|\widehat{\phi}(t)\right\|^2_{H^2_{\Gamma_0}(\Omega)}+\left\|\widehat{\phi}_t(t)\right\|^2_{H^1_{\Gamma_0}(\Omega)}\right]_{t=s}^{t=\tau}$$
$$\leq C\int_0^T\left\|\nabla\widehat{\phi}_t\right\|^2_{L^2(\Gamma_1)}dt+\text{l.o.t.}\left(\widehat{\psi},\widehat{\phi},\widehat{\phi}_t\right),$$

where l.o.t $\left(\widehat{\psi},\widehat{\phi},\widehat{\phi}_t\right)$ denotes, as usual, "lower order terms" (below the energy level) of $\widehat{\psi},\widehat{\phi}$, and $\widehat{\phi}_t$.

Step 3. Taking a radial vector field $\overline{h}\in\mathbb{R}^2$ which meets the requirement in (1.2), one can derive the following inequality which is an analogue to that demonstrated in [9]. In deriving this estimate the trace result in [11] is critically invoked.

Lemma 5 *For all $\epsilon_0\in(0,T)$, the solution $\left[\widehat{\phi},\widehat{\phi}_t\right]$ to (2.4) satisfies*

$$\int_{\epsilon_0}^{T-\epsilon_0}\left[\left\|\widehat{\phi}\right\|^2_{H^2_{\Gamma_0}(\Omega)}+\left\|\widehat{\phi}_t\right\|^2_{H^1_{\Gamma_0}(\Omega)}\right]dt$$
$$\leq C_T\int_0^T\left\|\nabla\widehat{\phi}_t\right\|^2_{L^2(\Gamma_1)}dt$$
$$+C\sum_{i=1}^2\left\|\left[\widehat{\phi}(s_i),\widehat{\phi}_t(s_i)\right]\right\|^2_{H^2_{\Gamma_0}(\Omega)\times H^1_{\Gamma_0}(\Omega)},$$

where $s_1=T-\epsilon_0$, and $s_2=\epsilon_0$.

Step 4. A standard energy argument shows that the heat component $\widehat{\psi}$ satisfies the estimate

$$\int_0^T \left\|\widehat{\psi}\right\|^2_{H^1(\Omega)} dt \;\le\; C\left(\int_0^T \left[\left\|\frac{\partial\widehat{\phi}_t}{\partial\nu}\right\|^2_{L^2(\Gamma_1)} + \left\|\widehat{\phi}\right\|^2_{H^2_{\Gamma_0}(\Omega)}\right] dt \right.$$
$$\left. + \int_0^T \left\|\widehat{\phi}_t\right\|^2_{H^1_{\Gamma_0}(\Omega)} dt\right) + \text{l.o.t.}\left(\widehat{\phi},\widehat{\phi}_t,\widehat{\psi}\right).$$

This inequality, in combination with **Lemmas** 4 and 5, eventually give (again the full details are in [4]),

Lemma 6 *For $T > 0$ large enough, the solution $\left[\widehat{\phi},\widehat{\phi}_t,\widehat{\psi}\right]$ of (2.4) satisfies the following estimate:*

$$\int_0^T \left[\left\|\widehat{\phi}\right\|^2_{H^2_{\Gamma_0}(\Omega)} + \left\|\widehat{\phi}_t\right\|^2_{H^1_{\Gamma_0}(\Omega)} + \left\|\widehat{\psi}\right\|^2_{H^1(\Omega)}\right] dt$$
$$+\left\|\left[\widehat{\phi}(T),\widehat{\phi}_t(T)\right]\right\|^2_{H^2_{\Gamma_0}(\Omega)\times H^1_{\Gamma_0}(\Omega)}$$
$$\le\; C_T\int_0^T \left\|\left|\nabla\widehat{\phi}_t\right|\right\|^2_{L^2(\Gamma_1)} dt + \text{l.o.t.}\left(\widehat{\phi},\widehat{\phi}_t,\widehat{\psi}\right). \tag{2.6}$$

Step 5. Note that in the inequality (2.6), there is no boundary trace term $\psi|_{\Gamma_2}$, reflecting the contribution of the control u_3; the observability estimate (2.6) is independent of thermal control. However, to remove the corrupting lower order terms in this estimate so as to have the desired inequality (2.1), the thermal control now comes directly into play. Indeed, a compactness–uniqueness argument is now to be employed, with this argument making critical use of the Holmgren's–type uniqueness result derived in [7] for overdetermined (in both the mechanical and thermal variables) thermoelastic systems. The correct use of this uniqueness theorem necessitates the appearance of the thermal control u_3. With such control in place, we then have

Lemma 7 *For $T > 0$ large enough, the existence of the inequality (2.6) implies that there exists a C_T such that the following estimate holds true:*

$$\text{l.o.t.}\left(\widehat{\phi},\widehat{\phi}_t,\widehat{\psi}\right) \le C_T\left(\int_0^T \left\|\left|\nabla\widehat{\phi}_t\right|\right\|^2_{L^2(\Gamma_1)} dt + \left\|\widehat{\psi}\right\|^2_{[H^s(\Sigma_{2,T})]'}\right). \tag{2.7}$$

The inequalities (2.6) and (2.7), and the transformation $\widehat{\phi}(t) = e^{-\xi t}\phi(t)$ and $\widehat{\psi}(t) = e^{-\xi t}\psi(t)$ give the desired inequality (2.1), thereby completing the proof of **Theorem 3**. By the remarks made at the beginning of this section, with this partial exact controllability result in hand, the exact–approximate controllability statement **Theorem 1** follows.

References

[1] G. Avalos and I. Lasiecka, "Exponential stability of a free thermoelastic system without mechanical dissipation", SIAM Journal Math. Anal., Vol. 29, No. 1, (January 1998), pp. 155–182.

[2] G. Avalos and I. Lasiecka, "Exponential stability of a thermoelastic system without mechanical dissipation", Rendiconti Di Instituto Di Matematica Dell'Universitá di Trieste, Supp. Vol. XXVIII, 1–28 (1997).

[3] G. Avalos, "Exact controllability of a thermoelastic system with control in the thermal component only", IMA Preprint Series #1532 (University of Minnesota), and submitted to Advances in Differential Equations.

[4] G. Avalos and I. Lasiecka, "Boundary controllability of thermoelastic plates with free boundary conditions", IMA Preprint Series #1567 (University of Minnesota).

[5] L. de Teresa and E. Zuazua, "Controllability for the linear system of thermoelastic plates", Advances in Differential Equations, Vol. 1, Number 3 (1996), pp. 369–402.

[6] S. Hansen and B. Zhang, "Boundary control of a linear thermoelastic beam", J. Math. Anal. Appl., 210, (1997), 182–205.

[7] V. Isakov, "On the uniqueness of the continuation for a thermoelasticity system", preprint (1998).

[8] J. Lagnese, Boundary stabilization of thin plates, SIAM Stud. Appl. Math., 10 (1989).

[9] J. Lagnese, "The reachability problem for thermoelastic plates", Arch. Rational Mech. Anal., 112 (1990), pp. 223–267.

[10] I. Lasiecka and R. Triggiani, "Exact controllability of the wave equation with Neumann boundary control", Appl. Math. Optim., 19 (1989), 243–290.

[11] I. Lasiecka and R. Triggiani, "Sharp trace estimates of solutions to Kirchoff and Euler–Bernoulli equations", Appl. Math. Optim., 28 (1993), 277–306

[12] I. Lasiecka and R. Triggiani,"Structural decomposition of thermoelastic semigroups with rotational forces", to appear in Semigroup Forum.

[13] G. Lebeau and E. Zuazua, "Null–controllability of a system of linear thermoelasticity", Arch. Rational Mech., Vol. 141 (1998), 297–329.

[14] W. Liu, "Partial exact controllability and exponential stability in higher–dimensional linear thermoelasticity", ESAIM: Control, Optimisation and Calculus of Variations, Vol. 3 (1998), 23–48.

[15] W. Liu, "Erratum on partial exact controllability and exponential stability in higher–dimensional linear thermoelasticity", to appear.

[16] D. Russell, "Controllability and stabilizability theory for linear partial differential equations: recent progress and open questions, SIAM Review, vol. 20, No. 4 (1978), 639–739.

A LINEAR PARABOLIC BOUNDARY CONTROL PROBLEM WITH MIXED CONTROL-STATE CONSTRAINT

Maïtine Bergounioux

Département de Mathématiques
Faculty of Mathematics UMR 6628 -
Université d'Orléans, France

Fredi Tröltzsch

Fakultät für Mathematik
Technische Universität Chemnitz-Zwickau
D-09107 Chemnitz, Germany

Abstract:

A simple class of linear optimal control problems for parabolic equations with mixed control-state inequality constraints is investigated. The constraints are formulated pointwise in L^∞. It is shown how to obtain associated Lagrange multipliers in L^p-spaces.

1 INTRODUCTION

We discuss the following linear optimal boundary control problem for the heat equation:

$$\max \int_\Omega \alpha_\Omega \, y(T) \, dx + \int_Q \alpha_Q \, y \, dxdt + \int_\Sigma \alpha_\Sigma \, y \, d\sigma dt + \int_\Sigma \alpha_u u \, d\sigma dt$$

subject to the state equation

$$\begin{aligned} y_t - \Delta y + d\,y &= 0 && \text{in } Q \,, \\ \partial_\nu y + b\,y &= u && \text{on } \Sigma \,, \\ y(0) &= 0 && \text{in } \Omega \,, \end{aligned} \tag{1.1}$$

and to the mixed control-state constraints

$$\begin{aligned} u(x,t) &\leq c(x,t) + y(x,t), \\ u(x,t) &\geq 0, \end{aligned}$$

which are required a.e. on Σ. The heat equation is defined in a bounded domain $\Omega \subset \mathbb{R}^N$, $N \geq 2$, with sufficiently smooth boundary Γ. In the fixed time interval $(0,T)$ we put $Q := \Omega \times (0,T)$, and $\Sigma := \Gamma \times\ (0,T)$. Moreover, functions $\alpha_\Omega \in L^\infty(\Omega)$, d, $\alpha_Q \in L^\infty(Q)$, and c, b, α_Σ, $\alpha_u \in L^\infty(\Sigma)$ are given. The control function u is assumed to be bounded and measurable. In this way, the feasible set of the control problem belongs to $L^\infty(\Sigma)$, and the mixed control-state constraint $u \le c + y$ must be regarded in the same space. Therefore, one might expect that an associated Lagrange multiplier has to be found in $L^\infty(\Sigma)^*$. In contrast to this, we shall verify the existence of at least one multiplier in $L^\infty(\Sigma)$. For linear programming problems in L^p-spaces with constraints of bottleneck type this surprising fact is known since long time, see for instance [1], [2], and the references cited therein. In this short note, we extend these early ideas to the parabolic boundary control problem defined above. Asscociated distributed control problems have been discussed extensively in our recent paper [3]. In the forthcoming paper [4], a class of nonlinear parabolic control problems with pointwise mixed control-state inequality constraints will be discussed on using these results.

We assume that Γ is so smooth that a Green's function $G = G(x,\xi,t)$, $G : \bar{\Omega}^2 \times \mathbb{R}_+ \to \mathbb{R}$ exists, which satisfies the known estimate

$$|G(x,\xi,t)| \le k_1 t^{-\frac{N}{2}} \exp(-k_2 \frac{|x-\xi|^2}{t}) \tag{1.2}$$

with positive real numbers k_1, k_2. Then the (weak) solution of (1.1) is given by

$$y(x,t) = \int_0^t \int_\Sigma G(x,\xi,t-s)\, u(\xi,s)\, d\sigma(\xi) ds, \tag{1.3}$$

where $d\sigma$ denotes the surface measure on Γ.

2 COMPARISON PRINCIPLES FOR AN INTEGRAL EQUATION

First we discuss the integral equation

$$u(x,t) = f(x,t) + \int_0^t \int_\Gamma G(x,\xi,t-s)\ u(\xi,s) d\sigma(\xi) ds \tag{2.1}$$

for given $f \in L^\infty(\Sigma)$. Introduce the integral operator K:

$$(Ku)(x,t) = \int_0^t \int_\Gamma G(x,\xi,t-s)\ u(\xi,s)\ d\sigma(\xi) ds. \tag{2.2}$$

K is continuous in $L^\infty(\Sigma)$, also from $L^p(\Sigma)$ to $L^\infty(\Sigma)$ for $p > N+1$, see [5], p. 138, Lemma 5.6.6. K is also continuous from $L^p(\Sigma)$ to $C(\bar{Q})$. In this way, $y(\cdot,T)$ is well defined in $C(\bar{\Omega})$. Endow the space $L^\infty(\Sigma)$ with the equivalent norm $||u||_\lambda$,

$$||u||_\lambda = \operatorname{vrai\,max}_{(x,t)\in\Sigma} |e^{-\lambda t} u(x,t)|\ , \tag{2.3}$$

where $\lambda > 0$. Then it is an easy and standard exercise to show the following result:

Lemma 2.1 *K is a contraction in $L^\infty(\Sigma)$, provided that $\lambda > \lambda_o$ is sufficiently large.*

Corollary 2.1 *For all $f \in L^\infty(\Sigma)$, the equation (2.1) has a unique solution $u \in L^\infty(\Sigma)$. The mapping $f \mapsto u$ is continuous in $L^\infty(\Sigma)$. If $G(x,\xi,t) \geq 0$ and $f(x,t) \geq 0$ a.e. in Σ, then $u(x,t) \geq 0$ a.e. in Σ.*

Proof: We have

$$u = (I-K)^{-1} f = (\sum_{n=0}^{\infty} K^n) f \tag{2.4}$$

by well known results on Neumann series. Moreover, $||(I-K)^{-1}||_\lambda \leq 1/(1 - ||K||_\lambda)$, where we have used for convenience the symbol $||\cdot||_\lambda$ to denote also the norm of the operator K induced by $||\cdot||_\lambda$. The first part of the lemma follows from the equivalence of the norms $||\cdot||_\lambda$ and $||\cdot||_{L^\infty(\Sigma)}$. If G is a nonnegative function, then K is a nonnegative operator, that is $f \geq 0 \Rightarrow Kf \geq 0$. The second result follows immediately from this and (2.4). ∎

If f_1, f_2 are two bounded and measurable right hand sides for (2.1) and G is nonnegative, then $u_1 \geq u_2$ holds for the associated solutions. Now we verify these facts for the space $L^p(\Sigma)$. To see this, regard K as an operator from $L^p(\Sigma)$ to $L^\infty(\Sigma)$. This smoothing property of K is basic for the next result.

Corollary 2.2 (Comparison principle) *Suppose $p > N+1$. Then the integral equation (2.1) has for each $f \in L^p(\Sigma)$ a unique solution $u \in L^p(\Sigma)$, and the mapping $f \mapsto u$ is continuous in $L^p(\Sigma)$. If $G(x,\xi,t) \geq 0$, $u_i \in L^p(\Sigma)$, $i = 1,2$, are solutions of (2.1) associated to $f_i \in L^p(\Sigma)$, $i = 1,2$, and $f_1(x,t) \geq f_2(x,t)$ holds a.e. in Σ, then $u_1(x,t) \geq u_2(x,t)$ a.e. in Σ.*

Proof: Put $v := u - f$, then (2.1) reads $v = Kf + Kv$. By the smoothing property of K, the right hand side Kf of this transformed equation is bounded and measurable. The last corollary implies that this equation admits a unique solution $v \in L^\infty(\Sigma)$ depending continuously on Kf and hence on f. Clearly, $u = f + v$ is a solution of (2.1) in $L^p(\Sigma)$. The uniqueness of u in $L^p(\Sigma)$ is and easy consequence, since the difference of two solutions solves the equation with right hand side zero, which belongs to $L^\infty(\Sigma)$. The comparison part of Corollary 2.2 follows from the arguments after Corollary 2.1. ∎

In a dual problem, which is defined later, the (formal) adjoint integral operator $K^\top$,

$$(K^\top \mu)(x,t) = \int_t^T \int_\Gamma G(\xi, x, s-t)\, \mu(\xi, s)\, d\sigma(\xi) ds \tag{2.5}$$

appears. $K^\top$ has the same properties as K. In particular, for $\lambda > \lambda_o$ it is a contraction in the norm $||\cdot||_\lambda$.

Lemma 2.2 *For every function $a \in L^\infty(\Sigma)$, the equation*

$$\mu(x,t) = \max\{\, 0,\ a(x,t) + \int_t^T \int_\Gamma G(\xi, x, s-t)\, \mu(\xi, s)\, d\sigma(\xi) ds\} \tag{2.6}$$

has exactly one solution $\mu \in L^\infty(\Sigma)$.

The proof is an application of the contraction mapping principle. It can be applied, since the operator $(\Pi z)(x,t) = \max\{\ 0,\ z(x,t)\ \}$ is Lipschitz continuous with Lipschitz constant one in $L^\infty(\Sigma)$, and $K^\top$ is a contraction.

3 PRIMAL AND DUAL PROBLEM

By inserting the integral representation (1.3) of the state y in the objective functional of the control problem, we get after changing the order of integration

$$\int_\Omega \alpha_\Omega\, y(T)\, dx + \int_\Sigma \alpha_Q\, y\, dx dt + \int_\Sigma \alpha_\Sigma\, y\, d\sigma dt + \int_\Sigma \alpha_u u\, d\sigma dt = \int_\Sigma (-a(x,t))\, u(x,t)\, d\sigma\, dt$$

with a certain function $a \in L^\infty(\Sigma)$ (a concrete expression for a is given in section 5). In this way, the optimal control problem becomes a *continuous linear programming problem* with constraints of bottleneck type. This is our *Primal Problem*

$$(\mathcal{P}) \begin{cases} \max \int_\Sigma a\ u\ d\sigma dt \\ u(x,t) \le [c + Ku](x,t) & \text{a.e. in } \Sigma\ , \\ u(x,t) \ge 0 & \text{a.e. in } \Sigma\ . \end{cases}$$

Note that $(\mathcal{P})$ is defined in the space $L^\infty(\Sigma)$. From now on, we assume that $G(x,\xi,t-s) \ge 0$ and $p > N+1$. We define u_c by the equation

$$u_c = c + K u_c\ . \tag{3.1}$$

According to Corollary 2.1, u_c is bounded and measurable.

Theorem 3.1 $(\mathcal{P})$ *has a solution* $\bar{u}$ *if and only if* $u_c \ge 0$.

Proof: If $u_c \ge 0$, then the feasible set for $(\mathcal{P})$ is non-empty. All feasible elements $u \in L^p(\Sigma)$ satisfy $u \le u_c$ by the comparison principle of Corollary 2.2, and we have $u_c \in L^\infty(\Sigma)$. Moreover, $u \ge 0$ follows from the constraints. This implies the L^∞-boundedness of u by $\|u_c\|_{L^\infty(\Sigma)}$. So the feasible set is bounded, closed, and convex in all (reflexive) L^p-spaces for $1 + N < p < \infty$. The existence of $\bar{u} \in L^p(\Sigma)$ is an immediate conclusion. Of course $\bar{u}$ belongs to $L^\infty(\Sigma)$. On the other hand, if u_c is not greater or equal than 0, then the feasible set is empty by the comparison principle. ∎

To introduce the dual problem to $(\mathcal{P})$, we extend the feasible set of $(\mathcal{P})$ from $L^\infty(\Sigma)$ to $L^p(\Sigma)$ $(p > N+1)$. Any feasible solution u of $(\mathcal{P})$ satisfies $0 \le u \le u_c$, where u_c is defined in (3.1). Therefore, all feasible solutions of $L^p(\Sigma)$ belong automatically to $L^\infty(\Sigma)$, and the feasible set is not influenced by this extension to $L^p(\Sigma)$. From standard techniques to establish dual problems (see our discussion in section 6), in $L^q(\Sigma)$ we get the following *Dual Problem*:

$$(\mathcal{D}) \quad \begin{cases} \min \int_\Sigma c\,\mu\, d\sigma dt \\ \mu(x,t) \ge [a + K^\top \mu](x,t) & \text{a.e. in } \Sigma\ , \\ \mu(x,t) \ge 0 & \text{a.e. in } \Sigma\ , \end{cases}$$

where $\mu \in L^q(\Sigma)$, $\frac{1}{p} + \frac{1}{q} = 1$, and $K^\top$ is the adjoint integral operator defined by (2.5). The kernel of $K^\top$ satisfies the estimate (1.2), hence $K^\top : L^p(\Sigma) \to L^\infty(\Sigma)$ holds for $p > N + 1$, too. However, this is not true from $L^q(\Sigma)$ to $L^\infty(\Sigma)$. On the other hand, $K^\top$ represents the adjoint operator of $K : L^p(\Sigma) \to L^\infty(\Sigma) \subset L^p(\Sigma)$. Therefore, $K^\top$ is continuous in $L^q(\Sigma) \sim L^p(\Sigma)^*$, too.
In view of (1.2) and Lemma 2.1, $K^\top$ is a contraction in $L^\infty(\Sigma)$, hence the equation

$$\mu = \beta + K^\top \mu \tag{3.2}$$

has a unique solution $\mu \in L^\infty(\Sigma)$ for each $\beta \in L^\infty(\Sigma)$. Moreover, we have uniqueness for (3.2) in $L^q(\Sigma)$. This can be shown by duality, since by Corollary (2.2) the range of $v - Kv$ is $L^p(\Sigma)$. The next theorem contains the basic idea of this paper.

Theorem 3.2 *If $c \ge 0$, then $(\mathcal{D})$ has at least one bounded and measurable optimal solution $\bar{\mu}$.*

Proof: We know by Lemma 2.2 that equation (2.6)

$$\mu(x,t) = \max\{\ 0,\ a(x,t) + (K^\top\ u)(x,t)\}$$

has in $L^\infty(\Sigma)$ exactly one solution $\bar{\mu}$. Let $\mu \in L^q(\Sigma)$ be any other feasible element for $(\mathcal{D})$, which is different from $\bar{\mu}$. Then $\mu \ge a + K^\top \mu$ a.e. in Σ and, of course, $\mu \ge 0$. Next, we construct a sequence $\mu_1 \ge \mu_2 \ge \cdots$ as follows : $\mu_1 = \mu \in L^q(\Sigma)$ and $\mu_2 = \max\{\ 0,\ a + K^\top \mu_1\ \}$. A simple discussion yields $\mu_2 \le \mu_1$ a.e. in Σ. Then by positivity, $K^\top \mu_2 \le K^\top \mu_1$, and we get

$$\mu_2 \ge a + K^\top \mu_1 \ge a + K^\top \mu_2 \text{ a.e. on } \Sigma\ .$$

So μ_2 is feasible and $\mu_2 \le \mu_1$ on Σ. Repeating this process, one constructs a non-increasing feasible sequence $\{\mu_n\}$ which has to be pointwise convergent towards some $\tilde{\mu} \ge 0$, that is

$$\lim_{n\to+\infty} \mu_n(x,t) = \tilde{\mu}(x,t) \text{ a.e. on } \Sigma\ .$$

An application of the Lebesgue dominated convergence theorem yields that μ_n tends to $\tilde{\mu}$ in $L^q(\Sigma)$. Passing to the limit in $\mu_n = \max\{\ 0,\ a + K^\top \mu_{n-1}\ \}$ gives in $L^q(\Sigma)$

$$\tilde{\mu} = \max\{\ 0,\ a + K^\top \tilde{\mu}\ \}\ .$$

Finally, by simple arguments, we are able to conclude $\tilde{\mu} \in L^\infty(\Sigma)$. By uniqueness in $L^\infty(\Sigma)$ we have $\tilde{\mu} = \bar{\mu}$; moreover $\mu \geq \tilde{\mu}$. Therefore $\mu \geq \bar{\mu}$ holds for all feasible solutions and, since $c \geq 0$,

$$\int_\Sigma c(x,t)\ \mu(x,t)\ d\sigma dt \geq \int_\Sigma c(x,t)\ \bar{\mu}(x,t)\ d\sigma dt.$$

∎

4 THE DUALITY RELATION

The discussion of the duality between $(\mathcal{P})$ and $(\mathcal{D})$ is not yet complete. We have only shown that under certain assumptions the dual problem admits a solution $\bar{\mu}$. To make sure that $\bar{\mu}$ is a Lagrange multiplier associated to a solution of $(\mathcal{P})$, we need additionally the *strong duality relation*, that is the equality of primal and dual optimal value. To this aim, we briefly sketch some main ideas of duality for linear programs.

Let $X = L^p(\Sigma)$ with its natural partial ordering $\geq$, and define $A := I - K$ having the adjoint operator A'. By $\langle \cdot\,, \cdot \rangle$ we denote the pairing between X and its dual space $X' = L^q(\Sigma)$. Then the *primal problem* is

$$(\mathcal{P}) \qquad \begin{array}{rl} \max & \langle a, x \rangle \\ A\,x & \leq c \\ x & \geq 0. \end{array}$$

On using the *Lagrange function* $\mathcal{L}(x,\mu) := \langle a, x\rangle + \langle \mu, c - Ax\rangle$, the primal problem can be written in sup-inf form. Reversing the order of supremum and infimum we arrive at the dual problem

$$(\mathcal{D}) \qquad \begin{array}{rl} \min & \langle \mu, c \rangle \\ A'\,\mu & \geq a \\ \mu & \geq 0. \end{array}$$

Let $\bar{x}$ be optimal for $(\mathcal{P})$. Then $\bar{\mu}$ is an associated Lagrange multiplier if and only if the pair $(\bar{x}, \bar{\mu})$ is a saddle point of $\mathcal{L}$, i.e. $\mathcal{L}(x,\bar{\mu}) \leq \mathcal{L}(\bar{x},\bar{\mu}) \leq \mathcal{L}(\bar{x},\mu)$ for all $x \geq 0$, $\mu \geq 0$. A necessary and sufficient condition for $(\bar{x},\bar{\mu})$ to be a saddle point is that $\bar{x}$ solves $(\mathcal{P})$, $\bar{\mu}$ solves $(\mathcal{D})$, and the *strong duality relation* $v = v'$ holds true. Let us define by $P(c)$ and $D(a)$ the feasible sets of $(\mathcal{P})$ and $(\mathcal{D})$, respectively, i.e.

$$P(c) = \{x \in X | x \geq 0,\ A\,x \leq c\}, \quad D(a) = \{\mu \in X' | \mu \geq 0,\ A'\,\mu \geq a\}.$$

It is easy to verify that the *weak duality relation*

$$\sup_{x \in P(c)} \langle a, x \rangle \leq \inf_{\mu \in D(a)} \langle \mu, c \rangle \tag{4.1}$$

holds always true. To show the *strong* duality relation, we need the convex cone

$$K(A) := \{(\alpha, d) \in \mathbb{R} \times X | \exists x \geq 0 : Ax \leq d, \langle a, x \rangle \geq \alpha\}.$$

Theorem 4.1 *If $K(A)$ is closed and $(\mathcal{P})$ admits at least one solution, then the strong duality relation holds, that is*

$$\max_{x\in P(c)} \langle a, x\rangle = \inf_{\mu\in D(a)} \langle \mu, c\rangle. \tag{4.2}$$

This is a standard result of linear programming theory. We should mention that (4.2) implies the *complementary slackness conditions*

$$\langle A\,\bar{x} - c, \bar{\mu}\rangle = \langle A'\,\bar{\mu} - a, \bar{x}\rangle = 0. \tag{4.3}$$

It turns out that the assumptions of the theorem are fulfilled for our problem:

Lemma 4.1 *$K(A)$ is closed for $(\mathcal{P})$.*

We know from the proof of the Theorem 3.1 that the norm of $u \in P(c)$ is bounded by the norm of u_c, which is bounded by Corollary 2.2. Therefore, this Lemma is easy to prove.

5 NECESSARY OPTIMALITY CONDITIONS

Finally, we prove that the solution $\bar{\mu}$ of $(\mathcal{D})$ is a Lagrange multiplier for the state-constraint of the parabolic control problem. By definition, $\bar{\mu}$ is an associated *Lagrange multiplier*, if $\bar{u}$, together with $\bar{\mu}$ and an *adjoint state* $\bar{p}$, satisfies the *first order necessary optimality conditions*: They consist of the *adjoint equation*

$$\left\{\begin{array}{rcll} -\bar{p}_t - \Delta\,\bar{p} + d\,\bar{p} &=& \alpha_Q & \text{in } Q \\ \partial_\nu \bar{p} + b\,\bar{p} &=& \alpha_\Sigma & \text{on } \Sigma\,, \\ \bar{p}(T) &=& \alpha_\Omega & \text{in } \Omega\,, \end{array}\right. \tag{5.1}$$

the *variational inequality*

$$\int_\Sigma (\alpha_u + \bar{p} + \bar{\mu})(u - \bar{u})\, d\sigma\, dt \geq 0 \quad \forall u \geq 0, \tag{5.2}$$

the *complementary slackness condition*

$$(\bar{u} - c - \bar{y})\,\bar{\mu} = 0 \qquad \text{a.e. in } \Sigma, \tag{5.3}$$

and the *nonnegativity condition* $\bar{\mu}(x,t) \geq 0$, which must be satisfied a.e. on Σ.

To verify these conditions, we introduce some auxiliary functions: Put $a_\Omega = -\alpha_\Omega$, $a_\Sigma = -\alpha_\Sigma$, $a_Q = -\alpha_Q$, $a_u = -\alpha_u$. Moreover, we define Ψ and φ by

$$\begin{array}{rcl} -\Psi_t - \Delta\,\Psi + d\,\Psi &=& a_Q \\ \partial_\nu \Psi + b\,\Psi &=& a_\Sigma \\ \Psi(T) &=& a_\Omega \end{array} \qquad \begin{array}{rcl} -\varphi_t - \Delta\,\varphi + d\,\varphi &=& 0 \\ \partial_\nu \varphi + b\,\varphi &=& \bar{\mu} \\ \varphi(T) &=& 0\;. \end{array}$$

The function φ plays the role of the *dual state*. It holds $K^{\mathsf{T}}\bar{\mu} = \varphi$, hence the constraints of $(\mathcal{D})$ admit the form $\mu \geq a + \varphi$. Moreover, (4.3) yields the complementary slackness condition for the dual problem,

$$(-a - \varphi + \bar{\mu})\,\bar{u} = 0 \qquad \text{a.e. on } \Sigma. \tag{5.4}$$

Ψ was defined to satisfy $a = \Psi + a_u = \Psi - \alpha_u$.

It is clear that we have to put $\bar{p} := -(\Psi + \varphi)$. Then p solves (5.1). Moreover, the complementary slackness condition (5.3) is satisfied by (4.3). Finally, the variational inequality follows from

$$\begin{aligned}\int_\Sigma(\alpha_u + \bar{p} + \bar{\mu})(u - \bar{u}) &= \int_\Sigma(-a_u + \bar{p} + \bar{\mu})(u - \bar{u}) = \int_\Sigma(\Psi - a + \bar{p} + \bar{\mu})(u - \bar{u}) \\ &= \int_\Sigma(-a - \varphi + \bar{\mu})(u - \bar{u}) = \int_\Sigma(-a - \varphi + \bar{\mu})u - \int_\Sigma(-a - \varphi + \bar{\mu})\bar{u} \geq 0.\end{aligned}$$

In the last estimate, the relations $-a - \varphi + \mu \geq 0$ and (5.4) were used. The optimality conditions are verified. Altogether, we have found our final result:

Theorem 5.1 *The dual problem (D) has at least one bounded and measurable solution $\bar{\mu}$. Let $\bar{u}$ be optimal for the linear parabolic boundary control problem. Then $\bar{\mu}$ is a Lagrange multiplier associated with the mixed control-state constraint $u \leq c + y$.*

References

[1] E.J. Anderson, A review of duality theory for linear programming over topological vector spaces, *Journal of Mathematical Analysis and Applications* 97, 380-392 (1983).

[2] E.J. Anderson and P. Nash, *Linear programming in infinite-dimensional spaces*, J. Wiley & Sons, Chichester etc. 1987.

[3] M. Bergounioux and F. Tröltzsch, Optimal control of linear bottleneck problems, *ESAIM: Control, Optimisation and Calculus of Variations*, Vol. 3, n° 5&6, pp.235–250 (1998).

[4] M. Bergounioux and F. Tröltzsch, Optimal control of semilinear parabolic equations with state-constraints of bottleneck type, Faculty of Math., TU Chemnitz, Preprint 98-19, to appear.

[5] F. Tröltzsch, Optimality conditions for parabolic control problems and applications, *Teubner Texte zur Mathematik*, Leipzig (1984).

MEMBRANE SHELL EQUATION: CHARACTERIZATION OF THE SPACE OF SOLUTIONS*

Michel C. Delfour

Centre de recherches mathématiques, Université de Montréal
C. P. 6128, succ. centre-ville, Montréal QC, Canada H3C 3J7
delfourCRM.UMontreal.CA

Abstract: The existence of solution to *membrane shell equation* is studied in a bounded open connected domain ω (Lipschitzian when ω has a boundary γ) in a $C^{1,1}$ midsurface for homogeneous Neumann boundary conditions or homogeneous Dirichlet boundary conditions on a part γ_0 of γ. It is proved that its tangential part is solution of the *reduced membrane shell equation* in $H^1(\omega)^N$ (resp. $H^1_{\gamma_0}(\omega)^N$) unique up to an element of a finite dimensional subspace, while its normal component belongs to a weighed $L^2(\omega)$ space by the pointwise norm of the second fundamental form. It is also shown that the reduced equation is equivalent to the equation for the projection onto the linear subspace of vector functions whose *linear change of metric tensor* is orthogonal to the second fundamental form of the midsurface.

1 INTRODUCTION

In recent papers ([8, 9, 11, 12]) it was established that the polynomial $P(2,1)$ model is both pertinent and basic in the theory of *thin shells*. It was shown in [8] that its solution converges to the solution of a coupled system of variational equations. For the plate and the bending dominated shell it yields (as the thickness $2h$ goes to zero) the *membrane shell equation* and the *asymptotic bending equation*.

The first variational equation of the asymptotic coupled system coincides with the variational equation characterizing the *asymptotic* $P(0,1)$ *model*. It was shown in [8] that this equation decomposes into two equations: a first equation containing the *Love-Kirchhoff* group of terms and a second equation which coincides with the classical *membrane shell equation*. The detailed cor-

*This research has been supported by National Sciences and Engineering Research Council of Canada research grant A–8730 and by a FCAR grant (Ministère de l'Éducation du Québec).

respondence with the covariant form of the classical membrane shell equation is given in [10]. The decomposition is achieved by variable elimination which results in the introduction of an *effective compliance* C_{eP} associated with the initial three dimensional compliance C.

In this paper the *membrane shell equation* is studied in a bounded open connected domain ω (Lipschitzian when ω has a boundary γ) in a $C^{1,1}$ submanifold of codimension one for homogeneous Neumann boundary conditions or homogeneous Dirichlet boundary conditions on a part γ_0 of γ when γ_0 has nonzero H_{N-2} Hausdorff measure. This paper is a companion paper to [8] where the spaces of solution E^0 and E^P corresponding to the respective asymptotic $P(0,1)$ model and the membrane shell equation are defined as completions of appropriate quotient spaces. It gives a complete characterization of the space E^P without extra condition on the second fundamental form. Such a characterization is currently available for the *plate* and *uniform strong elliptic shells* in [13, 5, 6, 7]. It also shows that we can always associate with the vector functions of the space E^P a class of tangential components which turns out to be solutions of the *reduced membrane shell equation*. This reduced equation is also connected with a projection onto a linear subspace of elements of E^P whose *linear change of metric tensor* is orthogonal to the second fundamental form. Another consequence of the characterization of E^P is the fact that in the asymptotic convergence of the solution of the $P(2,1)$ model we now know that the tangential component of the displacement of the midsurface strongly converges in $H^1(\omega)^N$ and the normal component in a weighed $L^2(\omega)$ space by the pointwise norm of the second fundamental form. The characterization given in this paper and the one of E^{01} given in [9] for the $P(2,1)$ model sharpen the abstract results of [8].

For $N = 3$, this extends to arbitrary $D^2 b$ the available existence of solutions obtained by [1, 3] for $g^0 = 0$, homogeneous Dirichlet boundary conditions on the whole boundary, the special constitutive law $C^{-1}\varepsilon = 2\,\mu\,\varepsilon + \lambda\,\mathrm{tr}\,\varepsilon\, I$ and the *uniform ellipticity* of the 2-dimensional C^2-midsurface ω. However in the case of uniform elliptic shells uniqueness does not so far follows directly in an obvious way. The first existence and uniqueness result seems to be due to [13] under relatively strong conditions. For a domain ω with a C^3 boundary γ in an analytic midsurface, the existence and uniqueness of solutions $(\hat v^0_\Gamma, \hat v^0_n)$ in $H^1_0(\omega)^3 \times L^2(\omega)$ was established by [6, 7]. The conditions were relaxed by [1, 3]: the midsurface is of class C^2 and the boundary γ is Lipschitz for the existence (midsurface C^5 and the boundary γ of class C^4 for existence and uniqueness).

Notation and Background Material. The inner product in $\mathbf{R}^N$ and the double inner product in $\mathcal{L}(\mathbf{R}^N;\mathbf{R}^N)$ (space of $N \times N$ matrices or tensors) are denoted as

$$x \cdot y = \sum_{i=1}^N x_i\, y_i, \quad A \cdot\cdot B = \sum_{i=1}^N \sum_{j=1}^N A_{ij}\, B_{ij}.$$

${}^*M$ denotes the transpose of of an arbitrary $k \times m$ matrix M.

In this paper the submanifold Γ of codimension one is specified as the boundary of a subset Ω of $\mathbf{R}^N$. It is assumed that ω is a bounded open subset of Γ

and that Γ is of class $C^{1,1}$ in a neighbourhood of ω. This is equivalent to say that the algebraic distance function b of Ω is $C^{1,1}$ in that neighborhood. Its gradient ∇b coincides with the unit exterior normal n to Γ on ω and its Hessian matrix $D^2 b$ to the second fundamental form. Finally $P \stackrel{\text{def}}{=} I - n\,{}^*n$ will denote the orthogonal projection onto the tangent plane to ω ($[n\,{}^*n]_{ij} = n_i\, n_j$). For a detailed account of the intrinsic differential calculus on a $C^{1,1}$-submanifold, the reader is referred to the now available lecture notes [10, 8]. Finally it will be convenient to introduce the following notation for the decompositions of an $N \times N$ matrix τ into its tangential and normal parts along ω

$$\tau^P \stackrel{\text{def}}{=} P\tau P, \quad \tau_{nn} \stackrel{\text{def}}{=} \tau n \cdot n, \quad [t]\tau = \tau^P + (P\tau n)\,{}^*n + n\,{}^*(P\tau n) + \tau_{nn}\, n\,{}^*n$$

and the spaces of symmetrical matrices

$$\mathrm{Sym}_N \stackrel{\text{def}}{=} \{\tau \in \mathcal{L}(\mathbf{R}^N; \mathbf{R}^N) : {}^*\tau = \tau\}$$
$$\mathrm{Sym}_N^P \stackrel{\text{def}}{=} \{\tau \in \mathrm{Sym}_N : \tau n = 0\} \Leftrightarrow \forall \tau \in \mathrm{Sym}_N,\ \tau^P \in \mathrm{Sym}_N^P.$$

2 MEMBRANE SHELL EQUATION

It was shown in [8] that the membrane shell equation can be obtained by decomposition of the variational equation of the asymptotic $P(1,0)$ model which also yields the typical group of terms occurring in the *Love-Kirchhoff condition*. It involves an effective compliance C_{eP} which retains the properties of the three-dimensional compliance C. So for the purposes of this paper it is convenient to start with the following assumption on the effective compliance.

Assumption 2.1 *The* effective compliance *is a linear bijective and symmetrical transformation of* Sym_N^P *such that there exists a constant* $\alpha > 0$ *for which*

$$\forall \tau^P \in \mathrm{Sym}_N^P, \quad C_{eP}^{-1}\tau^P \cdot\cdot\, \tau^P \geq \alpha\, \|\tau^P\|^2$$

The *membrane shell variational equation* is given by: for all $v^0 \in H^1(\omega)^N$

$$\int_\omega C_{eP}^{-1} \varepsilon_\Gamma^P(\hat{v}^0) \cdot\cdot\, \varepsilon_\Gamma^P(v^0)\, d\Gamma = \ell^P(v^0) \tag{1.1}$$

where the right-hand side is specified by a linear functional ℓ^P. Associate with ε_Γ^P the space

$$V \stackrel{\text{def}}{=} \{v \in L^2(\omega)^N : v_\Gamma \in H^1(\omega)^N\} \subset H \stackrel{\text{def}}{=} L^2(\omega)^N \tag{1.2}$$

and define E^P as the *completion* of the quotient space $V/\ker \varepsilon_\Gamma^P$ with respect to the norm associated with the inner product

$$\int_\omega \varepsilon_\Gamma^P(u) \cdot\cdot\, \varepsilon_\Gamma^P(v)\, d\Gamma. \tag{1.3}$$

Similarly for homogeneous Dirichlet boundary conditions on a part γ_0 of γ, denote by $E^P_{\gamma_0}$ the completion of the quotient space

$$V_{\gamma_0}/\ker \varepsilon^P, \quad V_{\gamma_0} \stackrel{\text{def}}{=} \{v \in L^2(\omega)^N : v_\Gamma \in H^1_{\gamma_0}(\omega)^N\}$$

with respect to the norm generated by the scalar product (1.3). By Assumption 2.1 on C_{eP}, the bilinear term in (1.1) is continuous and coercive in E^P.

Theorem 2.2 *Let Assumption 2.1 on C_{eP} be verified.*

(i) Given $\ell^P \in (E^P)'$, that is there exists $c > 0$ such that for all $v^0 \in H^1(\omega)^N$

$$|\ell^P(v^0)| \le c\,\|\varepsilon^P_\Gamma(v^0)\|_{L^2(\omega)} \tag{1.4}$$

the variational equation: to find $\hat{v}^0 \in E^P$ such that for all $v^0 \in H^1(\omega)^N$

$$\int_\omega [C^{-1}_{eP}\varepsilon^P_\Gamma(\hat{v}^0)]\cdot\cdot\,\varepsilon^P_\Gamma(v^0)\,d\Gamma = \ell^P(v^0) \tag{1.5}$$

has a unique solution $\hat{v}^0$ in E^P.

(ii) Assume that ω is connected and that γ_0 is a subset of γ with strictly positive H_{N-2} measure. Given $\ell^P \in (E^P_{\gamma_0})'$, that is there exists $c > 0$ such that for all $v^0 \in H^1_{\gamma_0}(\omega)^N$

$$|\ell^P(v^0)| \le c\,\|\varepsilon^P_\Gamma(v^0)\|_{L^2(\omega)} \tag{1.6}$$

the variational equation: to find $\hat{v}^0 \in E^P_{\gamma_0}$ such that for all $v^0 \in H^1_{\gamma_0}(\omega)^N$

$$\int_\omega [C^{-1}_{eP}\varepsilon^P_\Gamma(\hat{v}^0)]\cdot\cdot\,\varepsilon^P_\Gamma(v^0)\,d\Gamma = \ell^P(v^0) \tag{1.7}$$

has a unique solution $\hat{v}^0$ in $E^P_{\gamma_0}$.

3 REDUCED MEMBRANE SHELL EQUATION

The membrane shell equation can be further decomposed into a system of two equations. For test functions $v \in V$, that is $(v^0_\Gamma, v^0_n) \in H^1(\omega)^N \times L^2(\omega)$,

$$\begin{gathered}\varepsilon^P_\Gamma(v^0) = \varepsilon^P_\Gamma(v^0_\Gamma) + v^0_n\, D^2 b \\ \exists c > 0,\ \forall v^0_n \in H^1(\omega), \quad |\ell^P(v^0_n\, n)| \le c\,\|v^0_n\, D^2 b\|_{L^2(\omega)} \le c'\,\|v^0_n\|_{L^2(\omega)} \\ \Rightarrow \exists f^P \in L^2(\omega) \text{ such that } \ell^P(v^0_n\, n) = \textstyle\int_\omega f^P\, v^0_n\, d\Gamma\end{gathered}$$

It will be convenient to define the function

$$f^P_n(X) \stackrel{\text{def}}{=} \begin{cases} f^P(X)/\|D^2 b(X)\|, & \text{if } \|D^2 b(X)\| \neq 0 \\ 0, & \text{if } \|D^2 b(X)\| = 0 \end{cases}$$

By construction $f^P_n\,\|D^2 b\| \in L^2(\omega)$.

Denote by $H^1_t(\omega)^N$ (resp. $H^1_{\gamma_0 t}(\omega)^N$) the subspace $\{v \in H^1(\omega)^N$ (resp. $H^1_{\gamma_0}(\omega)^N) : v \cdot n = 0\}$ of tangential vectors. The decomposition yields the two equations

$$\begin{gathered}[C^{-1}_{eP}\varepsilon^P_\Gamma(\hat{v}^0)]\cdot\cdot D^2 b = f^P = f^P_n \|D^2 b\| \\ \forall v^0_\Gamma \in H^1_t(\omega)^N, \quad \int_\omega [C^{-1}_{eP}\varepsilon^P_\Gamma(\hat{v}^0)]\cdot\cdot\varepsilon^P_\Gamma(v^0_\Gamma)\, d\Gamma = \ell^P(v^0_\Gamma)\end{gathered} \tag{1.8}$$

where by condition (1.4) on ℓ^P

$$\exists c > 0, \forall v^0_\Gamma \in H^1_t(\omega)^N, \quad |\ell^P(v^0_\Gamma)| \le c\,\|\varepsilon^P_\Gamma(v^0_\Gamma)\|_{L^2(\omega)}.$$

In the case of the plate ($D^2 b = 0$), $\varepsilon^P_\Gamma(v^0) = \varepsilon^P_\Gamma(v^0_\Gamma) + v^0_n\, D^2 b = \varepsilon^P_\Gamma(v^0_\Gamma)$ and there is only the variational equation

$$\forall v^0_\Gamma \in H^1(\omega)^N, \quad \int_\omega [C^{-1}_{eP}\varepsilon^P_\Gamma(\hat{v}^0_\Gamma)]\cdot\cdot\varepsilon^P_\Gamma(v^0_\Gamma)\, d\Gamma = \ell^P(v^0_\Gamma)$$

which completely specifies $\hat{v}^0_\Gamma \in H^1_t(\omega)^N / \ker \varepsilon^P_\Gamma$ (resp. $H^1_{\gamma_0 t}(\omega)^N$) and $\hat{v}^0_n$ is arbitrary. There is a generalization of this result without adding new conditions on $D^2 b$. The second equation (1.8) specifies the *tangential part* $\hat{v}^0_\Gamma$ of $\hat{v}^0$ up to an element of some appropriate equivalence class providing a natural decomposition of the membrane shell equation into an equation for the equivalence class of $\hat{v}^0_\Gamma$ and an equation for $\hat{v}^0_n$ again modulo another equivalence class. In the case of the plate the corresponding equivalence class for $\hat{v}^0_n$ is so big that there is no information on $\hat{v}^0_n$ and we have uniqueness for $\hat{v}^0_\Gamma$ in the case of homogeneous Dirichlet boundary conditions on a part of the boundary.

Denote by $[v]_E$ the equivalence class of v in E^P (resp. $E^P_{\gamma_0}$). Let

$$\omega_0 \stackrel{\text{def}}{=} \{x \in \omega : D^2 b(x) = 0\} \text{ and } \omega_+ \stackrel{\text{def}}{=} \omega \backslash \omega_0.$$

For $v \in V$ (resp. V_{γ_0}) define the function

$$\pi_S(v) \stackrel{\text{def}}{=} \begin{cases} v - \dfrac{C^{-1}_{eP}\varepsilon^P_\Gamma(v)\cdot\cdot D^2 b}{C^{-1}_{eP} D^2 b\cdot\cdot D^2 b}\, n, & \text{in } \omega_+ \\ v, & \text{in } \omega_0 \end{cases} \tag{1.9}$$

Using the identity $\varepsilon^P_\Gamma(v) = \varepsilon^P_\Gamma(v_\Gamma) + v_n\, D^2 b$, it is easy to verify that for all $v \in V$ (resp. V_{γ_0})

$$\varepsilon^P_\Gamma(\pi_S(v)) = \begin{cases} \varepsilon^P_\Gamma(v) - \dfrac{C^{-1}_{eP}\varepsilon^P_\Gamma(v)\cdot\cdot D^2 b}{C^{-1}_{eP} D^2 b\cdot\cdot D^2 b}\, D^2 b, & \text{in } \omega_+ \\ \varepsilon^P_\Gamma(v), & \text{in } \omega_0 \end{cases} \tag{1.10}$$

For each $v_\Gamma \in H^1_t(\omega)^N$ (resp. $H^1_{\gamma_0 t}(\omega)^N$) define the tensor

$$\tilde{\varepsilon}^P_\Gamma(v_\Gamma) \stackrel{\text{def}}{=} \varepsilon^P_\Gamma(\pi_S(v_\Gamma)) = \begin{cases} \varepsilon^P_\Gamma(v_\Gamma) - \dfrac{C^{-1}_{eP}\varepsilon^P_\Gamma(v_\Gamma)\cdot\cdot D^2 b}{C^{-1}_{eP} D^2 b\cdot\cdot D^2 b}\, D^2 b, & \text{in } \omega_+ \\ \varepsilon^P_\Gamma(v_\Gamma), & \text{in } \omega_0 \end{cases} \tag{1.11}$$

the quotient space

$$V^P \stackrel{\text{def}}{=} H^1_t(\omega)^N / \ker \tilde{\varepsilon}^P_\Gamma \quad (\text{resp. } V^P_{\gamma_0} \stackrel{\text{def}}{=} H^1_{\gamma_0 t}(\omega)^N / \ker \tilde{\varepsilon}^P_\Gamma) \tag{1.12}$$

and the space

$$U \stackrel{\text{def}}{=} \pi_S(H_t^1(\omega)^N) \quad (\text{resp. } U_{\gamma_0} \stackrel{\text{def}}{=} \pi_S(H_{\gamma_0 t}^1(\omega)^N)) \tag{1.13}$$

Consider the *reduced membrane shell equation*: to find $v_\Gamma \in V^P$ (resp. $V_{\gamma_0}^P$) such that for all $w_\Gamma \in H_t^1(\omega)^N$ (resp. $H_{\gamma_0 t}^1(\omega)^N$)

$$\int_\omega C_{eP}^{-1}\tilde{\varepsilon}_\Gamma^P(v_\Gamma)\cdot\cdot\, \tilde{\varepsilon}_\Gamma^P(w_\Gamma)\, d\Gamma = \ell^P(\pi_S(w_\Gamma)) \tag{1.14}$$

By condition (1.4) on ℓ^P there exists $c > 0$ such that for all $w_\Gamma \in H_t^1(\omega)^N$

$$|\ell^P(\pi_S(w_\Gamma)| \le c\,||\varepsilon_\Gamma^P(\pi_S(w_\Gamma))||_{L^2(\omega)} = c\,||\tilde{\varepsilon}_\Gamma^P(w_\Gamma)||_{L^2(\omega)}$$

and equation (1.14) has a unique solution in the completion of the quotient space $V^P = V/\ker\tilde{\varepsilon}_\Gamma^P$ with respect to the topology generated by the norm $||\tilde{\varepsilon}_\Gamma^P(v_\Gamma)||_{L^2(\omega)}$. We now give a sharper existence theorem for the reduced membrane shell equation and the membrane shell equation. This theorem is based on a characterization of the elements of the spaces E^P and $E_{\gamma_0}^P$.

Theorem 3.1 *Let Assumption 2.1 on C_{eP} and* (1.4) on ℓ^P be verified.

(i) There exists a solution $\hat{v}_\Gamma$ in $H_t^1(\omega)^N$ *(resp. $H_{\gamma_0 t}^1(\omega)^N$)* to the *reduced membrane shell equation* (1.14) unique up to an element of $\ker\tilde{\varepsilon}_\Gamma^P$ and

$$[\pi_S(\hat{v}_\Gamma)]_E = [\pi_S(\hat{v})]_E \tag{1.15}$$

where $[\hat{v}]_E$ is the solution of the membrane shell equation (1.5) *(resp.* (1.7)*)* in E^P *(resp. $E_{\gamma_0}^P$).*

(ii) There exists a solution $\hat{u}$ such that $\hat{u}_\Gamma \in H_t^1(\omega)^N$ *(resp. $H_{\gamma_0 t}^1(\omega)^N$)* and $\hat{u}_n||D^2 b|| \in L^2(\omega)$ to the membrane shell equation (1.5) *(resp.* (1.7)*)* which is unique up to an element of $\ker\varepsilon_\Gamma^P$.

(iii) $\ker\tilde{\varepsilon}_\Gamma^P$ is finite dimensional. When $D^2 b \neq 0$ almost everywhere in ω, $\ker\varepsilon_\Gamma^P$ is also finite dimensional and

$$\ker\varepsilon_\Gamma^P = \left\{ v \,:\, v_\Gamma \in \ker\tilde{\varepsilon}_\Gamma^P \text{ and } v_n = -\frac{C_{eP}^{-1}\varepsilon_\Gamma^P(v_\Gamma)\cdot\cdot\, D^2 b}{C_{eP}^{-1}D^2 b\cdot\cdot\, D^2 b} \right\} \tag{1.16}$$

This theorem necessitates the following theorem on the structure of the spaces spaces E^P and $E_{\gamma_0}^P$ which follows from a sequence of lemmas. The proofs will be omitted for lack of space and will be given in a subsequent paper.

Theorem 3.2 *Let Assumption 2.1 on C_{eP} be verified.*

(i) $\ker\tilde{\varepsilon}_\Gamma^P$ *is finite dimensional and the space* $V^P = V/\ker\tilde{\varepsilon}_\Gamma^P$ (resp. $V_{\gamma_0}^P = V_{\gamma_0}/\ker\tilde{\varepsilon}_\Gamma^P$) *is complete for the norm* $||\tilde{\varepsilon}_\Gamma^P(v_\Gamma)||_{L^2(\omega)}$.

(ii) The space E^P (resp. $E^P_{\gamma_0}$) *is equal to*

$$\left\{ u_\Gamma + u_n\, n : \begin{array}{l} u_\Gamma \in H^1_t(\omega)^N \,(\text{resp. } H^1_{\gamma_0 t}(\omega)^N) \\ \text{and } u_n \, \|D^2 b\| \in L^2(\omega) \end{array} \right\} / \ker \varepsilon^P_\Gamma \quad (1.17)$$

Specifically for each $[v]_E \in E^P$, *there exists a unique* $[v_\Gamma]_V \in V^P = H^1_t(\omega)^N / \ker \tilde{\varepsilon}^P_\Gamma$ (resp. $V^P_{\gamma_0} = H^1_{\gamma_0 t}(\omega)^N / \ker \tilde{\varepsilon}^P_\Gamma$) *such that*

$$[\pi_S(u_\Gamma)]_E = [\pi_S(v)]_E \quad (1.18)$$

and for each u_Γ *in the equivalence class* $[v_\Gamma]_V$ *the* normal component

$$u_n \stackrel{\text{def}}{=} \begin{cases} \dfrac{C^{-1}_{eP}\varepsilon^P_\Gamma(v-u_\Gamma)\cdot\cdot D^2 b}{C^{-1}_{eP} D^2 b \cdot\cdot D^2 b}, & \text{in } \omega_0 \\ 0, & \text{in } \omega_+ \end{cases}$$

is such that $u_n \, \|D^2 b\| \in L^2(\omega)$ *and*

$$[u_\Gamma + u_n\, n]_E = [v]_E.$$

Conversely for all $u_\Gamma \in H^1_t(\omega)^N$ *and* $u_n \, \|D^2 b\| \in L^2(\omega)$

$$[u_\Gamma + u_n\, n]_E \in E^P.$$

(iii) When $D^2 b \neq 0$ *almost everywhere in* ω, *then* $\ker \varepsilon^P_\Gamma$ *is finite dimensional.*

Define the closed linear subspace

$$\begin{array}{c} S^P \stackrel{\text{def}}{=} \{ v \in E^P : C^{-1}_{eP}\varepsilon^P_\Gamma(v)\cdot\cdot D^2 b = 0 \} \\ (\text{resp } S^P_{\gamma_0} \stackrel{\text{def}}{=} \{ v \in E^P_{\gamma_0} : C^{-1}_{eP}\varepsilon^P_\Gamma(v)\cdot\cdot D^2 b = 0 \}) \end{array} \quad (1.19)$$

of E^P (resp. $E^P_{\gamma_0}$). We first make sense of the map π_S on E^P.

Lemma 3.3 *The map*

$$[v]_E \mapsto \pi_S([v]_E) \stackrel{\text{def}}{=} [\pi_S(v)]_E : E^P \to S^P \quad (1.20)$$

is well-defined, linear and continous. Moreover

$$\overline{\pi_S(V)/\ker \varepsilon^P_\Gamma}^{E^P} = \overline{\pi_S(V/\ker \varepsilon^P_\Gamma)}^{E^P} = S^P \quad (1.21)$$

$$\forall v \in S^P, \quad \varepsilon^P_\Gamma(\pi_S(v)) = \varepsilon^P_\Gamma(v). \quad (1.22)$$

and π_S *is a projection, that is* $[\pi_S(v)]_E = [\pi_S(\pi_S(v))]_E$ *in* E^P.

Lemma 3.4 *The map* $\pi_S : H^1_t(\omega)^N \to U$ *is a continuous linear bijection and* U *is closed for the topology generated by the norm*

$$\|u\|_U = \left\{ \|\varepsilon^P_\Gamma(u)\|^2_{L^2(\omega)} + \|u_\Gamma\|^2_{L^2(\omega)} + \|u_n\, \|D^2 b\|\|^2_{L^2(\omega)} \right\}^{1/2}.$$

The space V^P *is complete for the norm* $\|\tilde{\varepsilon}_\Gamma^P(v_\Gamma)\|_{L^2(\omega)}$ *and* $\ker \tilde{\varepsilon}_\Gamma^P$ *is finite dimensional.*

Lemma 3.5 *The map*

$$\begin{aligned} [v_\Gamma]_V \mapsto \quad & \pi_S([v_\Gamma]_V) \stackrel{\text{def}}{=} [\pi_S(v)]_E \\ & : V^P = H_t^1(\omega)^N / \ker \tilde{\varepsilon}_\Gamma^P \to U / \ker \varepsilon_\Gamma^P \end{aligned} \tag{1.23}$$

is a well-defined isomorphism, where $[v]_V$ *denotes the equivalence class of* v *in* V^P, V^P *is endowed with the topology generated by the norm* $\|\tilde{\varepsilon}_\Gamma^P(v_\Gamma)\|$ *and* $U/\ker \varepsilon_\Gamma^P$ *by the norm* $\|\varepsilon_\Gamma^P(v)\|$ *on* E^P. *Moreover*

$$S^P = U/\ker \varepsilon_\Gamma^P = \pi_S(H_t^1(\omega)^N)/\ker \varepsilon_\Gamma^P = \pi_S(V^P).$$

Lemma 3.6 *For each* $v \in E^P$, *the projection* $[\pi_S(v)]_E$ *is the unique solution in* S^P *of the variational equation: for all* $w \in H^1(\omega)^N$

$$\int_\omega C_{eP}^{-1} \varepsilon_\Gamma^P(\pi_S(v) - v) \cdot\cdot\, \varepsilon_\Gamma^P(w)\, d\Gamma = 0 \tag{1.24}$$

and there is a solution $v_\Gamma \in H_t^1(\omega)^N$ *unique up to an element of* $\ker \tilde{\varepsilon}_\Gamma^P$ *to the variational equation: for all* w_Γ *in* $H_t^1(\omega)^N$

$$\int_\omega C_{eP}^{-1} \tilde{\varepsilon}_\Gamma^P(v_\Gamma) \cdot\cdot\, \tilde{\varepsilon}_\Gamma^P(w_\Gamma)\, d\Gamma = \int_\omega C_{eP}^{-1} \varepsilon_\Gamma^P(v) \cdot\cdot\, \varepsilon_\Gamma^P(\pi_S(w_\Gamma))\, d\Gamma. \tag{1.25}$$

Moreover

$$[\pi_S(v)]_E = [\pi_S(v_\Gamma)]_E. \tag{1.26}$$

All this remains true for Dirichlet boundary conditions on a part γ_0 *of the boundary with* $H_{\gamma_0 t}^1(\omega)^N$ *in place of* $H_t^1(\omega)^N$ *and* $E_{\gamma_0}^P$ *in place of* E^P.

Lemma 3.7 *(i) The space* E^P *is equal to*

$$\left\{u_\Gamma + u_n\, n \;:\; u_\Gamma \in H_t^1(\omega)^N \text{ and } u_n \|D^2 b\| \in L^2(\omega)\right\} / \ker \varepsilon_\Gamma^P \tag{1.27}$$

Specifically for each $[v]_E \in E^P$, *there exists a unique* $[v_\Gamma]_V \in V^P = H_t^1(\omega)^N / \ker \tilde{\varepsilon}_\Gamma^P$ (resp. $V_{\gamma_0}^P = H_{\gamma_0 t}^1(\omega)^N / \ker \tilde{\varepsilon}_\Gamma^P$) *such that*

$$[\pi_S(v_\Gamma)]_E = [\pi_S(v)]_E \tag{1.28}$$

and for each u_Γ *in the equivalence class* $[u_\Gamma]_V$ *the* normal component

$$u_n \stackrel{\text{def}}{=} \begin{cases} \dfrac{C_{eP}^{-1} \varepsilon_\Gamma^P(\hat{v}^0 - u_\Gamma) \cdot\cdot\, D^2 b}{C_{eP}^{-1} D^2 b \cdot\cdot\, D^2 b}, & \text{in } \omega_0 \\ 0, & \text{in } \omega_+ \end{cases} \tag{1.29}$$

is such that $u_n \|D^2 b\| \in L^2(\omega)$ *and*

$$[u_\Gamma + u_n\, n]_E = [v]_E.$$

Conversely for all $u_\Gamma \in H^1_t(\omega)^N$ *and* $u_n \, \|D^2 b\| \in L^2(\omega)$

$$[u_\Gamma + u_n \, n]_E \in E^P.$$

(ii) When $D^2 b \neq 0$ *almost everywhere in* ω, *then* $\ker \varepsilon^P_\Gamma$ *is finite dimensional.*

The lemma remains true for Dirichlet boundary conditions on a part γ_0 *of the boundary with* $H^1_{t,\gamma_0}(\omega)^N$ *and* $E^P_{\gamma_0}$ *in place of* $H^1_t(\omega)^N$ *and* $H^1_{\gamma_0 t}(\omega)^N$.

References

[1] Ph. G. Ciarlet and V. Lods, *Ellipticité des équations membranaires d'une coque uniformément elliptique*, C. R. Acad. Sci. Paris Sér. I Math. **318** (1994), 195–200.

[2] —, *Analyse asymptotique des coques linéairement élastique. I. Coques membranaires*, C. R. Acad. Sci. Paris Sér. I Math. **318** (1994), 863–868.

[3] —, *On the ellipticity of linear membrane shell equations*, J. Math. Pures Appl. **75** (1996), 107–124.

[4] —, *Asymptotic analysis of linearly elastic shells. Generalized membrane shells*, J. Elasticity **43** (1996), 147–188.

[5] —, *Asymptotic analysis of linearly elastic shells. I. Justification of membrane shell equations*, Arch. Rational Mech. Anal. **136** (1996), 116–161.

[6] Ph. G. Ciarlet and É. Sanchez-Palencia, *Un théorème d'existence et d'unicité pour les équations des coques membranaires*, C. R. Acad. Sci. Paris Sér. I Math. **317** (1993), 801–805.

[7] —, *An existence and uniqueness theorem for the two-dimensional linear membrane shell equations*, J. Math. Pures Appl. **75** (1996), 51–67.

[8] M.C. Delfour, *Intrinsic differential geometric methods in the asymptotic analysis of linear thin shells*, Boundaries, interfaces and transitions (M.C. Delfour, ed.), CRM Proc. Lecture Notes, Amer. Math. Soc., Providence, RI, 1998, pp. 19-90.

[9] —, *Intrinsic $P(2,1)$ thin shell model and Naghdi's models without a priori assumption on the stress tensor*, in Proc International Conference on Optimal Control of Partial Differential Equations, K.H. Hoffmann, G. Leugering, F. Tröltzsch , eds., Birkhäuser, to appear.

[10] M.C. Delfour and J.-P. Zolésio, *Differential equations for linear shells: comparison between intrinsic and classical models*, Advances in Mathematical Sciences–CRM's 25 years (Luc Vinet, ed.), CRM Proc. Lecture Notes, Amer. Math. Soc., Providence, RI, 1997, pp. 42–124.

[11] —, *Convergence to the asymptotic model for linear thin shells*, Optimization Methods in Partial Differential Equations (S. Cox and I. Lasiecka, eds.), Contemp. Math., vol. 209, AMS, Providence, RI, 1977.

[12] —, *Convergence of the linear $P(1,1)$ and $P(2,1)$ thin shells to asymptotic shells*, Proc. Plates and Shells: from theory to practice (M. Fortin, ed.), CRM Proc. Lect. Notes ser., Amer. Math. Soc., Providence, RI, 1977, to appear

[13] Ph. Destuynder, *Sur la justification des modèles de plaques et de coques par les méthodes asymptotiques*, Doctoral dissertation, Univ. Pierre et Marie Curie, 1980.

[14] É. Sanchez-Palencia, *Statique et dynamique des coques minces. I. Cas de flexion pure non inhibée*, C. R. Acad. Sc. Paris Sér. I Math. **309** (1989), 411–417.

[15] —, *Statique et dynamique des coques minces. II. Cas de flexion pure inhibée - Approximation membranaire*, C. R. Acad. Sc. Paris Sér. I Math. **309** (1989), 531–537.

RENORMING FOR ELASTIC SYSTEMS WITH STRUCTURAL DAMPING

Richard H. Fabiano*
Department of Mathematical Sciences
340 Bryan Building
University of North Carolina at Greensboro
Greensboro, NC 27412
fabiano@uncg.edu

Abstract: We consider a model of an elastic system with so-called structural damping. The operator associated with such models is known to generate an analytic semigroup, but it is neither sectorial nor associated with a coercive sesquilinear form. We show that these properties can be obtained by using an appropriate inner product other than the usual energy inner product.

1 INTRODUCTION

We consider the following abstract model of an elastic system with damping, which is a second order differential equation on a Hilbert space X:

$$\begin{aligned} &\ddot{u}(t) + Au(t) + 2aA^{1/2}\dot{u}(t) = 0, \\ &u(0) = u_0, \quad \dot{u}(0) = v_0. \end{aligned} \tag{1.1}$$

Here we assume that A is a positive definite, self adjoint unbounded linear operator on X, with compact inverse, and with a positive definite, self adjoint square root $A^{1/2}$. It is standard to reformulate (1.1) on the energy space $H = \text{dom}A^{1/2} \times X$ equipped with the energy norm $\|(u,v)\|_H^2 = \|A^{1/2}u\|_X^2 + \|v\|_X^2$. If we identify $z(t) = (u(t), \dot{u}(t))$, then (1.1) can be reformulated as the first order system

$$\begin{aligned} \dot{z}(t) &= \mathcal{A}z(t) \\ z(0) &= (u_0, v_0). \end{aligned} \tag{1.2}$$

Here $\mathcal{A}$ is defined on the domain

$$\text{dom}\mathcal{A} = \{(u,v) \in H : v \in \text{dom}A^{1/2}, \ u \in \text{dom}A\}$$

*This research was supported in part by the National Science Foundation under grant DMS-9696239.

by $\mathcal{A}(u,v) = (v, -Au - 2aA^{1/2}v)$. It is well known that the closure of $\mathcal{A}$ is the infinitesimal generator of an analytic semigroup $T(t)$ on H (see [3], [4], [7], and [10]). However, the operator $-\mathcal{A}$ is neither sectorial nor associated with a coercive form. In this paper we show how to use an equivalent inner product to obtain these properties. Let us first recall some relevant definitions and results.

For an operator A defined in a Hilbert space H, the numerical range of A (also called the field of values of A), denoted by $\Theta(A)$, is the subset of the complex plane defined by $\Theta(A) = \{\langle Ax, x\rangle_H : x \in \text{dom}A, \ ||x||_H = 1\}$. The numerical range is inner product dependent, and sometimes this is indicated explicitly by the notation $\Theta(A, \langle\cdot,\cdot\rangle)$. The numerical range is convex [8] and always contains the eigenvalues, and hence contains the convex hull of the eigenvalues. The closed operator A is m-sectorial (with vertex at the origin) if the numerical range $\Theta(A)$ is contained in a sector of the form $|\arg(\lambda)| \leq \theta < \pi/2$ and if $||(\lambda I - A)^{-1}|| \leq 1/(-\text{Re}\,\lambda)$ for $\text{Re}\lambda < 0$. We say that A is quasi m-sectorial if a translation $A + kI$ is m-sectorial. Next, given Hilbert spaces $V \subset H$, with the embedding dense and continuous, the sesquilinear form $\sigma : V \times V \to \mathbb{C}$ is bounded if there exists $K > 0$ such that

$$|\sigma(x,y)| \leq K||x||_V||y||_V \quad \text{for all } x, y \in V,$$

is V-elliptic if there exists $c > 0$ such that

$$\text{Re}\,\sigma(x,x) \geq c||x||_V^2 \quad \text{for all } x \in V,$$

and is V-coercive if there exists $c, k \in \mathbb{R}$, $c > 0$, such that

$$\text{Re}\,\sigma(x,x) \geq c||x||_V^2 - k||x||_H^2 \quad \text{for all } x \in V.$$

In either case σ uniquely defines an operator $A : \text{dom}A \subset H \to H$ by

$$\sigma(x,y) = \langle Ax, y\rangle \tag{1.3}$$

for all $x \in \text{dom}A$ and $y \in V$, and

$$\text{dom}A = \{x \in V : |\sigma(x,y)| \leq K_x||y||_H \text{ for all } y \in V\}$$

where K_x depends on x. It is known that if σ is V-elliptic (resp. V-coercive), then the unique operator A defined by σ is m-sectorial (resp. quasi m-sectorial). Also, given any quasi m-sectorial operator A (whether or not it is defined by a sesquilinear form as above), it is known that $-A$ is the infinitesimal generator of an analytic semigroup. However, the converse is not true. Thus it is possible for $-A$ to generate an analytic semigroup even though A is not quasi m-sectorial and so does not 'come from' any V-coercive form. This is unfortunate, because there are some nice convergence results available for parameter estimation and LQR problems associated with coercive forms (see [2], [1], [9]), especially when the embedding $V \subset Z$ is compact. On the other hand, in such a case we are motivated to seek a new inner product in which A becomes m-sectorial and possibly associated with a V-coercive form. This is precisely the situation for the model considered in this paper. We refer the reader to ([8], [11], [12], [13]) for the definitions and results in this paragraph.

In the next section we show how the non-sectorial operator associated with the structural damping model can be made sectorial. Moreover we will see that the operator also becomes associated with a coercive sesquilinear form in the new inner product, which is not possible in the original inner product. Finally, we will see that Galerkin approximations constructed with the new inner product exhibit improved stability robustness compared to approximations in the original inner product.

2 ELASTIC SYSTEM WITH STRUCTURAL DAMPING

Consider again equation (1.1). Let $\{\phi_i\}_{i=1}^{\infty}$ denote the eigenvectors of A which form an orthonormal basis for X, with corresponding eigenvalues $\{\gamma_i\}_{i=1}^{\infty}$ satisfying $0 < \gamma_1 < \cdots < \gamma_n \to \infty$. We assume that the damping coefficient satisfies $0 < a < 1$. Physically this is equivalent to assuming that the system is not overdamped. In particular, when $a \geq 1$ then all eigenvalues of $\mathcal{A}$ are negative real numbers (every vibratory mode is overdamped), whereas for $0 < a < 1$ the spectrum of $\mathcal{A}$ consists of eigenvalues given by

$$\lambda_n^{\pm} = \alpha_{\pm}\gamma_n^{1/2}, \qquad \text{where } \alpha_{\pm} = -a \pm \sqrt{1-a^2}\, i.$$

Thus the eigenvalues lie on the lines $\mathrm{Im}\lambda = \frac{\pm\sqrt{1-a^2}}{a}\mathrm{Re}\lambda$, and also satisfy $\mathrm{Re}\lambda \leq -a\gamma_1^{1/2}$.

It is possible to associate $-\mathcal{A}$ with a sesquilinear form, and the standard way to do this is to set $V = \mathrm{dom}A^{1/2} \times \mathrm{dom}A^{1/2}$ with norm $||(u,v)||_V^2 = ||A^{1/2}u||_X^2 + ||A^{1/2}v||_X^2$. Then $V \subset H$ and the embedding is continuous. Also define the sesquilinear form $\sigma : V \times V \to \mathbb{C}$ by

$$\sigma((u,v),(f,g)) = -\langle A^{1/2}v, A^{1/2}f\rangle_X + \langle A^{1/2}u, A^{1/2}g\rangle_X + 2a\langle A^{1/2}v, g\rangle_X.$$

Then $\mathrm{dom}\mathcal{A} \subset V \subset H$ and we have

$$\langle -\mathcal{A}x, y\rangle_H = \sigma(x,y) \quad \text{for all } x \in \mathrm{dom}\mathcal{A},\ y \in V. \tag{2.1}$$

At this point we note that while σ is V-bounded, it is neither V-elliptic nor V-coercive, and the embedding $V \subset H$ is not compact. In fact, it is impossible to associate $-\mathcal{A}$ with a coercive form in this problem. That is, there does not exist a space V continuously embedded in H and a coercive form $\sigma : V \times V \to \mathbb{C}$ for which $\mathrm{dom}\mathcal{A} \subset V$ and (2.1) holds. The reason is that if there were such a space V and coercive form σ, then as noted in the introduction it would follow that the numerical range $\Theta(-\mathcal{A})$ is contained in a sector in the complex plane. However, in the energy norm it can be shown that $\Theta(-\mathcal{A})$ is the whole right half complex plane (see [5]). Thus we are motivated to seek a new inner product on H, hopefully with the property that the operator $-\mathcal{A}$ in this new inner product is m-sectorial. It turns out that we can achieve the best possible outcome - an inner product in which the numerical range $\Theta(-\mathcal{A})$ is *equal* to the convex hull of the eigenvalues. Furthermore, we can find a space V *compactly* embedded in H, on which there is defined a V-elliptic sesquilinear form associated with $-\mathcal{A}$.

To proceed, define the norm $\|\cdot\|_e$ on H by

$$\|(u,v)\|_e^2 = \frac{1}{2(1-a^2)}\left\{\|A^{1/2}u\|_X^2 + \|v\|_X^2 + 2a\mathrm{Re}\langle A^{1/2}u, v\rangle_X\right\}.$$

It is easy to verify that this is a norm which is equivalent to the energy norm, and which has a compatible inner product given by

$$\begin{aligned}\langle (u,v),(f,g)\rangle_e = \ & \tfrac{1}{2(1-a^2)}\left\{\langle A^{1/2}u, A^{1/2}f\rangle_X + \langle v,g\rangle_X + a\langle A^{1/2}u, g\rangle_X\right. \\ & \left. + a\langle v, A^{1/2}f\rangle_X\right\}. \end{aligned} \tag{2.2}$$

While it is true that similarly structured quadratic Liapunov functions have been used before for second order damped systems (for example, [6]), it is this precise form which is needed to shrink the numerical range down to the convex hull of the eigenvalues. This is shown in the following theorem.

Theorem *The numerical range $\Theta(\mathcal{A}, \langle\cdot,\cdot\rangle_e)$ is the convex hull of the eigenvalues of $\mathcal{A}$.*

Proof: We must show that $\mathrm{Re}\langle \mathcal{A}x, x\rangle_e \le -a\gamma_1^{1/2}\|x\|_e^2$ and $|\mathrm{Im}\langle \mathcal{A}x, x\rangle_e| \le -\frac{\sqrt{1-a^2}}{a}\mathrm{Re}\langle \mathcal{A}x, x\rangle_e$ for all $x \in \mathrm{dom}\mathcal{A}$. For $x = (u,v) \in \mathrm{dom}\mathcal{A}$ we have

$$\begin{aligned}\langle \mathcal{A}(u,v),(u,v)\rangle_e &= \langle (v, -Au - 2aA^{1/2}v),(u,v)\rangle_e \\ &= \frac{1}{2(1-a^2)}\left\{\langle A^{1/2}v, A^{1/2}u\rangle_X + \langle -Au - 2aA^{1/2}v, v\rangle_X\right. \\ &\quad \left. + a\langle A^{1/2}v, v\rangle_X + a\langle -Au - 2aA^{1/2}v, A^{1/2}u\rangle_X\right\} \\ &= \frac{1}{2(1-a^2)}\left\{2i\mathrm{Im}\langle A^{1/2}v, A^{1/2}u\rangle_X - a\langle A^{1/2}v, v\rangle_X\right. \\ &\quad - a\langle Au, A^{1/2}u\rangle_X - 2a^2\mathrm{Re}\langle A^{1/2}v, A^{1/2}u\rangle_X \\ &\quad \left. - 2a^2 i\mathrm{Im}\langle A^{1/2}v, A^{1/2}u\rangle_X.\right\}\end{aligned}$$

Now use the fact that $\gamma_1^{1/2}\|x\|_X^2 \le \|A^{1/4}x\|_X^2$ for all $x \in \mathrm{dom}A^{1/2}$ to get

$$\begin{aligned}\mathrm{Re}\langle \mathcal{A}(u,v),(u,v)\rangle_e &= \frac{-a}{2(1-a^2)}\left\{\|A^{1/4}v\|_X^2 + \|A^{3/4}u\|_X^2\right. \\ &\quad \left. + 2a\mathrm{Re}\langle A^{1/4}v, A^{3/4}u\rangle_X\right\} \\ &= \frac{-a}{2(1-a^2)}\left\{(1-a)\|A^{1/4}v\|_X^2 + (1-a)\|A^{3/4}u\|_X^2\right. \\ &\quad \left. + a\|A^{1/4}v + A^{3/4}u\|_X^2\right\}\end{aligned}$$

$$\begin{aligned}
&\le \frac{-a\gamma_1^{1/2}}{2(1-a^2)}\Big\{(1-a)\|v\|_X^2 + (1-a)\|A^{1/2}u\|_X^2 \\
&\qquad + a\|v + A^{1/2}u\|_X^2\Big\} \\
&= -a\gamma_1^{1/2}\|(u,v)\|_e^2.
\end{aligned}$$

Also,

$$\begin{aligned}
&\frac{\sqrt{1-a^2}}{a}\left(-\frac{\sqrt{1-a^2}}{a}\mathrm{Re}\langle \mathcal{A}(u,v),(u,v)\rangle_e - |\mathrm{Im}\langle \mathcal{A}(u,v),(u,v)\rangle_e|\right) \\
&= \frac{1}{2a}\Big(\|A^{1/4}v\|_X^2 + \|A^{3/4}u\|_X^2 + 2a\mathrm{Re}\langle A^{1/4}v, A^{3/4}u\rangle_X \\
&\qquad -2\sqrt{1-a^2}\left|\mathrm{Im}\langle A^{1/4}v, A^{3/4}u\rangle_X\right|\Big) \\
&\ge 0
\end{aligned}$$

This last inequality follows from the lemma below. Thus,

$$|\mathrm{Im}\langle \mathcal{A}(u,v),(u,v)\rangle_e| \le -\frac{\sqrt{1-a^2}}{a}\mathrm{Re}\langle \mathcal{A}(u,v),(u,v)\rangle_e$$

and the result follows. □

The following lemma, which applies in a general Hilbert space setting, was needed in the previous proof.

Lemma: *Let X be a Hilbert space, and $0 < a < 1$. Then*

$$\|x\|^2 + \|y\|^2 + 2a\mathrm{Re}\langle x,y\rangle - 2\sqrt{1-a^2}\,|\mathrm{Im}\langle x,y\rangle| \ge 0$$

for all $x, y \in X$

Proof: Consider $f(\theta) = 1 + a\cos\theta - \sqrt{1-a^2}|\sin\theta|$. An elementary argument shows that $f(\theta) \ge 0$ for $0 \le \theta \le 2\pi$, so that

$$2a\cos\theta - 2\sqrt{1-a^2}|\sin\theta| \ge -2.$$

Since $\langle x,y\rangle \in \mathbb{C}$, we have

$$\mathrm{Re}\langle x,y\rangle = r\cos\theta,\ \mathrm{Im}\langle x,y\rangle = r\sin\theta,\ r = |<x,y>| \ge 0,\ \text{and } 0 \le \theta \le 2\pi.$$

Thus

$$\begin{aligned}
2a\mathrm{Re}\langle x,y\rangle - 2\sqrt{1-a^2}|\mathrm{Im}\langle x,y\rangle| &= r(2a\cos\theta - 2\sqrt{1-a^2}|\sin\theta|) \\
&\ge -2r = -2|\langle x,y\rangle| \\
&\ge -2\|x\|\,\|y\|,
\end{aligned}$$

by the Cauchy-Schwarz inequality. So

$$\begin{aligned} ||x||^2 + ||y||^2 + 2a\text{Re}\langle x, y\rangle - 2\sqrt{1-a^2}\,|\text{Im}\langle x, y\rangle| &\geq ||x||^2 + ||y||^2 - 2||x||\,||y|| \\ &= (||x|| - ||y||)^2 \geq 0 \end{aligned}$$

and the result follows. □

Thus in our new inner product the numerical range $\Theta(\mathcal{A}, \langle\cdot,\cdot\rangle_e)$ is equal to the convex hull of the eigenvalues of $\mathcal{A}$. (On a related note, it is straightforward to verify that in the new inner product the operator $\mathcal{A}$ is normal, which is not true in the energy inner product). It remains for us to define a new space V_e and sesquilinear form σ_e. To this end, define $V_e = \text{dom}A^{3/4} \times \text{dom}A^{1/4}$ with norm

$$||(u,v)||_{V_e}^2 = \frac{1}{2(1-a^2)}\left\{||A^{3/4}u||_X^2 + ||A^{1/4}v||_X^2 + 2a\text{Re}\langle A^{3/4}u, A^{1/4}v\rangle_X\right\}$$

and the obvious compatible inner product. Also define $\sigma_e : V_e \times V_e \to \mathbb{C}$ by

$$\begin{aligned} \sigma_e((u,v),(f,g)) &= \frac{1}{2(1-a^2)}\left\{-a\langle A^{3/4}u, A^{3/4}f\rangle_X - a\langle A^{1/4}v, A^{1/4}g\rangle_X\right. \\ &\quad \left. -\langle A^{3/4}u, A^{1/4}g\rangle_X + (1-2a^2)\langle A^{1/4}v, A^{3/4}f\rangle_X\right\}. \end{aligned}$$

Without providing details, we observe that $\text{dom}\mathcal{A} \subset V_e \subset H$, the embedding $V_e \subset H$ is compact, the sesquilinear form σ_e is V_e-elliptic, and the operator $\mathcal{A}$ is associated with σ_e via (2.1). As mentioned previously, one can now apply results for 'coercive' systems to (1.1). Instead, we finish by comparing Galerkin approximations defined using σ_e and finite dimensional subspaces of V_e versus those defined using σ and subspaces of V. In particular we measure sensitivity of eigenvalues to unstructured perturbations by plotting the ϵ-pseudospectrum of the matrix representations for the respective Galerkin schemes. Recall that for any $\epsilon > 0$ the *ϵ-pseudospectrum* of A is the subset of the complex plane defined by (see [14] for details on definitions and results concerning pseudospectra)

$$\Sigma_\epsilon(A) = \{\lambda \in \mathbb{C} \,:\, \lambda \in \Sigma(A + \Delta A) \text{ for some } \Delta A \text{ with } ||\Delta A|| \leq \epsilon\}, \tag{2.3}$$

where $\Sigma(A)$ is the spectrum of A. It is always true that $\Sigma(A) \subset \Sigma_\epsilon(A)$ and it is clear that the pseudospectra give a reasonable measure of sensitivity of eigenvalues to unstructured perturbations.

We consider a specific example using $X = L^2(0,1)$ and $A = \frac{d^4}{dx^4}$ in (1.1), which corresponds to a structurally damped Euler-Bernoulli beam with pinned boundary conditions. In Figure 1 we plot the 2-pseudospectrum of Galerkin approximation matrix representations A^N in the energy norm for $N = 8$ and $N = 16$ subdivisions of $(0,1)$ and linear splines. In Figure 2 we do the same using the new inner product. However, for this case the pseudospectra are so well behaved that the 2-pseudospectrum cannot be distinguished from the

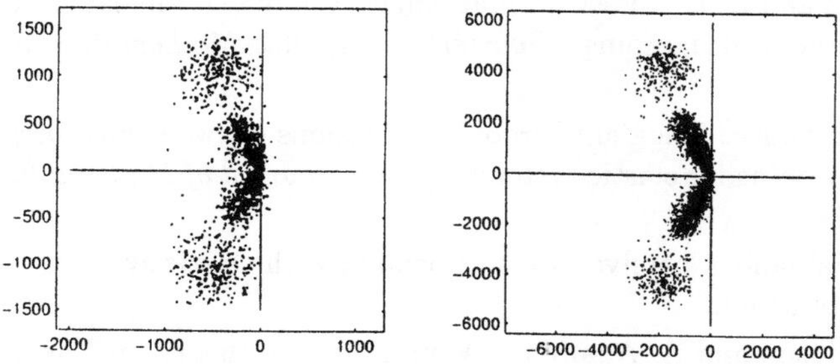

Figure 1 2-pseudospectrum of A^N for $N = 8$ and $N = 16$

eigenvalues. So instead we plot the 75-pseudospectrum(!) - that is, the eigenvalues of the matrix representation A_e^N (Galerkin approximation in the new inner product) are less sensitive to perturbations of magnitude 75 than are the eigenvalues of A^N to perturbations of magnitude 2.

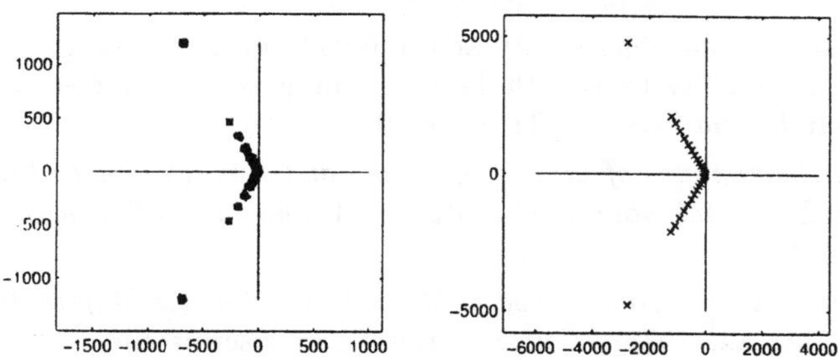

Figure 2 75-pseudospectrum of A_e^N for $N = 8$ and $N = 16$

Finally, we point out that the model (1.1) has been considered in more generality in [3], [4], [7], and [10]. In particular, the operator $A^{1/2}$ in (1.1) has been replaced by A^γ for $0 < \gamma \leq 1$ or more generally by an operator B 'similar' to A^γ. In [5] the ideas in this paper are considered in the more general case.

References

[1] H. T. Banks and K. Ito. A unified framework for approximation in inverse problems for distributed parameter systems. *Control: Theory and Advanced Technology*, 4(1):73–90, 1988.

[2] H. T. Banks and K. Kunisch. The linear regulator problem for parabolic systems. *SIAM J. Control and Optimization*, 22:684–699, 1984.

[3] G. Chen and D. L. Russell. A mathematical model for linear elastic systems with structural damping. *Quarterly of Applied Mathematics*, 39:433–454, 1982.

[4] S. Chen and R. Triggiani. Proof of extensions of two conjectures on structural damping for elastic systems. *Pacific Journal of Mathematics*, 136:15–55, 1989.

[5] R. H. Fabiano. Coercive forms associated with elastic systems with damping. submitted.

[6] J. S. Gibson and A. Adamian. Approximation theory for linear quadratic Gaussian control of flexible structures. *SIAM J. Control and Optimization*, 29:1–37, 1991.

[7] F. Huang. On the mathematical model for linear elastic systems with analytic damping. *SIAM Journal of Control and Optimization*, 26:714–724, 1988.

[8] T. Kato. *Perturbation Theory for Linear Operators*. Springer-Verlag, 1976.

[9] I. Lasiecka and R. Triggiani. *Differential and Algebraic Riccati Equations with Applications to Boundary/Point Control Problems: Continuous Theory and Approximation Theory*, volume 164 of *Lecture Notes in Control and Information Sciences*. Springer-Verlag, 1991.

[10] K. Liu and Z. Liu. Analyticity and differentiability of semigroups associated with elastic systems with damping and gyroscopic forces. *Journal of Differential Equations*, 141:340–355, 1997.

[11] A. Pazy. *Semigroups of Linear Operators and Applications to Partial Differential Equations*, volume 44 of *Applied Mathematical Sciences*. Springer-Verlag, 1983.

[12] R. E. Showalter. *Hilbert Space Methods for Partial Differential Equations*. Pitman, 1977. Out of print. A postscript file is available at http://ejde.math.swt.edu//mono-toc.html.

[13] H. Tanabe. *Equations of Evolution*. Pitman, 1979.

[14] L. N. Trefethen. Pseudospectra of linear operators. *SIAM Review*, 39:383–406, 1997.

STABILITY AND APPROXIMATIONS OF AN ACOUSTIC-STRUCTURE MODEL

Fariba Fahroo

Naval Postgraduate School
Mathematics Department, Code MA/Ff
Monterey, CA 93943-5216, USA*

fahroo@math.nps.navy.mil

Chunming Wang

University of Southern California
Department of Mathematics
Los Angeles, CA 90089-1113, USA†

cwang@optimize2.usc.edu

Abstract: We consider uniform exponential stability and preservation of this property under approximations for a coupled acoustic- structure system with "porous" damped boundary conditions. This system consists of a 2-D acoustic cavity whose portion of its boundary is modeled as a flexible beam. Using a multiplier technique we show that the "porous" boundary condition provides enough damping to yield uniform exponential stability of the model. We also present a polynomial-based Galerkin numerical scheme which results in approximate solutions with a uniform rate of decay.

Key Words: acoustic-structure models, multiplier technique, exponential stability, Galerkin method.

AMS subject classification: 35L05, 93D20, 65N12, 65N30.

1 INTRODUCTION

In this paper, we consider a two dimensional acoustic-structure system which consists of a rectangular acoustic cavity whose two sides are bounded by Euler-Bernoulli beams. The acoustic cavity is $\Omega = (-1,1) \times (-1,1)$ and the two beams are located at the sides $x = -1$ and $y = -1$.

*The research of this author was supported in part by the Naval Postgraduate School under a faculty development grant.

†The research of this author was supported in part by the Air Force Office of Scientific Research under grant AFOSR-90-0091

The model systems are given by the following partial differential equations (see [4]): for $t > 0$

$$\begin{aligned}
\partial_t^2\phi(t,x,y) &= \partial_x^2\phi(t,x,y) + \partial_y^2\phi(t,x,y), \quad (x,y)\in\Omega, && (1.1)\\
\partial_t^2 u_1(t,x) &= -\partial_x^4\left(Eu_1(t,x) + C\partial_t u_1(t,x)\right) + \partial_t\phi(t,x,-1), \quad x\in(-1,1), && (1.2)\\
\partial_t^2 u_2(t,y) &= -\partial_y^4\left(Eu_2(t,y) + C\partial_t u_2(t,y)\right) + \partial_t\phi(t,-1,y), \quad y\in(-1,1), && (1.3)
\end{aligned}$$

where ϕ is the velocity potential function for the acoustic cavity and u_1, u_2 are the transversal displacements of the beams numbered 1 and 2, respectively. The velocity transmission at the interface between the acoustic cavity and the beams is modeled by

$$\begin{aligned}
\partial_y\phi(t,x,-1) &= \partial_t u_1(t,x) + \alpha_1\partial_t\phi(t,x,-1), \quad x\in(-1,1), && (1.4)\\
\partial_x\phi(t,-1,y) &= \partial_t u_2(t,y) + \alpha_2\partial_t\phi(t,-1,y), \quad y\in(-1,1). && (1.5)
\end{aligned}$$

The coefficients α_i, $i = 1,2$ represent the velocity loss factor. We note that if $\alpha_i = 0$, the boundary conditions above correspond to the continuity of the velocity at the boundaries. Additional boundary conditions are given by

$$\begin{aligned}
\phi(t,x,1) = \phi(t,1,y) &= 0, \quad x,y\in(-1,1), && (1.6)\\
u_1(t,x) = \partial_x u_1(t,x) &= 0, \quad x = -1,1, && (1.7)\\
u_2(t,y) = \partial_y u_2(t,y) &= 0, \quad y = -1,1. && (1.8)
\end{aligned}$$

In this work we present a summary of the results obtained in [4] on well-posedness and uniform exponential stability of the solutions to the equations (1.1-1.8). We also present a polynomial-based Galerkin scheme that preserves the exponential stability of the solutions under approximations. The model presented here is a modification of the one presented first in [3]. The main difference lies in substituting the condition of continuity of normal velocities of the structure and the acoustic field at the boundary of the acoustic cavity used in [3] by boundary conditions (1.4)–(1.5) that allow for a difference in the two velocities proportional to the acoustic pressure. Recently, several studies have appeared in literature that analyze the stability properties of different models derived from the one in [3], (see [1, 2, 6, 7]). In the same vein the work here also studies the stability of a new model for acoustic-structure interactions, but in addition provides a framework that can be used for proving the existence of a uniform rate of decay for the approximate solutions obtained from a Galerkin method. Since the model equations presented here arise in active noise control problems, for implementation of the linear quadratic regulator control algorithm numerical approximation of the solutions and their exponential stability is of utmost importance. In this regard, this framework facilitates the study of convergence of the approximate Riccati operators and the feedback controls to the infinite-dimensional ones.

2 WELL-POSEDNESS OF THE MODEL

For these model equations the state space $\mathcal{H}$ is

$$\mathcal{H} = V_a \times H_a \times V_b \times H_b \times V_b \times H_b, \tag{1.9}$$

where the Hilbert spaces $H_a = L^2((-1,1)\times(-1,1))$, $H_b = L^2(-1,1)$ and the Hilbert spaces V_a and V_b are defined by

$$\begin{aligned} V_a &= \left\{\phi \in H^1((-1,1)\times(-1,1)) : \phi(x,1) = \phi(1,y) = 0, \forall\, x, y \in (-1,1)\right\}, \\ V_b &= \left\{u \in H^2(-1,1) : u(s) = u'(s) = 0, \quad \text{for} \quad s = -1, 1\right\}, \end{aligned}$$

with inner products defined by

$$\langle \phi, \tilde{\phi}\rangle_{V_a} = \int_\Omega \nabla\phi(x,y)\cdot\nabla\tilde{\phi}(x,y)dxdy, \quad \langle u, \tilde{u}\rangle_{V_b} = \int_{-1}^{1} Eu''(s)\tilde{u}''(s)ds.$$

A solution of the variational form of the model equations is a $\mathcal{H}$ valued function of time t, $\xi(t) = (\phi(t), \theta(t), u_1(t), v_1(t), u_2(t), v_2(t))$, which satisfies the equation

$$\tfrac{d}{dt}\langle \xi(t), \tilde{\xi}\rangle_{\mathcal{H}} = \sigma(\xi(t), \tilde{\xi}) \tag{1.10}$$

for all $\tilde{\xi} = (\tilde{\phi}, \tilde{\theta}, \tilde{u}_1, \tilde{v}_1, \tilde{u}_2, \tilde{v}_2) \in \mathcal{V}$ and where the Hilbert space $\mathcal{V}$ is defined as $\mathcal{V} = V_a \times V_a \times V_b \times V_b \times V_b \times V_b$ and the sesquilinear form σ on $\mathcal{V}\times\mathcal{V}$ is defined by

$$\begin{aligned} \sigma(\xi, txi) = \sigma_a(\theta, \tilde{\phi}) - \sigma_a(\phi, \tilde{\theta}) - \rho_1(v_1, \tilde{\theta}) - \rho_2(v_2, \tilde{\theta}) - \gamma_a(\theta, \tilde{\theta}) \\ +\sigma_b(v_1, \tilde{u}_1) - \sigma_b(u_1, \tilde{v}_1) + \rho_1(\tilde{v}_1, \theta) - \gamma_b(v_1, \tilde{v}_1) \\ +\sigma_b(v_2, \tilde{u}_2) - \sigma_b(u_2, \tilde{v}_2) + \rho_2(\tilde{v}_2, \theta) - \gamma_b(v_2, \tilde{v}_2). \end{aligned}$$

The above sesquilinear forms are defined by

$$\begin{aligned} &\sigma_a(\hat{\phi}, \tilde{\phi}) = \textstyle\int_\Omega \nabla\hat{\phi}(x,y)\cdot\nabla\tilde{\phi}(x,y)dxdy, \\ &\gamma_a(\hat{\phi}, \tilde{\phi}) = \textstyle\int_{-1}^{1} \alpha_1\hat{\phi}(x,-1)\tilde{\phi}(x,-1)dx + \int_{-1}^{1}\alpha_2\hat{\phi}(-1,y)\tilde{\phi}(-1,y)dy, \\ &\sigma_b(\hat{u}, \tilde{u}) = \textstyle\int_{-1}^{1} E\hat{u}''(s)\tilde{u}''(s)ds, \qquad \gamma_b(\hat{u}, \tilde{u}) = \int_{-1}^{1} C\hat{u}''(s)\tilde{u}''(s)ds, \\ &\rho_1(\tilde{u}, \tilde{\phi}) = \textstyle\int_{-1}^{1} \tilde{u}(x)\tilde{\phi}(x,-1)dx, \qquad \rho_2(\tilde{u}, \tilde{\phi}) = \int_{-1}^{1}\tilde{u}(y)\tilde{\phi}(-1,y)dy. \end{aligned} \tag{1.11}$$

We define an unbounded linear operator $\mathcal{A} : \mathcal{D}(\mathcal{A}) \subset \mathcal{H} \mapsto \mathcal{H}$ as follows:

$$\mathcal{D}(\mathcal{A}) = \left\{ \begin{array}{ll} \xi = (\phi, \theta, u_1, v_1, u_2, v_2) \in \mathcal{V} : \exists\hat{\xi} = (\hat{\phi}, \hat{\theta}, \hat{u}_1, \hat{v}_1, \hat{u}_2, \hat{v}_2) \in \mathcal{H}, & \text{such that} \\ \sigma(\xi, \tilde{\xi}) = \langle \hat{\xi}, \tilde{\xi}\rangle_{\mathcal{H}} \text{for all} \quad \tilde{\xi} = (\tilde{\phi}, \tilde{\theta}, \tilde{u}_1, \tilde{v}_1, \tilde{u}_2, \tilde{v}_2) \in \mathcal{V}. & \end{array} \right\},$$

and for all $\xi \in \mathcal{D}(\mathcal{A})$, $\mathcal{A}\xi = \hat{\xi}$ where $\hat{\xi} = (\hat{\phi}, \hat{\theta}, \hat{u}_1, \hat{v}_1, \hat{u}_2, \hat{v}_2)$ satisfies $\sigma(\xi, \tilde{\xi}) = \langle \hat{\xi}, \tilde{\xi}\rangle_{\mathcal{H}}$, for all $\tilde{\xi} = (\tilde{\phi}, \tilde{\theta}, \tilde{u}_1, \tilde{v}_1, \tilde{u}_2, \tilde{v}_2) \in \mathcal{V}$.

The main result of this section is the following theorem.

Theorem 2.1 *The operator $\mathcal{A}$ is the infinitesimal generator of a C_0-semigroup of contractions denoted by $e^{\mathcal{A}t}$ in $\mathcal{H}$ and for any $\xi_0 = (\phi_0, \theta_0, u_{1,0}, v_{1,0}, u_{2,0}, v_{2,0}) \in \mathcal{D}(\mathcal{A})$, the function $\xi(t) = e^{\mathcal{A}t}\xi_0$ is the unique weak solution of (1.10) with initial value ξ_0.*

The proof of the above theorem is based on the Lumer-Phillips theorem. The conditions in the Lumer-Phillips theorem are established through the following lemmas, (see [4]).

Lemma 2.1 *The operator $\mathcal{A}$ is dissipative in $\mathcal{H}$.*

Lemma 2.2 *For all real number $\lambda > 0$, the operator $\mathcal{A} - \lambda I : \mathcal{D}(\mathcal{A}) \subset \mathcal{H} \mapsto \mathcal{H}$ is onto.*

3 UNIFORM EXPONENTIAL STABILITY OF THE MODEL

In this section we outline the proof for existence of constants $M \geq 0$ and $\omega > 0$ such that

$$\|e^{\mathcal{A}t}\xi_0\|_{\mathcal{H}} \leq Me^{-\omega t}\|(\xi_0)\|_{\mathcal{H}}, \tag{1.12}$$

for all $\xi_0 = (\phi_0, \theta_0, u_{1,0}, v_{1,0}, u_{2,0}, v_{2,0}) \in \mathcal{H}$. Our approach is based on the use of a multiplier functional $Q(t)$ defined by

$$\begin{aligned} Q(t) &= \frac{t}{2}\|\xi(t)\|^2_{\mathcal{H}} + \int_{-1}^{1} u_1(t,x)v_1(t,x)dx + \int_{-1}^{1} u_2(t,y)v_2(t,y)dy \\ &+ \int_{\Omega} [2(x-1)\partial_x\phi(t,x,y) + 2(y-1)\partial_y\phi(t,x,y) + \phi(t,x,y)]\,\theta(t,x,y)dxdy. \end{aligned}$$

It is easy to see that there exists a constant T such that for all $t \geq T$,

$$\frac{t}{4}E(t) \leq Q(t) \leq tE(t),$$

where $E(t) = \|\xi(t)\|^2_{\mathcal{H}}$. The proof of the inequality (1.12) depends on showing that there exists a constant T' such that for all $t \geq T'$, $Q(t) \leq Q(T')$. Therefore, for all $t \geq \max(T, T')$, we have $\frac{t}{4}E(t) \leq Q(T') \leq T'E(T')$. As a result, we have $E(t) \leq \frac{4}{t}T'\|e^{\mathcal{A}T'}\|^2_{L(\mathcal{H})} \cdot \|\xi_0\|^2_{\mathcal{H}}$. Using the semigroup property of $e^{\mathcal{A}t}$, we obtain the exponential decay given in inequality (1.12). In order to show that Q is bounded, we shall show $Q'(t)$ is negative for all $t > T'$ for some constant T'. First, take $\xi_0 \in \mathcal{D}(\mathcal{A})$ and let $\xi(t) = e^{\mathcal{A}t}\xi_0$.

We consider the function $Q(t)$. Using the differentiability of the function $(\phi(\cdot), \theta(\cdot), u_1(\cdot), v_1(\cdot), u_2(\cdot), v_2(\cdot))$, we obtain

$$\begin{aligned} \frac{d}{dt}Q(t) &= \frac{1}{2}\|\xi(t)\|^2_{\mathcal{H}} + t\langle \mathcal{A}\xi(t), \xi(t)\rangle_{\mathcal{H}} \\ &+ \int_{-1}^{1} \partial_t u_1(t,x)v_1(t,x)dx + \int_{-1}^{1} u_1(t,x)\partial_t v_1(t,x)dx \\ &+ \int_{-1}^{1} \partial_t u_2(t,y)v_2(t,y)dy + \int_{-1}^{1} u_2(t,y)\partial_t v_2(t,y)dy \\ &+ \int_{\Omega} \partial_t\phi(t,x,y)\theta(t,x,y)dxdy + \int_{\Omega} \phi(t,x,y)\partial_t\theta(t,x,y)dxdy \\ &+ \int_{\Omega} [2(x-1)\partial_x\phi(t,x,y) + 2(y-1)\partial_y\phi(t,x,y)]\,\partial_t\theta(t,x,y)dxdy \\ &+ \int_{\Omega} \partial_t\,[2(x-1)\partial_x\phi(t,x,y) + 2(y-1)\partial_y\phi(t,x,y)]\,\theta(t,x,y)dxdy. \end{aligned}$$

From the definition of the operator $\mathcal{A}$, integration by parts and by rearranging the terms, we obtain

$$\begin{aligned}
\frac{d}{dt}Q(t) &= \frac{1}{2}\int_\Omega \nabla\phi(t,x,y)\cdot\nabla\phi(t,x,y)dxdy + \int_\Omega \phi(t,x,y)\Delta\phi(t,x,y)dxdy \\
&+ \int_\Omega \left[2(x-1)\partial_x\phi(t,x,y) + 2(y-1)\partial_y\phi(t,x,y)\right]\Delta\phi(t,x,y)dxdy \\
&+ \int_\Omega \left[2(x-1)\partial_x\theta(t,x,y) + 2(y-1)\partial_y\theta(t,x,y)\right]\theta(t,x,y)dxdy \\
&+ \frac{3}{2}\int_\Omega \theta^2(t,x,y)dxdy - t\int_{-1}^1 \alpha\theta^2(t,x,-1)dx - t\int_{-1}^1 \alpha\theta^2(t,-1,y)dy \\
&+ \frac{3}{2}\int_{-1}^1 v_1^2(t,x)dx + \frac{3}{2}\int_{-1}^1 v_2^2(t,y)dy \qquad (1.13) \\
&- \frac{1}{2}\int_{-1}^1 E(\partial_x^2 u_1(t,x))^2dx - \frac{1}{2}\int_{-1}^1 E(\partial_y^2 u_2(t,y))^2dy \\
&+ (-tC)\int_{-1}^1 (\partial_x^2 v_1(t,x))^2dx + (-tC)\int_{-1}^1 (\partial_y^2 v_2(t,y))^2dy \\
&- C\int_{-1}^1 \partial_x^2 v_1(t,x)\partial_x^2 u_1(t,x)dx - C\int_{-1}^1 \partial_y^2 v_2(t,y)\partial_y^2 u_2(t,y)dy \\
&+ \int_{-1}^1 \theta(t,x,-1)u_1(t,x)dx + \int_{-1}^1 \theta(t,-1,y)u_2(t,y)dy = \sum_{k=1}^{17} I_k,
\end{aligned}$$

where I_k is the k-th integral in the above equality. It is easy to see that there exists $T_1 > 0$ such that for all $t > T_1$,

$$\begin{aligned}
\sum_{k=8}^{15} I_k &\le -\frac{1}{4}\int_{-1}^1 E(\partial_x^2 u_1(t,x))^2dx - \frac{1}{4}\int_{-1}^1 E(\partial_y^2 u_2(t,y))^2dy \\
&\quad -\frac{t}{2}\int_{-1}^1 (\partial_x^2 v_1(t,x))^2dx - \frac{t}{2}\int_{-1}^1 (\partial_y^2 v_2(t,y))^2dy.
\end{aligned}$$

Using the divergence theorem, we can also show that there exists a constant T_2 such that for all $t > T_2$

$$I_4 + I_5 \le \frac{t\alpha}{2}\int_{-1}^1 \theta^2(t,x,-1)dx + \frac{t\alpha}{2}\int_{-1}^1 \theta^2(t,-1,y)dy.$$

On the other hand, for the terms I_{16} and I_{17}, it is easy to see that

$$\begin{aligned}
&\int_{-1}^1 \theta(t,x,-1)u_1(t,x)dx + \int_{-1}^1 \theta(t,-1,y)u_2(t,y)dy \\
&\le \epsilon\int_{-1}^1 u_1^2(t,x)dx + \epsilon\int_{-1}^1 u_2^2(t,y)dx \\
&\quad + \frac{1}{\epsilon}\int_{-1}^1 \theta^2(t,x,-1)dx + \frac{1}{\epsilon}\int_{-1}^1 \theta^2(t,-1,y)dy.
\end{aligned}$$

For the first 3 terms, we use a sequence of functions $\phi_n \in H^2(\Omega) \bigcap V_a$ that approximate ϕ and also have enough smoothness to justify application of the divergence theorem, (see [4]). Therefore, one can show there exists a constant T_3 such that for all $t > T_3$,

$$\begin{aligned} I_1 + I_2 + I_3 \quad &\leq \quad \frac{t\alpha}{2} \int_{-1}^{1} \theta^2(t, x, -1)dx + \frac{t\alpha}{2} \int_{-1}^{1} \theta^2(t, -1, y)dy \\ &+ \frac{t}{2} \int_{-1}^{1} (\partial_x^2 v_1(t, x))^2 dx + \frac{t}{2} \int_{-1}^{1} (\partial_y^2 v_2(t, y))^2 dy. \end{aligned} \tag{1.14}$$

By combining all these results, we get the desired uniform exponential stability.

4 GALERKIN APPROXIMATIONS

The Galerkin approximation of the variational form of the model equations requires the use of polynomial functions for the approximation of V_a. In particular, let H_a^N be a subspace of V_a given by

$$H_a^N = \left\{ \begin{array}{ll} \phi^N : \phi^N(x, y) \in \text{span}\{x^k y^j\}_{0 \leq k \leq N_x, 0 \leq j \leq N_y}, & \text{and} \\ \phi^N(x, 1) = \phi^N(1, y) = 0, \text{for all} \quad x, y \in (-1, 1) & \end{array} \right\}, \tag{1.15}$$

where N_x and N_y are positive integers. There are no special restriction on the selection of the subspaces H_b^N of V_b for the discussion of the exponential stability of solutions of the finite dimensional approximations of the model equations. A solution $\xi^N(t) = (\phi^N(t), \theta^N(t), u_1^N(t), v_1^N(t), u_2^N(t), v_2^N(t))$ to the Galerkin approximation on the subspace $\mathcal{H}^N$ defined by

$$\mathcal{H}^N = H_a^N \times H_a^N \times H_b^N \times H_b^N \times H_b^N \times H_b^N \tag{1.16}$$

is a $\mathcal{H}^N$ valued function of time t which satisfies the equation:

$$\tfrac{d}{dt} \langle \xi^N(t), \tilde{\xi}^N \rangle_{\mathcal{H}} = \sigma(\xi^N(t), \tilde{\xi}^N) \tag{1.17}$$

for all $\tilde{\xi}^N = (\tilde{\phi}^N, \tilde{\theta}^N, \tilde{u}_1^N, \tilde{v}_1^N, \tilde{u}_2^N, \tilde{v}_2^N) \in \mathcal{H}^N$. The equation above is a linear evolution equation in $\mathcal{H}^N$. As a consequence, there exists a semi-group $S^N(t)$ of linear operators on $\mathcal{H}^N$ such that for every given initial state vector $\xi_0^N = (\phi_0^N, \theta_0^N, u_{1,0}^N, v_{1,0}^N, u_{2,0}^N, v_{2,0}^N) \in \mathcal{H}^N$ the function defined by $\xi^N(t) = S^N(t)\xi_0^N$ is a solution of the dynamical equation in $\mathcal{H}^N$. The main result on the uniform exponential decay rate of the approximate solution semigroup $S^N(t)$ is given by the following theorem.

Theorem 4.1 *For the polynomial based Galerkin scheme there exist constants* $L \geq 1$ *and* $\beta > 0$ *such that*

$$||S^N(t)||_{L(\mathcal{H}^N)} \leq Le^{-\beta t}$$

for all $t \geq 0$ *and for all* N.

For the proof of this theorem, see [5].

For computational purposes, we represent the model equations in $\mathcal{H}^N$ in their coordinate vector form with respect to selected bases for H_a^N and H_b^N which are subspaces of polynomials. For this purpose, we let $\{w_i^n\}_{i=1}^n$ be a basis for H_b^N and let $\{\psi_k^m\}_{k=1}^m$ denote the 2-D polynomial basis functions which span H_a^N. By expanding the components of the approximate state vector ξ^N in terms of these basis functions, we obtain

$$\phi^N(t,x,y) = \sum_{k=1}^m \phi_k^N(t)\psi_k^m(x,y), \qquad \theta^N(t,x,y) = \sum_{k=1}^m \theta_k^N(t)\psi_k^m(x,y),$$

$$u_j^N(t,x) = \sum_{i=1}^n u_{j,i}^N(t)w_i^n(x), \qquad v_j^N(t,x) = \sum_{i=1}^n v_{j,i}^N(t)w_i^n(x), j=1,2$$

From equation (1.17), we can write the following first order approximate system

$$\mathcal{M}^N\dot{\bar{\xi}}^N(t) = \mathcal{A}^N\bar{\xi}^N(t), \tag{1.18}$$

where $\bar{\xi}^N(t) = (\bar{\phi}^N(t), \bar{\theta}^N(t), \bar{u}_1^N(t), \bar{v}_1^N(t), \bar{u}_2^N(t), \bar{v}_2^N(t)) \in R^N$, with $N = 2(m+2n)$ is the coordinate vector representation of the approximate state.

We represent the linear system (1.18) more explicitly as the following equation:

$$\begin{bmatrix} K_a^N & 0 & 0 & 0 & 0 & 0 \\ 0 & M_a^N & 0 & 0 & 0 & 0 \\ 0 & 0 & K_b^N & 0 & 0 & 0 \\ 0 & 0 & 0 & M_b^N & 0 & 0 \\ 0 & 0 & 0 & 0 & K_b^N & 0 \\ 0 & 0 & 0 & 0 & 0 & M_b^N \end{bmatrix} \frac{d}{dt} \begin{bmatrix} \bar{\phi}^N(t) \\ \bar{\theta}^N(t) \\ \bar{u}_1^N(t) \\ \bar{v}_1^N(t) \\ \bar{u}_2^N(t) \\ \bar{v}_2^N(t) \end{bmatrix} \tag{1.19}$$

$$= \begin{bmatrix} 0 & K_a^N & 0 & 0 & 0 & 0 \\ -K_a^N & -A_a^N & 0 & -A_{ab}^N & 0 & -A_{ab}^N \\ 0 & 0 & 0 & K_b^N & 0 & 0 \\ 0 & A_{ab}^{N\,T} & -K_b^N & -A_b^N & 0 & 0 \\ 0 & 0 & 0 & 0 & 0 & K_b^N \\ 0 & A_{ab}^{N\,T} & 0 & 0 & -K_b^N & -A_b^N \end{bmatrix} \begin{bmatrix} \bar{\phi}^N(t) \\ \bar{\theta}^N(t) \\ \bar{u}_1^N(t) \\ \bar{v}_1^N(t) \\ \bar{u}_2^N(t) \\ \bar{v}_2^N(t) \end{bmatrix}.$$

The component matrices can be computed from the sesquilinear forms given by (1.11):

$$[\ K_a^N\ ]_{k,\ell} = \sigma_a(\psi_k^m, \psi_\ell^m) \qquad [\ K_b^N\ ]_{i,j} = \sigma_b(w_i^n, w_j^n)$$

$$[\ M_a^N\ ]_{k,\ell} = \int_\Omega \psi_k^m \psi_\ell^m\, dxdy \qquad [\ M_b^N\ ]_{i,j} = \int_{-1}^1 w_i^n(s)w_j^n(s)\, ds$$

$$[\ A_a^N\ ]_{k,\ell} = \gamma_a(\psi_k^m, \psi_\ell^m) \quad [\ A_{ab}^N\ ]_{i,\ell} = \rho_1(w_i^n, \psi_\ell^m) \quad [\ A_b^N\ ]_{i,j} = \gamma_b(w_i^n, w_j^n) \tag{1.20}$$

5 NUMERICAL RESULTS

In this section, we demonstrate that the approximate solutions obtained from the Galerkin method described in the previous section have a uniform rate of decay. In carrying out this goal, our approach is to calculate the eigenvalues of the matrix

$$A^N = (\mathcal{M}^N)^{-1}\mathcal{A}^N,$$

for increasing values of N, and observe the trend in the location of the eigenvalues, and the margin of stability which is defined as the magnitude of the maximum of the real part of the eigenvalues. These computations are performed using MATLAB on a SPARC 10 workstation.
For the parameters, we chose $\alpha_1 = \alpha_2 = 0.1$, and E and C to be 1. We calculated the eigenvalues and the margin of stability using the *eig* routine in MATLAB. The results summarized in the table below, clearly indicate that a uniform margin of stability is preserved as the dimension of approximation is increased.

$N_a = N_b$	$\max\{Re\lambda, \lambda \in \sigma(A^N)\}$
10	-0.07786
14	-0.05948
18	-0.05372
22	-0.05174
26	-0.05092

References

[1] Avalos, G. The exponential stability of a coupled hyperbolic/parabolic system arising in structural acoustics. *Abstract Appl. Analysis*, to appear.

[2] Avalos, G. and Lasiecka, I. The strong stability of a semigroup arising from a coupled hyperbolic/parabolic system. *Semigroup Forum*, to appear.

[3] Banks, H. T., Fang, W., Silcox, R.J., and Smith, R. C. (1993). Approximation methods for control of structural acoustic models with piezoceramic actuators. *Journal of Intelligent Material Systems and Structures*, 4:98-116.

[4] Fahroo, F. and Wang, C. (1996) A new model for acoustic-structure interaction and its exponential stability. *The Applied Mathematics Quarterly*, to appear.

[5] Fahroo, F. and Wang, C. (1998). Convergence and uniform exponential stability of the approximations of acoustic-structure model equations. in preparation.

[6] Micu, S. and Zuazua, E. (1996). Stabilization and periodic solutions of a hybrid system arising in the control of noise. Proceedings of the IFIP TC7/WG-7.2 International Conference, Laredo, España, Lecture Notes in Pure and Applied Mathematics. Marcel Dekker, New York, 174:219-230.

[7] Micu, S. and Zuazua, E. (1998). On a weakly damped system arising in the control of noise. International Series of Numerical Mathematics, Birkhäuser Verlag, Basel. 126:207-222.

ANALYTICITY OF SEMIGROUP ASSOCIATED WITH A LAMINATED COMPOSITE BEAM

Scott Hansen*
Department of Mathematics
Iowa State University, Ames, IA 50011, USA
shansen@math.iastate.edu

Zhuangyi Liu
Department of Mathematics & Statistics
University of Minnesota, Duluth, MN 55812, USA.
zliu@d.umn.edu

Abstract: We consider a system of coupled partial differential equations that describe the vibrations of laminated beam in which the layers are bonded together by a medium that dissipates energy at a rate proportional to the shear. We show that for the simplest model, in which only transverse inertial energy is accounted for, the associated semigroup is analytic.

1 INTRODUCTION

In this paper we study the damping characteristics of a laminated beam model introduced (for plates) in Hansen [3]. The model is derived under the assumption that the laminated beam consists of $2n$ layers of Euler-Bernoulli beams bonded to one another by $2n - 1$ "adhesive layers" which resist shear, but otherwise have negligible physical characteristics. We suppose that damping is also included in the adhesive layers so that a force opposing shear, proportional to the rate of shear exists within the adhesive layers. No damping is included elsewhere in the model.

It was noticed in Hansen and Spies [5], that in the case of one adhesive layer, with constant coefficients and simply supported boundary conditions, the spectrum associated with the generator of the semigroup exhibits *frequency-proportional* damping characteristics (see Russell [16]). Many systems with this

*Research partially supported the National Science Foundation under grant DMS-9504822

type of damping have been shown to be associated with analytic semigroups (e.g., Chen and Russell [1], Huang [8], Chen and Triggiani [2], Liu and Renardy [11], Liu and Yong [13], Liu and Liu [9]-[10], Hansen and Zhang [6], Lasiecka and Triggiani [12]).

We prove that analyticity indeed holds for the case of symmetrically layered laminated beams with an arbitrary number of layers in the case of simply supported boundary conditions. The proof is based on a contradiction argument similar to the one used in [13] to establish analyticity of another model. Our choice of simply supported boundary conditions allows us to apply the contradiction argument with the least amount of technical difficulties. However the basic approach applies to certain other boundary conditions as well.

This paper is organized as follows. Background on the modeling and well-posedness are given in Section 2. The analyticity of the multilayer case with simply-supported boundary conditions is proved in Section 3.

2 LAMINATED BEAMS AND PLATES

The laminated beam model we consider are the beam-analogs of the laminated plate models described in Hansen [3] with viscous damping included in the adhesive layers. In the case of one adhesive layer this reduces to the beam model considered in Hansen and Spies [5].

We will consider the *symmetric case* in which the layers (both the beam layers and damping layers) of the laminated beam have material properties that are symmetric with respect to the centerline of the laminated beam. This allows for a decoupling of a set of "bending equations" that completely decouple from a set of "stretching equations". Without any symmetry assumptions, this decoupling does not exist and stronger damping would be needed to obtain energy decay due to longitudinal motions that are not damped.

2.1 Modeling Assumptions

In order to describe the meaning of the variables and physical constants it is necessary to describe briefly the modeling assumptions used in deriving the laminated beam model.

The laminated beam is assumed to consist of $2n$ beam layers and $2n-1$ adhesive layers that occupy the region $\Omega \times (-h/2, h/2)$ at equilibrium, where $\Omega = (0,1)$ for simplicity. Thus, since we are considering the symmetric case where all the material properties of the layers are symmetric with respect to the centerline of the laminated beam, this centerline must occur in an adhesive layer. (The case of an odd number of beam layers is similar.)

It is assumed in the modeling that (i) within each layer the longitudinal displacements vary linearly as a function of x_3, (ii) within each layer the transverse displacements are constant with respect to x_3, (iii) no slip is allowed on the interfaces, (iv) the Kirchhoff hypothesis (that normal sections remain normal after deformation) applies to the beam layers, (v) the adhesive layers resist shear but bending stresses and inertial forces are negligible, (vi) longitudinal kinetic evergy terms are assumed to be negligible.

To be more precise on assumption (v), if ϕ_i denotes the shear variable for the ith adhesive layer of thickness d_i, then $\phi_i = \delta_i/d_i + w_x$ where w_x is the partial of the transverse displacement w in the x_1 direction and δ_i is the change in longitudinal displacement of from the top to the bottom of the ith adhesive layer. (δ_i/d_i is also the rotation angle of the ith adhesive layer.) The strain energy associated with this ith layer is

$$\mathcal{P}_i(K_i, d_i) = \frac{1}{2}\int_\Omega E_i d_i^3 w_{xx}^2 + 12 E_i d_i (v_i)_x^2 + K_i d_i (\phi_i)^2 \, dx,$$

where K_i is the shear modulus of the ith adhesive layer, E_i is the Young's modulus of the ith adhesive layer, and v_i is the longitudinal displacement of this layer.

We formally take the limit as $K_i \to 0$ and $d_i \to 0$, but with

$$\gamma_i := \frac{K_i}{d_i} \quad \text{fixed.}$$

What remains of the energy of the ith adhesive layer is

$$\mathcal{P}_i = \frac{1}{2}\int_\Omega \gamma_i \delta_i^2 \, dx.$$

Due to the symmetry of motion, the energy in the ith layer (either adhesive or beam) above the centerline equal to that of the ith one below the centerline. This also applies to the center adhesive layer, provided we regard this layer as two consecutive adhesive layers with half the original thickness. This convention does not introduce additional degrees of freedom since we are restricting our interest to bending solutions, which are antisymmetric with respect to the centerline of the beam.

Define the following n by n matrices

$$\begin{aligned} \mathbf{g} &= \operatorname{diag}(\gamma_1, \gamma_2, \ldots, \gamma_n) & \mathbf{p} &= \operatorname{diag}(\rho_1, \rho_2, \ldots, \rho_n) \\ \mathbf{h} &= \operatorname{diag}(h_1, h_2, \ldots, h_n) & \mathbf{D} &= \operatorname{diag}(D_1, D_2, \ldots, D_n) \end{aligned}$$

where $\{\gamma_i\}$ are the previously described elastic parameters, h_i is the thickness of the i-th beam layer, $D_i h_i^3$ is the *modulus of flexural rigidity* for the ith beam layer, ρ_i is the density of the ith beam layer.

If M and N are matrices in $\mathbf{R}^{mn}$, by $M \cdot N$ we mean the natural scalar product in $\mathbf{R}^{mn}$. We denote $(\theta, \xi)_\Omega = \int_\Omega \theta \cdot \xi \, dx_1$.

Also define M to be the $n \times n$ matrix with 1's on and below the main diagonal and 0's elsewhere and let $\vec{1}$ denote the column vector consisting of n 1's. We can express v is terms of ψ and δ as

$$v = M\delta - (M - I/2)\mathbf{h}\vec{1} w_x =: M\delta - \mathbf{c} w_x. \tag{2.1}$$

The energy we use is $\mathcal{E}(t) = c(\dot{w}) + a(w, \delta)$ where

$$\begin{aligned} c(w; w) &= (mw, w)_\Omega \\ a(w, \delta; w, \delta) &= (D_{tot} w_{xx}, w_{xx})_\Omega + 12(\mathbf{Dh} v_x, v_x)_\Omega + (\mathbf{g}\delta, \delta)_\Omega . \end{aligned} \tag{2.2}$$

where

$$m = \sum_{i=1}^{n} \rho_i h_i, \qquad D_{tot} = \sum_{i=1}^{n} D_i h_i^3$$

and v is defined as in (2.1). Hamilton's principle can now be used to derive a conservative system of equations. (The system one obtains is the same as equations (23)–(24) of [3], but without the rotational inertia terms in (23).) Once this is done, damping can be included by the correspondence principle. Since we wish to include shear damping in the adhesive layers, to each γ_i is a corresponding damping parameter β_i, and hence corresponding to the diagonal stiffness matrix $\mathbf{g}$ is a corresponding viscosity matrix $\mathbf{b}$ =diag$(\beta_1, \beta_2, \ldots, \beta_n)$ that we include in the system of equations via the correspondence $\mathbf{g}\delta \to \mathbf{g}\delta + \mathbf{b}\delta_t$.

The final system of equations obtained is the following.

$$m\ddot{w} + (D_{tot} w_{xx})_{xx} - 12\mathbf{c}^T(\mathbf{Dh} v_{xx})_x = 0 \qquad x \in \Omega,\ t > 0 \tag{2.3}$$

$$-12M^T(\mathbf{Dh} v_x)_x + \mathbf{g}\delta + \mathbf{b}\dot{\delta} = 0 \qquad x \in \Omega,\ t > 0 \tag{2.4}$$

where v is defined in (2.1).

The simply supported boundary conditions are

$$w = 0, \quad w_{xx} = 0, \quad \delta_x = 0, \qquad x = 0, 1. \tag{2.5}$$

2.2 *Well-posedness*

We let

$$\mathcal{V} = (H_0^1 \cap H^2)(\Omega) \times (H^1(\Omega))^n, \qquad \mathcal{H} = L^2(\Omega)$$

with norms defined by

$$||\{w, \delta\}||_{\mathcal{V}}^2 = a(w, \delta; w, \delta), \quad ||y||_{\mathcal{H}}^2 = c(y; y).$$

The forms $a(\cdot;\cdot)$ and $c(\cdot;\cdot)$ can be shown to be continuous, symmetric, bilinear and positive definite on $\mathcal{V} \times \mathcal{V}$ and $\mathcal{H} \times \mathcal{H}$ respectively. (See [4] for proof of a similar situation.) Therefore these forms extend to complex inner products on $\mathcal{V}$ and $\mathcal{H}$ respectively. We define the *energy space* $\mathcal{E}$ by

$$\mathcal{E} = \{(w, y, \delta) : (w, \delta) \in \mathcal{V},\ y \in \mathcal{H}\}$$

with the inner product $< \cdot, \cdot >_{\mathcal{E}}$ that corresponds to the forms $a(\cdot;\cdot)$ and $c(\cdot;\cdot)$.

We let $y = \dot{w}$ and rewrite the equations of motion (2.3)–(2.4) as

$$\frac{d}{dt}\begin{pmatrix} w \\ y \\ \delta \end{pmatrix} = \mathcal{A}\begin{pmatrix} w \\ y \\ \delta \end{pmatrix} := \begin{pmatrix} y \\ \frac{1}{m}(-(D_{tot} w_{xx})_{xx} + 12(\mathbf{c}^T \mathbf{h} \mathbf{D} v_x)_{xx}) \\ \mathbf{b}^{-1}(12M^T \mathbf{h}(\mathbf{D} v_x)_x - \mathbf{g}\delta) \end{pmatrix} \tag{2.6}$$

For the moment, let use $\mathcal{A}(\mathbf{b})$ to denote the dependence of $\mathcal{A}$ on the damping vector $\mathbf{b}$. A calculation reveals that the adjoint $\mathcal{A}^*$ of $\mathcal{A}$ is given by

$$\mathcal{A}^* = -\mathcal{A}(-\mathbf{b}) \tag{2.7}$$

and that for $(w, y, \delta) \in \mathcal{D}(\mathcal{A})$, (for brevity, we omit a precise description of $\mathcal{D}(\mathcal{A})$ here)

$$\text{Re} < \mathcal{A}\begin{pmatrix} w \\ y \\ \delta \end{pmatrix}, \begin{pmatrix} w \\ y \\ \delta \end{pmatrix} >= \text{Re} < \mathcal{A}^*\begin{pmatrix} w \\ y \\ \delta \end{pmatrix}, \begin{pmatrix} w \\ y \\ \delta \end{pmatrix} >= -< \mathbf{b}^{-1}\xi, \xi >_{L^2},$$

where $\xi = M^T\mathbf{h}(\mathbf{D}v_x)_x - g\delta$.

It follows that both $\mathcal{A}$ and $\mathcal{A}^*$ are dissipative. Since A is easily seen to be densely defined and closed, by a corollary to the Lumer-Phillips theorem ([14], p.15), we have the following.

Proposition 2.1 *The operator $\mathcal{A}$ is the generator of a C_0 semigroup of contractions on $\mathcal{V} \times \mathcal{H}$. Consequently, given any initial data $\{w_0, \delta_0\} \in \mathcal{V}$ and $w_1 \in \mathcal{H}$, there is a unique solution $\{w, \delta\}$ for the equations of motion with*

$$\{w, \delta\} \in C([0, \infty); \mathcal{V}), \quad w_t \in C([0, \infty); \mathcal{H})$$

3 DECAY RESULTS

We show that $\mathcal{A}$ is the generator of an analytic semigroup.

Proposition 3.1 *Assume that each b_i, $i = 1, 2, \ldots n$ is positive. Then the operator $\mathcal{A}$ is the generator of an analytic semigroup.*

Proof: We employ a contradiction argument for the analyticity of a C_0 semigroup of contractions proposed in [13].

From Pazy [14], we have that $\{e^{t\mathcal{A}}\}$ is analytic if and only if $||\lambda(A - \lambda I)^{-1}||$ is uniformly bounded for all λ on some vertical line. We will show that this in fact holds on the imaginary axis. Equivalently, we will show that there exists $\epsilon > 0$ such that

$$\inf_{||z||=1} ||(\mathcal{A} - isI)z|| \geq \epsilon|is| \qquad \forall s \in \mathbf{R}.$$

If, on the contrary, this condition does not hold then there exists $(s_n) \in \mathbf{R}$ and $z_n = \{w_n, \dot{w}_n, \delta_n\}$ with $||z_n||_{\mathcal{E}} = 1$ such that

$$\lim_{n\to\infty} ||iz_n - \frac{1}{s_n}\mathcal{A}z_n||_{\mathcal{E}} = 0. \tag{3.8}$$

Due to the fact that the norm of the resolvent is symmetric with respect to the real axis we can assume that each s_n is positive. Furthermore, one can directly show (using the dissipativity of $\mathcal{A}$ and compactness of the resolvent) that $\mathcal{A}$ has no spectrum on the imaginary axis. Thus we may assume that $s_n \to +\infty$.

Let us use the notation $\phi' = \frac{d}{dx}\phi$, $\phi'' = \frac{d^2}{dx^2}\phi$ and so forth. Explicitly writing out (3.8) gives

$$i\begin{pmatrix} w_n \\ y_n \\ \delta_n \end{pmatrix} - s_n^{-1}\begin{pmatrix} y_n \\ m^{-1}[-(D_{tot}w_n'')'' + 12(\mathbf{c}^T\mathbf{h}\mathbf{D}v_n')''] \\ \mathbf{b}^{-1}[-12M^T\mathbf{h}(\mathbf{D}v_n')' + \mathbf{g}\delta] \end{pmatrix} =: \begin{pmatrix} \xi_1(n) \\ \xi_2(n) \\ \xi_3(n) \end{pmatrix} \longrightarrow 0 \tag{3.9}$$

as $n \to \infty$. The convergence is in the energy space, which implies

$$\xi_2 \to 0 \qquad \text{in } L^2(\Omega) \tag{3.10}$$
$$\xi_1'' \to 0 \qquad \text{in } L^2(\Omega) \tag{3.11}$$
$$\xi_3 \to 0 \qquad \text{in } L^2(\Omega) \tag{3.12}$$
$$(M\xi_3 - \mathbf{c}\xi_1')' \to 0 \qquad \text{in } L^2(\Omega) \tag{3.13}$$

On the other hand, from (3.8) we have

$$\operatorname{Re}\langle (i - s_n^{-1}\mathcal{A})z_n, z_n\rangle_{\mathcal{E}} \to 0 \qquad \text{as } n \to \infty$$

By a direct calculation one finds that

$$\begin{aligned} \operatorname{Re}\langle (i - s_n^{-1}\mathcal{A})z_n, z_n\rangle_{\mathcal{E}} &= -s_n^{-1}\operatorname{Re}\langle \mathcal{A}z_n, z_n\rangle_{\mathcal{E}} \\ &= -s_n^{-1}\|\mathbf{b}^{-1/2}(-12M^T\mathbf{h}(\mathbf{D}v_n')' + \mathbf{g}\delta_n)\|_{L^2}^2, \end{aligned}$$

and hence since $\mathbf{b}$ is invertible we obtain

$$s_n^{-1}\| - 12M^T\mathbf{h}(\mathbf{D}v_n')' + \mathbf{g}\delta_n\|_{L^2}^2 \to 0. \tag{3.14}$$

Substituting this into (3.12) gives

$$\|\delta_n\|_{L^2} \to 0 \tag{3.15}$$

and since also $\|\mathbf{g}\delta\|_{L^2} \to 0$, (3.14) also implies (using the invertibility of M, $\mathbf{h}$) that

$$s_n^{-1}\|(\mathbf{D}v_n')'\|_{L^2}^2 \to 0. \tag{3.16}$$

Since $\|\xi_1''\|_{L^2} \to 0$, (3.13) is equivalent to $\xi_3' \to 0$ in L^2, or, since (3.12) already holds,

$$i\mathbf{b}\delta_n - s_n^{-1}(-12M^T\mathbf{h}(\mathbf{D}v_n')' + \mathbf{g}\delta_n) \to 0 \ \text{ in } H^1(\Omega). \tag{3.17}$$

Note that (w_n'') is bounded in L^2 and therefore by (3.11)

$$(s_n^{-1}y_n'') \text{ bounded in } L^2. \tag{3.18}$$

Since (y_n) is also bounded, by interpolation (or integration by parts and Schwartz inequality) we have

$$(s_n^{-1/2}y_n') \qquad \text{bounded in } L^2(\Omega) \tag{3.19}$$

We take the inner product of (3.13) with $\mathbf{D}v_n'$ in $L^2(\Omega)$ to obtain

$$i < \mathbf{D}v_n', \mathbf{D}v_n' > + < M\mathbf{b}^{-1}[-12M^T\mathbf{h}(\mathbf{D}v_n')' + \mathbf{g}\delta] - \mathbf{c}y_n', s_n^{-1}(\mathbf{D}v_n')' > \quad (3.20)$$
$$= i < \mathbf{D}v_n', \mathbf{D}v_n' > + < M\mathbf{b}^{-1}[-12M^T\mathbf{h}(\mathbf{D}v_n')'], s_n^{-1}(\mathbf{D}v_n')' > \quad (3.21)$$
$$+ < M\mathbf{b}^{-1}\mathbf{g}\delta_n, s_n^{-1}(\mathbf{D}v_n')' > - < s_n^{-1/2}\mathbf{c}y_n', s_n^{-1/2}(\mathbf{D}v_n')' > \quad (3.22)$$

where the last three terms go to zero by (3.15),(3.16) and (3.19). Therefore

$$||v_n'||_{L^2} \to 0. \quad (3.23)$$

Now substitute (3.23) in (3.13) to obtain

$$s_n^{-1}(M\mathbf{b}^{-1}[-12M^T\mathbf{h}(\mathbf{D}v_n')' + \mathbf{g}\delta] - \mathbf{c}y_n')' \to 0. \quad (3.24)$$

Next we take the inner product of (3.24) with $\vec{1}y_n$, using the facts that by (3.15), (3.16), (3.19)

$$< s_n^{-1/2}(Dv_n')', s_n^{-1/2}\vec{1}y_n' > \to 0, \qquad < \delta_n, s_n^{-1}\vec{1}y_n' > \to 0 \quad (3.25)$$

to obtain that $s_n^{-1}||\mathbf{c}y_n'||_{L^2}^2 \to 0$ and hence

$$s_n^{-1/2}||y_n'||_{L^2} \to 0. \quad (3.26)$$

Take the inner product of (3.10) with my_n to obtain

$$im||y_n||^2 + s_n^{-1}12 < \mathbf{c}^T\mathbf{h}(\mathbf{D}v_n')', y_n' > + s_n^{-1} < D_{tot}w_n'', y_n'' > \to 0. \quad (3.27)$$

The second term goes to zero due to (3.16), (3.26) and the third term goes to zero due to (3.18) and the boundedness of $(D_{tot}w'')$ in L^2. Hence

$$||y_n||_{L^2}^2 \to 0. \quad (3.28)$$

Now calculate the inner product of (3.11) with $D_{tot}w_n''$ and subtract from this the inner product of (3.10) with y_n. After some integrations by parts we obtain

$$i < D_{tot}w_n'', w_n'' > -i < my_n, y_n > -s_n^{-1} < 12(\mathbf{c}^T\mathbf{h}\mathbf{D}v_n')', y_n' > \to 0$$

where the last term goes to zero by (3.16) and (3.26). It follows from (3.28) that

$$||w_n''||_{L^2} \to 0. \quad (3.29)$$

Combining (3.29) with (3.15),(3.23), and (3.28) we have that $||z_n||_{\mathcal{E}} \to 0$ as $n \to \infty$, which is a contradiction. □

In the preceeding proof we have shown that $||\lambda(\mathcal{A} - \lambda I)^{-1}||$ is uniformly bounded on the imaginary axis. Hence, since zero is in the resolvent set, we also have that $||(\mathcal{A} - \lambda I)^{-1}||$ is bounded on the imaginary axis. Therefore, using [7], as a corollary of Proposition 3.1 we have the following result.

Corollary 3.1 *The semigroup $e^{t\mathcal{A}}$ of Proposition 3.1 is exponential stable, i.e., there exists $\epsilon > 0$ and $M > 1$ such that*

$$||e^{t\mathcal{A}}z||_{\mathcal{E}} \leq Me^{-\epsilon t}||z||_{\mathcal{E}}.$$

References

[1] Chen, G. and Russell, D. L. (1981/1982). A mathematical model for linear elastic systems with structural damping. *Quart. Appl. Math.*, 39:433-454.

[2] Chen, S. and Triggiani, R. (1989). Proof of extensions of two conjectures on structural damping for elastic systems, the case $\frac{1}{2} \leq \alpha \leq 1$. *Pacific J. Math.*, 39:15-55.

[3] Hansen, S. (1998). Modeling and analysis of multilayer laminated plates. to appear: *ESAIM* Proc. "Control and Partial Differential" Conference at CIRM, Marseille-Luminy, 16-20 jun 1997.

[4] Hansen, S. (1997). A dynamical model for multilayered plates with independent shear deformations *Quart. Appl. Math.* LV:601-621

[5] Hansen, S. and Spies, R. (1997) Structural damping in laminated beams due to interfacial slip. *J. Sound Vib.*, 204:183-202

[6] Hansen, S. and Zhang, B-Y. (1997) Boundary control of a linear thermoelastic beam. *JMAA*, 210:182-205

[7] Huang, F. (1985) Characteristic conditions for the exponential stability of linear dynamical systems in Hilbert spaces. *Ann. of Diff. Eqs.*, 1:43-56

[8] Huang, F. (1988). On the mathematical medel for linear elastic systems with analytic damping. *SIAM J. Control and Optimization*, 26:714-724.

[9] Liu, K. and Liu, Z. (1997). Exponential Stability and Analyticity of the abstract thermoelastic evolution equation. *Z. agnew Math. Phys*, 48:885-904.

[10] Liu, K. and Liu, Z. (1997). Analyticity and differentiability of semigroups associated with elastic systems with damping and gyroscopic forces. *J. Diff. Eqns.*, Vol. 141, 2:340-355.

[11] Liu, Z. and Renardy, M. (1995). A note on the equations of a thermoelastic plate. *Appl. Math. Lett.*, Vol. 8, 3:1-6.

[12] Lasiecka, I. and Triggiani, R. (1998). Two direct proofs on the analyticity of the S.C. semigroup arising in abstract thermo-elastic equation. *Adv. Diff. Eqns.*, to appear.

[13] Liu, Z. and Yong, J. (1998) Qualitative properties of certain C_0 semigroups arising in elastic systems with various dampings. *Adv. Diff. Eqns.*, to appear.

[14] Pazy, A. (1983). *Semigroups of Linear Operators and Applications to Partial Differential Equations.* Springer, New York.

[15] Russell, D. L. (1991). A comparison of certain elastic dissipation mechanisms via decoupling and projection techniques. *Quarterly of Appl. Math.*, Vol. XLIX, 2:373-396.

[16] Russell, D. L. (1992) On mathematical models for the elastic beam with frequency- proportional damping. *Control and Estimation of Distributed Parameter Systems*, H. T. Banks ed., SIAM, Philadelphia.

A PRACTICAL ESTIMATION TECHNIQUE FOR SPATIAL DISTRIBUTION OF GROUNDWATER CONTAMINANT

Sungkwon Kang
Department of Mathematics
Chosun University, Kwangju 501-759, Korea and
Institute for Scientific Computation
Texas A&M University, College Station, TX 77843-3404, U.S.A.
skang@chonnam.chonnam.ac.kr

Thomas B. Stauffer
AFRL/MLQR, Tyndall AFB, FL 32403, U.S.A.
tom_stuffer@ccmail.aleq.tyndall.af.mil

Kirk Hatfield
Department of Civil Engineering
University of Florida, Gainsville, FL 32611, U.S.A.
khatf@ce.ufl.edu

Abstract: To predict the fate of groundwater contaminants, accurate spatially continuous information is needed. Because most field sampling of groundwater contaminants are not conducted spatially continuous manner, a special estimation technique is required to interpolate/extrapolate concentration distributions at unmeasured locations. A practical three-dimensional estimation method for *in situ* groundwater contaminant concentrations is introduced.

1 INTRODUCTION

Groundwater contamination is an important environmental issue. Researchers have conducted extensive field experiments to analyze geophysical, chemical and biological processes that control the fate and movement of groundwater contaminants [1,3,4,9,11,12]. In order to describe and predict underlying physical, chemical and biological processes affecting chemical fate and transport, accurate spatially continuous information is needed. Because most field sampling of groundwater contaminants are not conducted spatially continuous manner, a special estimation technique is required to interpolate/extrapolate contaminant concentration distributions at unmeasured locations. These interpolations/extrapolations are complicated by uncertainties often associated with an unknown distribution of contaminant fluxes in space and time reflected in a complex velocity field within a heterogeneous aquifer.

Many geostatistical techniques have been developed for estimating geophysical/chemical parameters and groundwater contaminant concentrations (see, for example, [2,5,6,7,10] and references there in). But, application of these estimation methods to groundwater contaminant data often fail to obtain satisfactory results because methods are based on the *geostatistical intrinsic assumptions* [10], and because field data behave irratically or contain "outliers." The geostatistical intrinsic hypothesis (or stationary assumptions) is that spatial correlationship between data points depend only on the separation vector (modulus and direction) and not on the individual sample location. But, the global behavior of contaminant plume follows dynamical process governed by groundwater flow; consequently, the concentration of contaminant strongly depends on sample location. Questions concerning the locations or regions of high pollutant concentrations and contaminant plume dimensions are key issues in enviromental concerns. The first obstacle, the intrinsic geostatistical hypothesis, can be overcome by extracting the *plume macroscopic behavior* from the field data. The concept of macroscopic plume behavior is essentially similar to that of *drift* or *trend* in geostatistics in the sense of nonstationarity. In geostatistics, the general profile of most regionalized variables is assumed to be stationary, and, hence, the slowly varying minor nonstationary components (drift or trend) observed in the field data may be approximated by lower order polynomials. However, the greater portion of a goundwater contaminant plume exhibits nonstationary characteristics. Thus, a major component of the plume should be estimated from dynamical processes and measured in a large region. On the other hand,=0Dthe drift should be estimated in a "small neighborhood" of the point where kriging to be performed. Additional problems arise when concentrations are estimated. For example, conventional semivariograms are too sensitive to obtain correct spatial correlations for data exhibiting a wide variance. These correlations are needed for determining variogram models in a kriging procedure. The log transformation commonly used to compress data variance contains a logical conflict between original data structure and application of kriging algorithm. Consequently, to make spatial interpolations of data exhibiting a large variance, there is a need to develop a new robust estimator. This paper introduces a new robust estimator which is consistent and robust.

The general procedure of our proposed estimation method is following:

(1) Divide the field site into several subregions based on all available information. (2) In each subregion, macroscopic plume behaviors (or deterministic transport components) are estimated from the field data. These estimated values are subtracted from the field data to obtain residuals. (3) Based on complexities of spatial distribution of the residuals, divide each subregion into

several small blocks. (4) In each block, calculate experimental variograms using a robust estimator and determine mathematical variogram models. (5) Perform kriging to estimate residual at each desired location. (6) Finally, combine kriged residual values with the estimated macroscopic transport components.

The purpose of this paper are to provide a systematic methodology for estimating *in situ* groundwater contaminant concentrations, to introduce the $\mathcal{R}_p$-*estimator* for producing correlations between data points characterized by large variance, and to address some of mathematical problems related to estimation of goundwater contaminants. The method can be used generally to estimate space and time dependent geophysical, chemical and biological parameters; thus it may be useful to those developing numerical models for capturing the main feature of the groundwater contaminant distributions.

2 ESTIMATION OF MACROSCOPIC PLUME BEHAVIOR

As the first step for estimating the global plume behavior represented in the field data, the field site is divided into several subregions. Adjacent subregions may be overlapped to obtain spatially continuous information. The size of each subregion strongly depends on the geological structure of the field site and the global characteristics of data distributions. Each region showing distinctive distribution behavior is contained in a separated subregion. All available field data are visualized and analyzed. Also, any information related to the field site such as geological aquifer history are incorporated.

The macroscopic behavior is a spatially continuous large scale behavior describing the main profile of the plume movement. This step is difficult because the parameters in the transport equations, such as, dispersion/diffusion and seepage velocity share a highly nonlinear interdependence in space and time. Also, the measured contaminant concentration themselves add uncertainties due to spatial variabilities and unequal analytical confidence intervals. The criteria on how to choose basis functions and some specific approximation functions to estimate the macroscopic plume behavior are proposed. The approximating functions are chosen based on the global characteristics of the solute transport process in porous media. Nonlinear optimization techniques are needed for estimating parameters appeared in the approximating basis functions.

The basis functions are chosen using the following criteria:

(1) "Simple" functions are preferred because they are easily evaluated. At the same time, a "low order" approximation should capture the main profile of the plume movement. Here, the order of the approximation refers to the number of basis functions needed for the approximation.

(2) The approximation functions must capture the global plume behavior outside sampling network. In many practical applications, sampling networks often don't cover the entire extent of the contaminant plume; however, extrapolation of plume movement outside the sampling network is often desired.

(3) Basis functions should be "robust" in the sense that they are not very sensitive to unevenly spaced data points.

According to the selection criteria described above, specific basis functions are chosen based on the available field data and the process to be described (e.g., transport of a tracer or the spatial distribution of aquifer permeability). Focusing on the problem at hand, advective, dispersive/diffusive solute transport in porous media depends on space, time, and solute concentration. If, in addition, recharge, chemical, biological, and other reactive processes are considered, then the solute transport may be approximated [8], in Cartesian coordinates, using

$$\frac{\partial C}{\partial t} = \nabla \cdot (D \cdot \nabla C) - \nabla \cdot (VC) + f \tag{2.1}$$

with appropriate initial and boundary conditions, where $C = C(x, y, z; t)$ is the concentration of the solute, i.e., the mass of solute per unit volume of fluid, $D = D(x, y, z; t; C)$ is the dispersion tensor, $V = V(x, y, z; t; C)$ is the pore water velocity vector, $f = f(x, y, z; t; C)$ is a "forcing function" related to recharge, chemical, and biological activities. Note that the coefficients D and V depend on space, time and concentration itself.

As a simple case, assume that the porous medium is homogeneous, isotropic, saturated, the flow is steady-state, and that there is no external source. Then the transport equation (2.1) can be simplified as

$$\frac{\partial C}{\partial t} = \left[D_x \frac{\partial^2 C}{\partial x^2} + D_y \frac{\partial^2 C}{\partial y^2} + D_z \frac{\partial^2 C}{\partial z^2}\right] - \left[\bar{v}_x \frac{\partial C}{\partial x} + \bar{v}_y \frac{\partial C}{\partial y} + \bar{v}_z \frac{\partial C}{\partial z}\right], \tag{2.2}$$

where D_x, D_y and D_z are dispersion coefficients in the x, y and z-directions, $\bar{v}_x$, $\bar{v}_y$ and $\bar{v}_z$ are the average linear pore water velocities in each coordinate direction defined by $\bar{v}_x = v_x/\phi$, $\bar{v}_y = v_y/\phi$, $\bar{v}_z = v_z/\phi$, in which v_x, v_y, and v_z are specific discharge components, and ϕ is the porosity of the medium. If a contaminant is released instantaneously at the origin $(x, y, z) = (0, 0, 0)$, the mass distribution of the contaminant at time t is given by

$$C(x, y, z; t) = \frac{M}{8(\pi t)^{3/2}\phi\sqrt{D_x D_y D_z}} \exp\left(-\frac{\bar{X}^2}{4D_x t} - \frac{\bar{Y}^2}{4D_y t} - \frac{\bar{Z}^2}{4D_z t}\right), \tag{2.3}$$

where M is the mass of contaminant introduced at the point source, $\bar{X} = x - \bar{v}_x t$, $\bar{Y} = y - \bar{v}_y t$ and $\bar{Z} = z - \bar{v}_z t$ [8]. The averaged pore water velocities $\bar{v}_x$, $\bar{v}_y$ and $\bar{v}_z$ contribute movement of the center of mass of the contaminant plume (the propagation process). D_x, D_y, and D_z contribute to the longitudinal and transverse spreading of the plume around the plume centroid (the dispersion/diffusion process). The solution (2.3) of the ideal equation (2.2) is a simple representation of these two processes throughout two parameter sets $V = (\bar{v}_x, \bar{v}_y, \bar{v}_z)$ and $D = (D_x, D_y, D_z)$.

To approximate the gross distribution of contaminant concentrations in space, we propose the following linear combination of exponential functions.

$$F(x, y, z; a, b, c) = \sum_{i=1}^{m} c_i \exp\left(-\left(\frac{x - a_i^x}{b_i^x}\right)^2 - \left(\frac{y - a_i^y}{b_i^y}\right)^2 - \left(\frac{z - a_i^z}{b_i^z}\right)^2\right), \tag{2.4}$$

where m is the number of basis functions, a_i^x, a_i^y, a_i^z, b_i^x, b_i^y, b_i^z, and c_i, $1 \leq i \leq m$, are parameters to be determined. In the same context as transport equations (2.1) and (2.2), the parameter set $a = \{(a_i^x, a_i^y, a_i^z) : 1 \leq i \leq m\}$ represents the propagation or advection process, the set $b = \{(b_i^x, b_i^y, b_i^z) : 1 \leq i \leq m\}$ represents the dispersion/diffusion process, and the set $c = \{c_i : 1 \leq i \leq m\}$ is related to the magnitude of the source load.

3 $\mathcal{R}_P$-ESTIMATOR

The residuals are obtained by subtracting the macroscopic transport portion from field data. The experimental semivariogram or variogram is used to describe the pattern of spatial correlation displayed by the residuals. A mathematical model is fitted to this experimental variogram, and this model is used in kriging to estimate the residuals at unmeasured locations. Some of mathematical models commonly used in practice can be found in [10]. In this section, we introduce the "$\mathcal{R}_p$*-estimator*," where $\mathcal{R}$ stands for "robust" and $p > 0$ indicates the order of robustness. For $0 < p \leq 1$, the estimator is robust; whereas for $p > 1$ the estimator becomes sensitive to apparent outliers. The estimator satisfies the following properties:

(1) It is consistent such that spatial correlations among the original data are preserved under linear transformation. Thus, the original data structure, estimation of correlations, and kriging procedures are consistent. (2) It is robust. It reduces outlier effects on estimated correlations between data points. (3) It is systematic. Depending on the distribution of the data, the order p of robustness can be adjusted systematically.

Let $Z(\mathbf{x})$ be a regionalized function on a domain Ω in three dimiensional space, and $Z(\mathbf{x}_i)$ be the realization of the function $Z(\mathbf{x})$ at $\mathbf{x}_i = (x_i, y_i, z_i) \in \Omega$, $i = 1, 2, \cdots, n$. Let $p > 0$ be a positive real number. For any vector $\mathbf{h} = (h_x, h_y, h_z)$, we define the $\mathcal{R}_p$*-estimator* as

$$\mathcal{R}_p(\mathbf{h}) = \left[\frac{1}{n(\mathbf{h})} \sum_{i=1}^{n(\mathbf{h})} |Z(\mathbf{x}_i) - Z(\mathbf{x}_i + \mathbf{h})|^p \right]^{\frac{1}{p}}, \tag{3.1}$$

where $n(\mathbf{h})$ is the number of data pairs separated by the vector $\mathbf{h}$.

For any positive integer k, and for any positive real number $p > 0$, the following inequalities hold:

$$\begin{aligned} a_1^p + a_2^p + \cdots + a_k^p &\leq (a_1 + a_2 + \cdots + a_k)^p, \quad p \geq 1, \\ a_1^p + a_2^p + \cdots + a_k^p &\geq (a_1 + a_2 + \cdots + a_k)^p, \quad 0 < p \leq 1, \end{aligned} \tag{3.2}$$

where $a_i \geq 0$, $i = 1, 2, \cdots, k$. Thus, the function

$$(a_1, a_2, \cdots, a_k) \mapsto \frac{a_1^p + a_2^p + \cdots + a_k^p}{k} \tag{3.3}$$

is a convex function for $p \geq 1$ and a concave function for $0 < p \leq 1$. As $p > 0$ approaches 0, robust effects are increased and the estimator $\mathcal{R}_p$ defined by equation (3.1) reduces effects of outliers for $0 < p \leq 1$.

Moreover, it is easy to see that, for any positive constant $c > 0$,

$$\left[\frac{1}{n(\mathbf{h})}\sum_{i=1}^{n(\mathbf{h})}|cZ(\mathbf{x}_i) - cZ(\mathbf{x}_i+\mathbf{h})|^p\right]^{\frac{1}{p}} = c\left[\frac{1}{n(\mathbf{h})}\sum_{i=1}^{n(\mathbf{h})}|Z(\mathbf{x}_i) - Z(\mathbf{x}_i+\mathbf{h})|^p\right]^{\frac{1}{p}}, \tag{3.4}$$

and, hence, the estimator $\mathcal{R}_p$ preserves any scaling factor. Therefore, if the original data set has a large range of values, then the range can be scaled by multiplying by a fixed positive constant without destroying any correlation structure found within the original data. The Cressie-Hawkins robust estimator [6]

$$\gamma_{ch}(\mathbf{h}) = \frac{1}{2}\frac{\left[\frac{1}{n(\mathbf{h})}\sum_{i=1}^{n(\mathbf{h})}|Z(\mathbf{x}_i) - Z(\mathbf{x}_i+\mathbf{h})|^{\frac{1}{2}}\right]^4}{[0.457 + 0.494/n(\mathbf{h})]} \tag{3.5}$$

which is commonly used in practice, the squared median of the absolute deviations estimator [7] $\gamma_{smad}(\mathbf{h}) = 2.198 \times [median\,|Z(\mathbf{x}_i) - Z(\mathbf{x}_i+\mathbf{h})|]^2$, and the conventional semivariogram [10]

$$\gamma(\mathbf{h}) = \frac{1}{2\,n(\mathbf{h})}\sum_{i=1}^{n(\mathbf{h})}|Z(\mathbf{x}_i) - Z(\mathbf{x}_i+\mathbf{h})|^2 \tag{3.6}$$

are essentially similar to the $\mathcal{R}_p$-estimator with $p = 1/2$, $p = 1$, and $p = 2$, respectively. However, the semivariogram γ is not robust. The influence of outliers on the semivariogram γ increases by the square $|Z(\mathbf{x}) - Z(\mathbf{x}+\mathbf{h})|^2$ as the difference $|Z(\mathbf{x}) - Z(\mathbf{x}+\mathbf{h})|$ increases. The Cressie-Hawkins estimator γ_{ch} and the squared median of the absolute deviations estimator γ_{smad} are not robust enough so that they do not produce correct correlation between data points showing erratic behaviors which are commonly observed in field data. Moreover, they do not preserve scaling factors and are not systematic.

4 KRIGING

Kriging is to estimate variables at unmeasured locations. It uses the mathematical model variograms fit to experimental variograms. Many kriging methods are available. Among them, the universal kriging and the punctual kriging are simple and can be easily implemented. In this section, the punctual kriging is explained when the experimental variograms are obtained by the $\mathcal{R}_p$-estimator. The application of the estimator to the universal kriging can be done in a similar way.

For each i, $i = 1, 2, \cdots, m$, let $Z(\mathbf{x}_i)$ be a given value at location $\mathbf{x}_i = (x_i, y_i, z_i)$ that is selected for kriging. For a given location $\mathbf{x}_o = (x_o, y_o, z_o)$, assume that the value $Z(\mathbf{x}_o)$ at $\mathbf{x}_o$ can be approximated by a linear sum of known values $Z(\mathbf{x}_i)$, $i = 1, \cdots, m$. Let

$$Z(\mathbf{x}_o) = \sum_{i=1}^{m}\omega_i Z(\mathbf{x}_i), \tag{4.1}$$

where $\omega_i \geq 0$, $i = 1, \cdots, m$, are weights to be determined by the following kriging system:

$$\begin{aligned} \sum_{j=1}^{m} \omega_j \mathcal{R}_p(h_{ij}) + \lambda &= \mathcal{R}_p(h_{io}), \quad 1 \leq i \leq m, \\ \sum_{i=1}^{m} \omega_i &= 1, \end{aligned} \tag{4.2}$$

where $\mathcal{R}_p(h_{ij})$ is the correlation value estimated by the $\mathcal{R}_p$-estimator at lag h_{ij}, the subscript p is the order of robustness, h_{ij} is the "correlation lag" between two points $\mathbf{x}_i$ and $\mathbf{x}_j$, λ is the Lagrange multiplier, and $\sum_{i=1}^{m} \omega_i = 1$ is the optimality condition. Note that this kriging system (4.2) is "optimal" in the sense that the method produces the exact (original) value at the sampled location. However, the kriging system is optimal only inside the sampling network (convex) domain. Thus, the estimation procedure for points outside the sampling domain must consider macroscopic properties such as trend, drift, etc., of the original data structure together with the kriging system because the optimality condition in equation (4.2) is no longer valid outside the (convex) domain. The optimality constraint described above is independent of the choice of variogram; $\mathcal{R}_p$-estimator, semivariogram γ in equation (3.6), or any other estimators.

With regard to the scaling factor, for each $p > 0$,

$$\mathcal{R}_p(ch) = c\mathcal{R}_p(h) \tag{4.3}$$

for any $c > 0$ and lag h. Thus, the scaling factor $c > 0$ of the original data set is preserved in the correlation estimation step. Moreover, for any constant $c > 0$, the following two kriging systems:

$$\begin{aligned} \sum_{j=1}^{m} \omega_j (c\mathcal{R}_p(h_{ij})) + \lambda &= c\mathcal{R}_p(h_{io}), \quad 1 \leq i \leq m, \\ \sum_{i=1}^{m} \omega_i &= 1, \end{aligned} \tag{4.4}$$

and

$$\begin{aligned} \sum_{j=1}^{m} \omega_j \mathcal{R}_p(h_{ij}) + \lambda/c &= \mathcal{R}_p(h_{io}), \quad 1 \leq i \leq m, \\ \sum_{i=1}^{m} \omega_i &= 1 \end{aligned} \tag{4.5}$$

are equivalent.

5 ACKNOWLEDGMENTS

This paper has not been subject to USAF review, however, and accordingly does not necessarily reflect the reviews of the USAF. The research of the first author was supported in part by the KOSEF 951-0104-024-2 and the BSRI-97-1426.

References

[1] Adams, E. E. and L. W. Gelhar (1992), Field study of dispersion in a heterogeneous aquifer, 2. Spatial moments analysis, *Water Resour. Res.*, 28(12): 3293-3307.

[2] ASCE Task Committee on Geostatistical Techniques in Geohydrology of the Ground Water Hydrology, Committee of the ASCE Hydraulics Division (1990), Review of geostatistics in geohydrology. I: Basic concepts; II. Applications, *J. Hydraulic Engineering*, 116(5): 612-632; 633-658.

[3] Boggs, J. M., L. M. Beard, W. R. Waldrop, T. B. Stauffer, W. G. MacIntyre and C. P. Antworth (1993), Transport of tritium and four organic compounds during a natural gradient experiment (MADE-2), *EPRI Report TR-101998*, Electric Power Research Institute, Palo Alto, CA 94304.

[4] Boggs, J. M., S. C. Young, L. M. Beard, L. W. Gelhar, K. R. Rehfeldt and E. E. Adams (1992), Field study of dispersion in a heterogeneous aquifer, 1. Overview and site description, *Water Resour. Res.*, 28(12): 3281-3291.

[5] Cooper, R. M. and J. D. Istok (1988), Geostatistics applied to groundwater contamination. I: Methodology; II: Application; III: Global estimates, *J. Environmental Engineering*, 114(2): 270-286; 287-299; 114(4): 915-928.

[6] Cressie, N. and D. M. Hawkins (1980), Robust estimation of the variogram: I, *Mathematical Geology*, 12(2): 115-125.

[7] Dowd, P. A. (1984), The variogram and kriging: Robust and resistant estimators, Geostatistics for Natural Resourses Characterization, Part I, *NATO ASI Ser., Ser. C*, 12: 91-107.

[8] Freeze, R. A. and J. A. Cherry (1979), *Groundwater*, Prentice-Hall, Englewood Cliffs, NJ.

[9] Garabedian, S. P., D. R. Leblanc, L. W. Gelhar, and M. A. Celia (1991), Large-scale natural gradient tracer test in sand and gravel, Cape Cod, Massachusetts, 2. Analysis of spatial moments for a nonreactive tracer, *Water Resour. Res.*, 27(5): 911-924.

[10] Journel, A. G. and J. C. Huijbregts (1991), *Mining Geostatistics, 5th ed.*,=20 Academic Press, San Diego.

[11] Leblanc, D. R., S. P. Garabedian, K. M. Hess, L. W. Gelhar, R. D. Quardri, K. G. Stollenwerk, and W. W. Wood (1991), Large-scale natural gradient tracer test in sand and gravel, Cape Cod, Massachussetts, 1. Experimental design and observed tracer movement, *Water Resour. Res.*, 27(5): 895-910.

[12] Mackay, D. M., D. L. Freyberg, P. V. Roberts, and J. A. Cherry (1986), A natural gradient experiment on solute transport in a sand aquifer, 1. Approach and overview of plume movement, *Water Resour. Res.*, 22(13): 2017-2079.

DOMAIN DECOMPOSITION IN OPTIMAL CONTROL OF ELLIPTIC SYSTEMS ON 2-D NETWORKS

John E. Lagnese

Department of Mathematics
Georgetown University
Washington, DC 20057-1233 USA

lagnesemath.georgetown.edu

Abstract: A method of decomposing the optimality system arising in optimal boundary control of elliptic systems of partial differential equations on 2-d networks into subproblems on individual elements of the network is presented.

1 ELLIPTIC SYSTEMS ON 2-D NETWORKS

Let $\mathcal{P}$ be a two-dimensional polygonal topological network in $\mathbb{R}^N$, $N \geq 3$. Thus $\mathcal{P}$ is a finite union of nonempty subsets $\mathcal{P}_I$, $I \in \mathcal{I}$, such that

(i) each $\mathcal{P}_I$ is a simply connected open polygonal subset of a plane Π_I in $\mathbb{R}^N$;

(ii) $\bigcup_{I \in \mathcal{I}} \overline{\mathcal{P}}_I$ is connected;

(iii) for all $I, J \in \mathcal{I}$, $\overline{\mathcal{P}}_I \bigcap \overline{\mathcal{P}}_J$ is either empty, a common vertex, or a whole common side.

For each $I \in \mathcal{I}$ we fix a system of Cartesian coordinates (x_1, x_2) in Π_I. Thus

$$\Pi_I = \{\mathbf{p}_{I0} + x_1 \boldsymbol{\eta}_{I1} + x_2 \boldsymbol{\eta}_{I2}\}, \quad x_1, x_2 \in \mathbb{R},$$

where $\boldsymbol{\eta}_{I1}, \boldsymbol{\eta}_{I2}$ are orthonormal vectors that span Π_I and $\mathbf{p}_{I0}$ denotes the origin of coordinates in Π_I. The linear segments that form the boundary $\partial\mathcal{P}_I$ of $\mathcal{P}_I$ are written $\overline{\Gamma}_{IJ}$, $J = 1, \ldots, N_I$. It is convenient to assume that Γ_{IJ} is open in $\partial\mathcal{P}_I$. The collection of all Γ_{IJ} are the *edges* of $\mathcal{P}$ and is denoted by $\mathcal{A}$. An edge Γ_{IJ} corresponding to an $A \in \mathcal{A}$ is denoted by Γ_{IA} or Γ_A and the *index set* $\mathcal{I}_A$ of A is $\mathcal{I}_A = \{I | \, A = \Gamma_{IA}\}$. The *degree* of an edge is the cardinality of $\mathcal{I}_A$ and is denoted by d_A. For each $I \in \mathcal{I}_A$ we denote by ν_{IA} the unit outer normal to $\mathcal{P}_I$ along Γ_{IA}. The coordinates of ν_{IA} in the given coordinate system of Π_I are written (ν_{IA}^1, ν_{IA}^2).

Let $m \geq 1$ be a given integer. For a function $\mathbf{W} : \mathcal{P} \mapsto \mathbb{R}^m$, $\mathbf{W}_I$ will denote the restriction of $\mathbf{W}$ to $\mathcal{P}_I$, that is $\mathbf{W}_I : \mathcal{P}_I \mapsto \mathbb{R}^m : x \mapsto \mathbf{W}(x)$. We introduce $m \times m$ matrices $A_I^{\alpha\beta}$, B_I^{β}, C_I, $I \in \mathcal{I}$, α, $\beta = 1, 2$, $A_I^{\alpha\beta} = (A_I^{\beta\alpha})^T$ and $C_I = C_I^T$, and where the superscript T indicates transpose, and we define the symmetric bilinear form

$$\mathcal{B}(\mathbf{W}, \mathbf{V}) = \sum_{I \in \mathcal{I}} \int_{\mathcal{P}_I} [A_I^{\alpha\beta}(\mathbf{W}_{I,\beta} + B_I^{\beta}\mathbf{W}_I) \cdot (\mathbf{V}_{I,\alpha} + B_I^{\alpha}\mathbf{V}_I) + C_I\mathbf{W}_I \cdot \mathbf{V}_I], \quad (1.1)$$

where repeated lower case Greek indices are summed over 1,2. A subscript following a comma indicated differentiation with respect to the corresponding variable, e.g., $\mathbf{W}_{I,\beta} = \partial \mathbf{W}_I / \partial x_\beta$. The matrices $A_I^{\alpha\beta}$, B_I^{β}, C_I may depend on $(x_1, x_2) \in \mathcal{P}_I$ and $\mathcal{B}$ is required to be $\mathcal{V}$-elliptic for an appropriate function space $\mathcal{V}$.

We shall consider the variational equation

$$\mathbf{W} \in \mathcal{V}, \quad \mathcal{B}(\mathbf{W}, \mathbf{V}) = \sum_{A \in \mathcal{N}^{\text{ext}}} \int_{\Gamma_{IA}} \mathbf{F}_A \cdot \mathbf{V}_I, \quad \forall \mathbf{V} \in \mathcal{V}, \quad (1.2)$$

where the space $\mathcal{V}$ is defined in terms of certain *geometric edge conditions*, $\mathcal{N}^{\text{ext}}$ is the set of *exterior Neumann edges* (see below), I is the unique index associated with an edge $A \in \mathcal{N}^{\text{ext}}$, and $\mathbf{F}_A \in [L^2(\Gamma_A)]^m$. To describe the space $\mathcal{V}$ we partition the edges of $\mathcal{A}$ into two disjoint subsets $\mathcal{D}$, the *Dirichlet edges*, and $\mathcal{N}$, the *Neumann edges*. Each edge $A \in \mathcal{D}$ is assumed to be an *exterior edge*, that is, $d_A = 1$. We denote by $\mathcal{N}^{\text{ext}}$ the exterior edges in $\mathcal{N}$ and set $\mathcal{N}^{\text{int}} = \mathcal{N} \backslash \mathcal{N}^{\text{ext}}$, the set of *interior Neumann edges*. We define

$$\mathcal{H} = \{\mathbf{V} : \mathcal{P} \mapsto \mathbb{R}^m |\, \mathbf{V}_I \in [L^2(\mathcal{P}_I)]^m\}$$

with the standard norm,

$$\mathcal{V} = \{\mathbf{V} \in \mathcal{H} |\, \mathbf{V}_I \in [H^1(\mathcal{P}_I)]^m,\ \mathbf{V}_I = 0 \text{ on } \Gamma_{IA} \text{ if } \Gamma_{IA} \in \mathcal{D},\ \mathbf{V}_I = \mathbf{V}_J \text{ on } \Gamma_A \text{ if } \Gamma_{IA} = \Gamma_{JA}\}.$$

We shall assume that the bilinear form $\mathcal{B}$ satisfies

$$\mathcal{B}(\mathbf{V}, \mathbf{V}) \geq c \|\mathbf{V}\|^2_{\mathcal{H}^1(\mathcal{P})}, \quad \forall \mathbf{V} \in \mathcal{V},$$

for some constant $c > 0$. This ellipticity condition will hold in examples 2.2 and 2.3 below if $\mathcal{D} \neq \emptyset$, and also in example 2.1 if, for instance, for each $I \in \mathcal{I}$ the 2×2 matrix $(a_I^{\alpha\beta})$ is positive definite and $b_I^{\alpha} = c_I = 0$.

Let

$$\mathcal{U} = \prod_{A \in \mathcal{N}^{\text{ext}}} \mathcal{L}^2(\Gamma_A), \quad \mathcal{L}^2(\Gamma_A) := [L^2(\Gamma_A)]^m.$$

For $\mathbf{F} \in \mathcal{U}$ and $A \in \mathcal{N}^{\text{ext}}$, $\mathbf{F}_A$ will denote the restriction of $\mathbf{F}$ to Γ_A. Consider the optimal control problem

$$\inf_{\mathbf{F} \in \mathcal{U}} \left\{ \sum_{I \in \mathcal{I}} \int_{\mathcal{P}_I} |\mathbf{W}_I - \mathbf{Z}_I|^2 + k \sum_{A \in \mathcal{N}^{\text{ext}}} \int_{\Gamma_A} |\mathbf{F}_A|^2 \right\}$$

where $\mathbf{Z} = \{\mathbf{Z}_I\}_{I\in\mathcal{I}} \in \mathcal{H}$ is given, $k > 0$ is the penalization parameter and $\mathbf{W}_I$, $I \in \mathcal{I}$ is the solution of (1.2). By standard theory, the optimality system is

$$\mathcal{B}(\mathbf{W},\mathbf{V}) + \sum_{A\in\mathcal{N}^{\text{ext}}} \frac{1}{k}\int_{\Gamma_{IA}} \mathbf{\Phi}_I \cdot \mathbf{V}_I = 0,$$

$$\mathcal{B}(\mathbf{\Phi},\mathbf{V}) = \sum_{I\in\mathcal{I}} \int_{\mathcal{P}_I} (\mathbf{W}_I - \mathbf{Z}_I)\cdot \mathbf{V}_I, \quad \forall \mathbf{V} \in \mathcal{V}.$$

This is the variational formulation of the boundary value problem

$$\mathcal{L}_I\mathbf{W}_I = 0, \quad \mathcal{L}_I\mathbf{\Phi}_I = \mathbf{W}_I - \mathbf{Z}_I \text{ in } \mathcal{P}_I,$$
$$\mathbf{W}_I = \mathbf{\Phi}_I = 0 \text{ on } \Gamma_{IA} \text{ if } \Gamma_{IA} \in \mathcal{D},$$
$$\mathbf{W}_I = \mathbf{W}_J, \quad \mathbf{\Phi}_I = \mathbf{\Phi}_J \text{ on } A \text{ if } \Gamma_{IA} = \Gamma_{JA}, \tag{1.3}$$
$$D_{\nu_{IA}}\mathbf{\Phi}_I = 0, \quad D_{\nu_{IA}}\mathbf{W}_I + \frac{1}{k}\mathbf{\Phi}_I = 0 \text{ on } \Gamma_{IA} \text{ if } \Gamma_{IA} \in \mathcal{N}^{\text{ext}},$$
$$\sum_{I\in\mathcal{I}_A} D_{\nu_{IA}}\mathbf{W}_I = \sum_{I\in\mathcal{I}_A} D_{\nu_{IA}}\mathbf{\Phi}_I = 0 \text{ on } A \text{ if } A \in \mathcal{N}^{\text{int}}, \tag{1.4}$$

where we have set

$$\mathcal{L}\mathbf{W}_I := -\frac{\partial}{\partial x_\alpha}[A_I^{\alpha\beta}(\mathbf{W}_{I,\beta} + B_I^\beta \mathbf{W}_I)] + (B_I^\alpha)^T A_I^{\alpha\beta}(\mathbf{W}_{I,\beta} + B_I^\beta\mathbf{W}_I) + C_I\mathbf{W}_I,$$

$$D_{\nu_{IA}}\mathbf{W}_I := \nu_{IA}^\alpha A_I^{\alpha\beta}(\mathbf{W}_{I,\beta} + B_I^\beta \mathbf{W}_I).$$

Conditions (1.3) and (1.4) are referred to as *geometric junction conditions* and *mechanical junction conditions*, respectively.

2 EXAMPLES

Example 2.1 (Scalar elliptic equations on 2-d networks) Suppose that $m = 1$. In this case the matrices $A_I^{\alpha\beta}$, B_I^α, C_I reduce to scalars $a_I^{\alpha\beta}$, b_I^α, c_I, where $a_I^{\alpha\beta} = a_I^{\beta\alpha}$. The geometric and mechanical junction conditions are respectively

$$W_I = W_J \text{ on } A \text{ if } \Gamma_{IA} = \Gamma_{JA}\ , A \in \mathcal{N}^{\text{int}},$$
$$\sum_{I\in\mathcal{I}_A} \nu_{IA}^\alpha = a_I^{\alpha\beta}(W_{I,\beta} + b_I^\beta W_I) = 0 \text{ on } A,\ A \in \mathcal{N}^{\text{int}}.$$

The reader is referred to the book of Nicaise (1993) for a detailed existence-regularity theory for scalar elliptic equations on 2-d networks.

Example 2.2 (Networks of homogeneous isotropic membranes in $\mathbb{R}^3$) In this case $m = N = 3$. For each $I \in \mathcal{I}$ we choose a basis $\{\boldsymbol{\xi}_{Ik}\}_{k=1}^3$ so that $\boldsymbol{\xi}_{I1} = \boldsymbol{\eta}_{I1}$, $\boldsymbol{\xi}_{I2} = \boldsymbol{\eta}_{I2}$, $\boldsymbol{\xi}_{I3} = \boldsymbol{\xi}_{I1} \times \boldsymbol{\xi}_{I2}$, where $\boldsymbol{\eta}_{I1}$, $\boldsymbol{\eta}_{I2}$ are the unit coordinate vectors in Π_I. We set $B_I^\alpha = C_I = 0$. With respect to the $\{\boldsymbol{\xi}_{Ik}\}_{k=1}^3$ basis the matrices

$A_I^{\alpha\beta}$ are given by

$$A_I^{11} = \operatorname{diag}(2\mu_I + \lambda_I, \mu_I, \mu_I),$$
$$A_I^{22} = \operatorname{diag}(\mu_I, 2\mu_I + \lambda_I, \mu_I),$$
$$A_I^{12} = \begin{pmatrix} 0 & \lambda_I & 0 \\ \mu_I & 0 & 0 \\ 0 & 0 & 0 \end{pmatrix}, \quad A_I^{21} = \begin{pmatrix} 0 & \mu_I & 0 \\ \lambda_I & 0 & 0 \\ 0 & 0 & 0 \end{pmatrix}$$

Write $\mathbf{W}_I = W_{Ik}\boldsymbol{\xi}_{Ik}$ (summation convention over k from 1 to m), $\mathbf{w}_I = W_{I\alpha}\boldsymbol{\xi}_{I\alpha}$,

$$\varepsilon_{\alpha\beta}(\mathbf{w}_I) = \frac{1}{2}(W_{I\alpha,\beta} + W_{I\beta,\alpha}),$$
$$S_I^{\alpha\beta}(\mathbf{w}_I) = 2\mu_I\varepsilon_{\alpha\beta}(\mathbf{w}_I) + \lambda_I\varepsilon_{\gamma\gamma}(\mathbf{w}_I)\delta^{\alpha\beta}.$$

The bilinear form (1.1) may be written

$$\mathcal{B}(\mathbf{W}, \mathbf{V}) = \sum_{I\in\mathcal{I}} \int_{\mathcal{P}_I} [S_I^{\alpha\beta}(\mathbf{w}_I)\varepsilon_{\alpha\beta}(\mathbf{v}_I) + \mu_I W_{I3,\alpha}V_{I3,\alpha}].$$

The geometric and mechanical junction conditions for $\mathbf{W}$ are respectively

$$\mathbf{W}_I = \mathbf{W}_J \text{ on } A \text{ if } \Gamma_{IA} = \Gamma_{JA},$$
$$\sum_{I\in\mathcal{I}_A} [S_I^{\alpha\beta}(\mathbf{w}_I)\boldsymbol{\xi}_{I\beta} + \mu_I W_{I3,\alpha}\boldsymbol{\xi}_{I3}]\nu_{IA}^{\alpha} = 0 \text{ on } A \text{ when } A \in \mathcal{N}^{\text{int}},$$

which have the interpretations of continuity of displacements and balance of forces, respectively, at the membrane junctions (see Lagnese, Leugering and Schmidt (1994)).

Example 2.3 (Interface problems for serial Reissner-Mindlin plates) We take $N = m = 3$ and assume that the regions $\mathcal{P}_I$ are coplanar. The region $\cup_{I\in\mathcal{I}}\overline{\mathcal{P}}_I$ represents the middle surface of a thin plate of uniform thickness h. Within the region $\mathcal{P}_I \times (-h/2, h/2)$ the plate is assumed homogeneous and elastically isotropic with Lamé parameters λ_I, μ_I and shear modulus K_I. Let $\boldsymbol{\eta}_1$, $\boldsymbol{\eta}_2$ be unit coordinate vectors in the plane Π containing the regions $\mathcal{P}_I$, and set $\boldsymbol{\xi}_1 = \boldsymbol{\eta}_2$, $\boldsymbol{\xi}_2 = -\boldsymbol{\eta}_1$ and $\boldsymbol{\xi}_3 = \boldsymbol{\xi}_1 \times \boldsymbol{\xi}_2$. The matrix $C_I = 0$ and, with respect to the $\{\boldsymbol{\xi}_k\}_{k=1}^3$ basis,

$$A_I^{11} = h\operatorname{diag}(\frac{h^2}{12}(2\mu_I + \sigma_I), \frac{h^2}{12}\mu_I, K_I),$$
$$A_I^{22} = h\operatorname{diag}(\frac{h^2}{12}\mu_I, \frac{h^2}{12}(2\mu_I + \sigma_I), K_I)$$

where $\sigma_I = 2\mu_I\lambda_I/(2\mu_I + \lambda_I)$,

$$A_I^{12} = \frac{h^3}{12}\begin{pmatrix} 0 & \sigma_I & 0 \\ \mu_I & 0 & 0 \\ 0 & 0 & 0 \end{pmatrix}, \quad A_I^{21} = \frac{h^3}{12}\begin{pmatrix} 0 & \mu_I & 0 \\ \sigma_I & 0 & 0 \\ 0 & 0 & 0 \end{pmatrix}$$

$$B_I^1 = \begin{pmatrix} 0 & 0 & 0 \\ 0 & 0 & 0 \\ 1 & 0 & 0 \end{pmatrix}, \quad B_I^2 = \begin{pmatrix} 0 & 0 & 0 \\ 0 & 0 & 0 \\ 0 & 1 & 0 \end{pmatrix}$$

Write $\mathbf{W}_I = W_{Ik}\boldsymbol{\xi}_k$, $\mathbf{w}_I = W_{I\alpha}\boldsymbol{\xi}_\alpha$ and

$$S_I^{\alpha\beta}(\mathbf{w}_I) = \frac{h^3}{12}(2\mu_I\varepsilon_{\alpha\beta}(\mathbf{w}_I) + \sigma_I\varepsilon_{\gamma\gamma}(\mathbf{w}_I)\delta^{\alpha\beta}) \tag{1.5}$$

With this notation the bilinear form (1.1) is

$$\mathcal{B}(\mathbf{W}, \mathbf{V}) = \sum_{I\in\mathcal{I}} \int_{\mathcal{P}_I} [S_I^{\alpha\beta}(\mathbf{w}_I)\varepsilon_{\alpha\beta}(\mathbf{v}_I) + K_I h(W_{I3,\alpha} + W_{I\alpha})(V_{I3,\alpha} + V_{I\alpha})].$$

The function $W_{I3}(x_1, x_2)$ represents the transverse displacement of the material particle situated at $(x_1, x_2) \in \mathcal{P}_I$ in the reference configuration, and W_{I1}, W_{I2} measure transverse shearing. The geometric junction conditions for $\mathbf{W}$ are

$$W_{Ik} = W_{Jk} \quad \text{on } A \text{ when } \Gamma_{IA} = \Gamma_{JA},\ k = 1, 2, 3,$$

(the degree $d_A = 2$ for each interior edge) while the mechanical junction conditions are

$$\sum_{I\in\mathcal{I}_A} \nu_{IA}^\alpha = S_I^{\alpha\beta}(\mathbf{w}_I) = 0, \quad \beta = 1, 2,$$

$$\sum_{I\in\mathcal{I}_A} K_I\nu_{IA}^\alpha(W_{I3,\alpha} + W_{I\alpha}) = 0 \text{ on } A \text{ when } A \in \mathcal{N}^{\text{int}}.$$

These have the interpretations of balance of angular and linear momenta, respectively (see Lagnese, Leugering and Schmidt (1994)).

3 DOMAIN DECOMPOSITION

In the present context the object of domain decomposition is to replace the above global optimality system on the network $\mathcal{P}$ by a sequence of local problems on the elements $\mathcal{P}_I$ whose solutions converge to a solution of the global optimality system in an appropriate sense. Each local problem is itself the optimality system for some local optimal control problem. In the decomposition, the geometric and mechanical junction conditions that couple adjacent regions in the global network are replaced by local Robin-type boundary conditions in which the effect of adjacent regions is modeled by certain inhomogeneities in the boundary conditions. This strategy was first employed in the context of optimal control problems for single scalar equations by Benamou (see, e.g., Benamou (1995), (1998) and references therein) and later extended by Leugering (1998) to construct domain decompositions for the optimal control of certain mechanical networks of 1-dimensional elements, where multiple nodes naturally appear.

For each index $I \in \mathcal{I}$ let $\mathcal{D}_I$ denote the Dirichlet edges in $\partial\mathcal{P}_I$ and set $\mathcal{N}_I = \partial\mathcal{P}_I \backslash \mathcal{D}_I$, which is partitioned into $\mathcal{N}_I^{\text{ext}} \cup \mathcal{N}_I^{\text{int}}$, where $\mathcal{N}_I^{\text{ext}} = \mathcal{N}_I \cap \mathcal{N}^{\text{ext}}$. We write $\mathcal{H}_I^s$ for $[H^s(\mathcal{P}_I)]^m$, $\mathcal{H}_I = \mathcal{H}_I^0$, and we define

$$\mathcal{V}_I = \{\mathbf{V}_I | \, \mathbf{V}_I \in \mathcal{H}_I^1, \ \mathbf{V}_I = 0 \text{ on } \Gamma_{IA} \text{ if } \Gamma_{IA} \in \mathcal{D}_I\}$$

A continuous, symmetric bilinear form $\mathcal{B}_I$ is defined on $\mathcal{V}_I$ by

$$\mathcal{B}_I(\mathbf{W}_I, \mathbf{V}_I) = \int_{\mathcal{P}_I} [A_I^{\alpha\beta}(\mathbf{W}_{I,\beta} + B_I^\beta \mathbf{W}_I) \cdot (\mathbf{V}_{I,\alpha} + B_I^\alpha \mathbf{V}_I) + C_I \mathbf{W}_I \cdot \mathbf{V}_I]$$

and we assume that

$$\mathcal{B}_I(\mathbf{V}_I, \mathbf{V}_I) \geq c_I \|\mathbf{V}_I\|^2_{\mathcal{H}_I^1}, \quad \forall \mathbf{V}_I \in \mathcal{V}_I,$$

for some constant $c_I > 0$.

For each $I \in \mathcal{I}$ and $n = 0, 1, \ldots$, we consider the following sequence of local problems for functions $\mathbf{W}_I^{n+1}$, $\mathbf{\Phi}_I^{n+1}$:

$$\begin{aligned} B_I(\mathbf{W}_I^{n+1}, \mathbf{V}_I) + \frac{1}{k} \sum_{A \in \mathcal{N}_I^{\text{ext}}} \int_{\Gamma_{IA}} \mathbf{\Phi}_I^{n+1} \cdot \mathbf{V}_I + \sum_{A \in \mathcal{N}_I^{\text{int}}} k_A \int_{\Gamma_{IA}} \mathbf{\Phi}_I^{n+1} \cdot \mathbf{V}_I \\ = \sum_{A \in \mathcal{N}_I^{\text{int}}} \int_{\Gamma_{IA}} \mathbf{V}_I \cdot \Lambda_{IA}(\mathbf{W}^n, \mathbf{\Phi}^n), \quad \forall \mathbf{V}_I \in \mathcal{V}_I, \end{aligned} \tag{1.6}$$

$$\begin{aligned} B_I(\mathbf{\Phi}_I^{n+1}, \mathbf{V}_I) - \sum_{A \in \mathcal{N}_I^{\text{int}}} k_A \int_{\Gamma_{IA}} \mathbf{W}_I^{n+1} \cdot \mathbf{V}_I - \int_{\mathcal{P}_I} (\mathbf{W}_I^{n+1} - \mathbf{Z}_I) \cdot \mathbf{V}_I \\ = \sum_{A \in \mathcal{N}_I^{\text{int}}} \int_{\Gamma_{IA}} \mathbf{V}_I \cdot \Pi_{IA}(\mathbf{W}^n, \mathbf{\Phi}^n), \quad \forall \mathbf{V}_I \in \mathcal{V}_I, \end{aligned} \tag{1.7}$$

where $k_A > 0$ is constant,

$$\begin{aligned} \Lambda_{IA}(\mathbf{W}^n, \mathbf{\Phi}^n) &= \frac{2k_A}{d_A} \sum_{J \in \mathcal{I}_A} \mathbf{\Phi}_J^n - k_A \mathbf{\Phi}_I^n - \frac{2}{d_A} \sum_{J \in \mathcal{I}_A} D_{\nu_{JA}} \mathbf{W}_J^n + D_{\nu_{IA}} \mathbf{W}_I^n, \\ \Pi_{IA}(\mathbf{W}^n, \mathbf{\Phi}^n) &= \Lambda_{IA}(\mathbf{\Phi}^n, -\mathbf{W}^n). \end{aligned}$$

The boundary value problem corresponding to (1.6), (1.7) is

$$\begin{gathered} \mathcal{L}_I \mathbf{W}_I^{n+1} = 0, \quad \mathcal{L}_I \mathbf{\Phi}_I^{n+1} = \mathbf{W}_I^{n+1} - \mathbf{Z}_I \text{ in } \mathcal{P}_I, \\ \mathbf{W}_I^{n+1} = \mathbf{\Phi}_I^{n+1} = 0 \text{ on } \Gamma_{IA} \text{ if } \Gamma_{IA} \in \mathcal{D}_I, \\ D_{\nu_{IA}} \mathbf{W}_I^{n+1} + \frac{1}{k} \gamma_{IA} \mathbf{\Phi}_I^{n+1} = 0, \quad D_{\nu_{IA}} \mathbf{\Phi}_I^{n+1} = 0 \text{ on } \Gamma_{IA} \text{ if } \Gamma_{IA} \in \mathcal{N}_I^{\text{ext}}, \end{gathered}$$

$$\left. \begin{aligned} D_{\nu_{IA}} \mathbf{W}_I^{n+1} + k_A \gamma_{IA} \mathbf{\Phi}_I^{n+1} = \Lambda_{IA}(\mathbf{W}^n, \mathbf{\Phi}^n), \\ D_{\nu_{IA}} \mathbf{\Phi}_I^{n+1} - k_A \gamma_{IA} \mathbf{W}_I^{n+1} = \Pi_{IA}(\mathbf{W}^n, \mathbf{\Phi}^n) \end{aligned} \right. \quad \text{on } \Gamma_{IA} \text{ if } \Gamma_{IA} \in \mathcal{N}_I^{\text{int}}$$

The inhomogeneities $\Lambda_{IA}(\mathbf{W}^n, \mathbf{\Phi}^n)$, $\Pi_{IA}(\mathbf{W}^n \mathbf{\Phi}^n)$ reflects the influence of the regions adjacent to Γ_{IA}, $A \in \mathcal{N}_I^{\text{int}}$, on the solution in $\mathcal{P}_I$ at level $n+1$. The choice of $\{\mathbf{W}_I^0, \mathbf{\Phi}_I^0\}_{I\in\mathcal{I}}$ is arbitrary save for a regularity requirement on $\Lambda_{IA}(\mathbf{W}^0, \mathbf{\Phi}^0)$ and $\Pi_{IA}(\mathbf{W}^0, \mathbf{\Phi}^0)$. One may verify that the above system for $\mathbf{W}_I^{n+1}$, $\mathbf{\Phi}_I^{n+1}$ is the optimality system for the local optimal control problem

$$\inf_{\mathbf{F}\in\mathcal{L}^2(\mathcal{N}_I)} \left\{ \int_{\mathcal{P}_I} |\mathbf{W}_I^{n+1} - \mathbf{Z}_I|^2 + k \sum_{A\in\mathcal{N}_I^{\text{ext}}} \int_{\Gamma_{IA}} |\mathbf{F}_{IA}|^2 \right.$$
$$\left. + \sum_{A\in\mathcal{N}_I^{\text{int}}} \int_{\Gamma_{IA}} (|\mathbf{F}_{IA}|^2 + |\mathbf{W}_I^{n+1} + \Pi_{IA}(\mathbf{W}^n, \mathbf{\Phi}^n)|^2) \right\}$$

subject to

$$\mathbf{W}_I^{n+1} \in \mathcal{V}_I, \quad \mathcal{B}_I(\mathbf{W}_I^{n+1}, \mathbf{V}_I) = \sum_{A\in\mathcal{N}_I^{\text{ext}}} \int_{\Gamma_{IA}} \mathbf{F}_{IA} \cdot \mathbf{V}_I$$
$$+ \sum_{A\in\mathcal{N}_I^{\text{int}}} \frac{1}{k_A} \int_{\Gamma_{IA}} (\mathbf{F}_{IA} + \Lambda_{IA}(\mathbf{W}^n, \mathbf{\Phi}^n)) \cdot \mathbf{V}_I, \quad \forall \mathbf{V}_I \in \mathcal{V}_I$$

4 CONVERGENCE

Suppose that $\Lambda_{IA}(\mathbf{W}^0, \mathbf{\Phi}^0)$, $\Pi_{IA}(\mathbf{W}^0, \mathbf{\Phi}^0) \in \mathcal{L}^2(\Gamma_{IA})$, $\forall A \in \mathcal{N}_I^{\text{int}}$. Then, for each $n \geq 0$ and $I \in \mathcal{I}$, $\Lambda_{IA}(\mathbf{W}^n, \mathbf{\Phi}^n)$, $\Pi_{IA}(\mathbf{W}^n, \mathbf{\Phi}^n) \in \mathcal{L}^2(\Gamma_{IA})$ and the local optimality system has a unique solution $\mathbf{W}_I^{n+1} \in \mathcal{V}_I$, $\mathbf{\Phi}_I^{n+1} \in \mathcal{V}_I$.

Theorem 1 *Suppose that the solution of the global optimality system has the regularity*

$$D_{\nu_{IA}}\mathbf{W}_I,\ D_{\nu_{IA}}\mathbf{\Phi}_I \in \mathcal{L}^2(\Gamma_{IA}), \quad I \in \mathcal{I},\ A \in \mathcal{N}_I^{int}.$$

Then for each $I \in \mathcal{I}$,

$$\mathbf{W}_I^{n+1} \to \mathbf{W}_I \text{ strongly in } \mathcal{H}_I,$$
$$\mathbf{\Phi}_I^{n+1}|_{\Gamma_{IA}} \to \mathbf{\Phi}_I|_{\Gamma_{IA}} \text{ strongly in } \mathcal{L}^2(\Gamma_{IA}),\ A \in \mathcal{N}_I^{ext}.$$

In words, the solution $\mathbf{W} = 3D\{\mathbf{W}_I\}_{I\in\mathcal{I}}$ of the global optimal control problem is the strong limit in $\mathcal{H}$ of the solutions $\{\mathbf{W}_I^{n+1}\}_{I\in\mathcal{I}}$ of the local optimal control problems, and the global optimal control $\{-\frac{1}{k}\mathbf{\Phi}_I|_{\Gamma_{IA}}\}_{A\in\mathcal{N}^{ext}}$ is the strong limit in $\mathcal{U}$ of the local optimal controls $\{-\frac{1}{k}\mathbf{\Phi}_I^{n+1}|_{\Gamma_{IA}}\}_{A\in\mathcal{N}^{ext}}$.

Space limitations do not permit a detailed proof of this theorem. The idea is to set

$$\mathbf{U}_I^{n+1} = \mathbf{W}_I^{n+1} - \mathbf{W}_I, \quad \mathbf{\Psi}_I^{n+1} = \mathbf{\Phi}_I^{n+1} - \mathbf{\Phi}_I,$$

and to introduce the quantity

$$E^{n+1} := \sum_{A\in\mathcal{N}^{\text{int}}} \sum_{I\in\mathcal{I}_A} \int_{\Gamma_{IA}} \{k_A(|\mathbf{U}_I^{n+1}|^2 + |\mathbf{\Psi}_I^{n+1}|^2)$$
$$+ \frac{1}{k_A}(|D_{\nu_{IA}}\mathbf{U}_I^{n+1}|^2 + |D_{\nu_{IA}}\mathbf{\Psi}_I^{n+1}|^2)\}.$$

It is then possible to prove the recursion

$$E^{n+1} = E^n - 2\sum_{I\in\mathcal{I}}\int_{\mathcal{P}_I}(|\mathbf{U}_I^{n+1}|^2 + |\mathbf{U}_I^n|^2) - \frac{2}{k}\sum_{A\in\mathcal{N}^{\text{ext}}}\int_{\Gamma_{IA}}(|\mathbf{\Psi}_I^{n+1}|^2 + |\mathbf{\Psi}_I^n|^2).$$

By iteration we obtain

$$E^{n+1} = E^1 - 4\sum_{\ell=1}^{n+1}{}'(\frac{2}{k}\sum_{A\in\mathcal{N}^{\text{ext}}}\int_{\Gamma_{IA}}|\mathbf{\Psi}_I^\ell|^2 + \sum_{I\in\mathcal{I}}\int_{\mathcal{P}_I}|\mathbf{U}_I^\ell|^2),$$

where $\sum_{\ell=1}^{n+1}{}' a_\ell = 1/2(a_1 + a_{n+1}) + \sum_{\ell=2}^{n} a_\ell$. It follows in particular that

$$\sum_{\ell=1}^{\infty}\sum_{I\in\mathcal{I}}\int_{\mathcal{P}_I}|\mathbf{U}_I^\ell|^2 < \infty, \quad \sum_{\ell=1}^{\infty}\sum_{A\in\mathcal{N}^{\text{ext}}}\int_{\Gamma_{IA}}|\mathbf{\Psi}_I^\ell|^2 < \infty.$$

Therefore

$$\{\mathbf{U}_I^n\}_{I\in\mathcal{I}} \to 0 \text{ strongly in } \mathcal{H}, \quad \mathbf{\Psi}_I^n|_{\Gamma_{IA}} \to 0 \text{ strongly in } L^2(\Gamma_{IA}), \forall A \in \mathcal{N}^{\text{ext}}.$$

References

Benamou, J.-D. (1995). “A domain decomposition method with coupled transmission conditions for the optimal control of systems governed by elliptic partial differential equations,” *SIAM J. Num. Anal.*, **36**, 2401-2416.

Benamou, J.-D. (1998). ”Domain decomposition, optimal control of systems governed by partial differential equations and synthesis of feedback laws,” *J. Opt. Theory Appl.*, **99**, to appear.

Lagnese, J., Leugering, G., and G. Schmidt. (1994). *Modeling, Analysis and Control of Dynamic Elastic Multi-Link Structures*, Boston, MA: Birkhäuser.

Leugering, G. (1998). “On domain decomposition of optimal control problems for dynamic networks of elastic strings,” *Computational Optimization and Applications*, to appear.

Nicaise, S. (1993). *Polygonal Interface Problems*, Series “Methoden und Verfahren der Mathematischen Physik,” vol. 39, Frankfurt: Peter Lang.

AN OBSERVABILITY ESTIMATE IN $L_2(\Omega) \times H^{-1}(\Omega)$ FOR SECOND-ORDER HYPERBOLIC EQUATIONS WITH VARIABLE COEFFICIENTS

Irena Lasiecka
and Roberto Triggiani
Department of Mathematics
University of Virginia, Charlottesville, VA 22903 USA*

Peng-Fei Yao
Institute of Systems Science
Academia Sinica, Beijing 100080 CHINA†

Abstract: For second-order hyperbolic equations with variable coefficient principal part, we establish a global observability-type estimate, whereby the energy of the solutions in $L_2(\Omega) \times H^{-1}(\Omega)$ is bounded by appropriate boundary traces, modulo lower-order terms. This extends the proof and the results of [L-T.2] from the constant coefficient to the variable coefficient case. The pseudo-differential analysis of [L-T.2] is combined with Riemann geometric multipliers [Y.1], which replace the Euclidean multipliers of [L-T.2] in the constant coefficient case.

1 INTRODUCTION. STATEMENT OF MAIN RESULT

Let Ω be an open bounded domain in $\mathbb{R}^n$, with smooth boundary $\Gamma = \partial\Omega$. Let Γ_0 be a (possibly empty) portion of Γ. We shall consider regular solutions of

$$\begin{cases} w_{tt} + \mathcal{A}w \equiv f \quad \text{in } Q = (0,\infty) \times \Omega, & (1.1a) \\ w|_{\Sigma_0} \equiv 0 \quad \text{in } \Sigma_0 = (0,\infty) \times \Gamma_0, & (1.1b) \end{cases}$$

*Research partially supported by the National Science Foundation under grant DMS-9504822 and by the Army Research Office under grand DAAH04-96-1-0059.
†Research partially supported by the National Science Foundation and the Pangdeng Project of China.

and establish an *a-priori* energy estimate one unit *below* the usual energy level of $H^1(\Omega) \times L_2(\Omega)$. In (1.1a) we have set

$$\mathcal{A}w = -\sum_{i,j=1}^{n} \frac{\partial}{\partial x_i}(a_{ij}(x))\frac{\partial w}{\partial x_j}, \quad x = [x_1, x_2, \ldots, x_n], \tag{1.2}$$

$a_{ij} = a_{ji} \in C^\infty(\mathbb{R}^n)$, to be a second-order differential operator, which is assumed throughout to satisfy the ellipticity condition
(H.1)

$$\sum_{i,j=1}^{n} a_{ij}(x)\xi_i\xi_j > 0, \quad \forall\, x \in \mathbb{R}^n,\ 0 \neq \xi = [\xi_1, \xi_2, \ldots, \xi_n] \in \mathbb{R}^n. \tag{1.3}$$

Let $G(x) = (g_{ij}(x)) = [A(x)]^{-1}$ be the inverse of the symmetric positive definite coefficient matrix $A(x) = (a_{ij}(x))$. For each $x \in \mathbb{R}^n$, the matrix $G(x)$ defines in a natural way a Riemann metric $g(X,Y) = \langle X, Y\rangle_g$ on the tangent space $\mathbb{R}^n_x = \mathbb{R}^n \ni X, Y$ [L-T-Y.1], [Y.1], so that $(\mathbb{R}^n, g)$ is a Riemann manifold. Let D be the Levi-Civita connection in the Riemann metric g, $D_X H$ be the covariant derivative of the vector field $H \in \mathbb{R}^n_x$ with respect to the vector field $X \in \mathbb{R}^n_x$; and let $DH(\,\cdot\,,\,\cdot\,)$ be the resulting covariant differential $DH(X,X) = \langle D_X H, X\rangle_g$. We may now state the main assumption, as in [Y.1], imposed throughout this paper upon the differential operator $\mathcal{A}$ in (1.2). Several nontrivial examples where this assumption is verified to hold true, by using Riemann geometry, are given in [L-T-Y.1], [Y.1].

Main Assumption (H.2). There exists a vector field H on the Riemann manifold $(\mathbb{R}^n, g)$ such that

$$DH(X,X) = \langle D_X H, X\rangle_g \geq a|X|_g^2, \quad \forall\, X \in \mathbb{R}^n_x,\ x \in \Omega, \tag{1.4}$$

where $a > 0$ is a constant. □

Main result. To state our main result, we first introduce the positive, self-adjoint operator on $L_2(\Omega)$:

$$\begin{cases} \mathcal{A}_0 w = \mathcal{A}w, \quad \mathcal{D}(\mathcal{A}_0) = H^2(\Omega) \cap H_0^1(\Omega); & (1.5a) \\ \mathcal{D}(\mathcal{A}_0^{\frac{1}{2}}) = H_0^1(\Omega),\ [\mathcal{D}(\mathcal{A}_0^{\frac{1}{2}})]' = H^{-1}(\Omega) \text{ (equivalent norms)} & (1.5b) \end{cases}$$

Next, we introduce the desired 'energy' of the solution w of Eqn. (1.1):

$$\begin{cases} \mathcal{E}_w(t) \equiv \|w(t)\|^2_{L_2(\Omega)} + \|\mathcal{A}_0^{-\frac{1}{2}} w_t(t)\|^2_{L_2(\Omega)} \equiv \|\{w(t), w_t(t)\}\|^2_Z & (1.6a) \\ \text{equivalent to } \|\{w(t), w_t(t)\}\|^2_{L_2(\Omega)\times H^{-1}(\Omega)}; & (1.6b) \end{cases}$$

$$Z \equiv L_2(\Omega) \times [\mathcal{D}(\mathcal{A}_0^{\frac{1}{2}})]' \equiv L_2(\Omega) \times H^{-1}(\Omega) \text{ (equivalent norms)}. \tag{1.7}$$

Our main result below refers to solutions of problem (1.1) within the following class:

$$\begin{cases} \{w, w_t\} \in C([0,T]; L_2(\Omega) \times H^{-1}(\Omega)); & (1.8a) \\ w|_{\Sigma_1} \in L_2(\Sigma_1); \ \left.\dfrac{\partial w}{\partial \nu_A}\right|_{\Sigma_1} \in H^{-1}(\Sigma_1), & (1.8b) \end{cases}$$

where $\Gamma_1 \equiv \Gamma \setminus \Gamma_0$, $\Sigma_1 = (0,T] \times \Gamma_1$, $L_2(\Sigma_1) \equiv L_2(0,T; L_2(\Gamma_1))$, and where $\frac{\partial}{\partial \nu_A} = \sum_{i,j=1}^n a_{ij} \frac{\partial}{\partial x_j} \nu_i$ is the co-normal derivative, $\nu = [\nu_1, \nu_2, \ldots, \nu_n]$ being the unit outward normal at Γ.

Theorem 1 *Assume (H.1) = (1.3), (H.2) = (1.4). Let* $f \in L_2(0,T; H^{-1}(\Omega)) \cap L_2(R^1_x; H^{-1}(\Sigma_T))$ *where here* x *denotes the space variable in the normal direction. Let* w *be a solution of problem (1.1) within the class (1.8a–b). Assume, moreover, that the vector field* H *of hypothesis (1.4) satisfies the condition:*
(H.3)

$$H(x) \cdot \nu(x) \le 0, \ x \in \Gamma_0, \ \nu = [\nu_1, \nu_2, \ldots, \nu_n]. \tag{1.9}$$

Then, the following inequality holds true: If T *is large enough, then*

$$\int_0^T \mathcal{E}_w(t)dt \le C_T \Bigg\{ \int_0^T \int_{\Gamma_1} w^2 d\Sigma_1 + \left\| \frac{\partial w}{\partial \nu_A} \right\|^2_{H^{-1}(\Sigma_{1T})} + \|w\|^2_{H^{-1}(Q)}$$
$$+ \left| \int_0^T \left(w|_{\Gamma_1}, \frac{\partial A_0^{-1} w_t}{\partial \nu_A} \right)_{L_2(\Gamma_1)} dt \right| + \int_0^T |f|^2_{H^{-1}(\Omega)} dt + \|\Lambda f\|^2_{L_2(Q_T)} \Bigg\}, \tag{1.10}$$

where Λ *is the operator in (2.2) below lifting by one unit in time and in the tangential space variables.* □

Remark 1 Inequality (1.10) is a global observability-type estimate, whereby the $L_2(\Omega) \times H^{-1}(\Omega)$-norm of the solution in the interior is controlled (bounded) by the indicated boundary traces, modulo lower-order terms. When $f \equiv 0$ and $\mathcal{A} = -\Delta$ this inequality (1.10) can be extracted from the 1989 results of [L-T.2; e.g., Eqn. (4.34)]. See Remark 2 below. A micro-local (as opposed to global) version of this inequality (1.10) was later given in the variable coefficient case in [B-H-L-R-Z] expanding upon [B-L-R], subject to geometric optics assumptions. Still, in the variable coefficient case, a *related* version of (1.10) was later given in [T.1], subject to the assumption of existence of a pseudo-convex function. Our Riemann geometric assumption (H.2) is more general than the existence of a pseudo-convex function [Y.1]. □

Consequence: Uniform stabilization of Eqn. (1.1a–b) with boundary feedback in the Dirichlet B.C. We next consider the uniform stabilization problem of the following (closed-loop) wave equation with an explicit dissipative feedback control in the Dirichlet B.C. of Γ_1:

$$\begin{cases} w_{tt} + \mathcal{A}w = 0 & \text{in } Q \equiv (0,\infty) \times \Omega; & (1.11a) \\ w(0,\,\cdot\,) = w_0;\; w_t(0,\,\cdot\,) = w_1 & \text{in } \Omega; & (1.11b) \\ w|_{\Sigma_0} \equiv 0 & \text{in } \Sigma_0 \equiv (0,\infty) \times \Gamma_0; & (1.11c) \\ w|_{\Sigma_1} = \dfrac{\partial(\mathcal{A}_0^{-1} w_t)}{\partial \nu_{\mathcal{A}}} & \text{in } \Sigma_1 \equiv (0,\infty) \times \Gamma_1, & (1.11d) \end{cases}$$

where $\frac{\partial}{\partial \nu_{\mathcal{A}}}$ is the co-normal derivative, defined below (1.8b). Moreover, $\mathcal{A}_0$ is the operator defined in (1.5a). In the case of the pure wave equation (constant coefficients $a_{ij}(x)$ = Kronecker δ_{ij}), the explicit, dissipative feedback control in (1.11d) was introduced in [L-T.1]: this reference then provided, for the first time (the semigroup well-posedness result in Theorem 2 below and) a uniform stabilization result of the wave equation on the space $L_2(\Omega) \times H^{-1}(\Omega)$ (which had been discovered a few years earlier by I. Lasiecka-R. Triggiani and J. L. Lions to be the space of optimal regularity for the solution of second-order hyperbolic equations under $L_2(0,T;L_2(\Gamma)) \equiv L_2(\Sigma)$-Dirichlet control), by means of an $L_2(0,\infty;L_2(\Gamma_1))$-feedback control as in (1.11d).

Theorem 2 (Well-posedness [L-T.1, Thm. 1.1, p. 345]) *Assume (H.1) = (1.3). Problem (1.11) defines a s.c. contraction semigroup $T_D(t)$ on Z, see (1.7).* □

Theorem 3 (Uniform stabilization) *Assume (H.1), (H.2), (H.3). Then, the s.c. contraction semigruop $T_D(t)$ guaranteed by Theorem 1 is uniformly stable on Z: there exist constants $M \geq 1$, $\delta > 0$ such that*

$$\|T_D(t)\|_{\mathcal{L}(Z)} \leq Me^{-\delta t},\; t \geq 0; \;\text{ or } \mathcal{E}_w(t) \leq M_1 e^{-\delta t} \mathcal{E}_w(0). \quad \square \qquad (1.12)$$

Theorem 3 may be obtained as a corollary of Theorem 2.

2 PROOF OF THEOREM 1 FOR $F \equiv 0$: ENERGY ESTIMATE

Orientation. The proof follows closely the strategy of [L-T.2], which consists of two main steps, as critically generalized in its second step to the variable coefficient case, by a Riemann geometric approach [Y.1]. These two main steps are:

(a) a preliminary pseudo-differential change of variable (for the corresponding half-space problem) from the original variable $w(t,x)$ of problem (1.1) to a new variable $v(t,x)$ of problem (2.4) below. This change of variable has the purpose of lifting by one unit upward the regularity of w in t and in the tangential space variables and thus, via the governing equation, of w in the normal space variable as well. Thus, the required topology $L_2(\Omega) \times H^{-1}(\Omega)$ for $\{w, w_t\}$ is lifted to the $H^1(\Omega) \times L_2(\Omega)$-topology for $\{v, v_t\}$, a level where energy methods apply.

(b) An energy (multiplier) analysis at the $H^1(\Omega) \times L_2(\Omega)$-level of the resulting v-problem (2.4), which in [L-T.2] was carried out via (by now) "classical" multipliers suitable for the constant coefficient case ($\mathcal{A} = -\Delta$) considered there (i.e., the multipliers $h \cdot \nabla v$, $v \operatorname{div}_0 h$, v_t in the Euclidean metric, where h is a smooth coercive vector field on Ω).

The present proof then uses step (a) verbatim and replaces the classical multiplier analysis for step (b) in the Euclidean metric with the Riemann geometry multipliers version recently introduced by [Y.1] to handle the variable coefficient case of an operator such as $\mathcal{A}$: i.e., $\langle H, \nabla_g v\rangle_g$, $v \operatorname{div}_0 H$, v_t in the Riemann metric g.

2.1 *From the original w-problem to the new v-problem*

For simplicity of exposition, we take here $f \equiv 0$ in order to fall into [L-T.2]. Keeping track of $f \neq 0$ is readily done. Let w be a solution of (1.1a–b), in the class (1.8a–b), subject to assumption (1.9). Starting with the original variable w of the problem in (1.1), we define the truncation of w by

$$\bar{w} = w, \quad 0 < t < T; \; \bar{w} = 0 \text{ elsewhere.} \tag{2.1}$$

From [L-T.2, p. 211–212] there is a bounded operator

$$\Lambda : H^s(R^1_t \times R^{n-1}_y) \to H^{s+1}(R^1_t \times R^{n-1}_y) \tag{2.2}$$

for $s \geq 0$, given explicitly by going over to the half-space problem in [L-T.2, Sec. 5], where y is the tangential variable, such that the change of variable

$$v = \Lambda \bar{w} \tag{2.3}$$

transforms the w-problem (1.1a–b) into the problem

$$\begin{cases} v_{tt} = \mathcal{A}v + K\bar{w}, & \text{in } (0,T) \times \Omega, \\ v|_{\Gamma_0} = 0, & \text{in } (-\infty, \infty) \times \Gamma_0, \end{cases} \tag{2.4}$$

K being a commutator operator, possessing the crucial property of Proposition 5 below. Denote $\Sigma_{iT} = (0,T) \times \Gamma_i$, $\Sigma_{i\infty} = (-\infty,\infty) \times \Gamma_i$, $\Sigma_T = (0,T) \times \Gamma$, $Q_T = (0,T) \times \Omega$, and $Q_\infty = (-\infty, \infty) \times \Omega$.

Proposition 4 (L-T.2, Lemma 3.1, p. 199) *Let w be a solution of problem (1.1) with $f \equiv 0$ in the class (1.8a–b), and let v be given by (2.3). Then*

$$\|v|_{\Sigma_1}\|^2_{H^1(\Sigma_{1,\infty})} + \left\|\frac{\partial v}{\partial \nu_{\mathcal{A}}}\right\|^2_{L_2(\Sigma_{1,\infty})} \leq C\left\{\|w|_{\Sigma_1}\|^2_{L_2(\Sigma_{1T})} + \left\|\frac{\partial w}{\partial \nu_{\mathcal{A}}}\right\|^2_{H^{-1}(\Sigma_{1T})}\right\}, \tag{2.5}$$

where the constant $C > 0$ does not depend on T.

Proposition 5 (L-T.2, Corollary 3.3, p. 200) *Let w be a solution of (1.1a–b) with $f \equiv 0$ in the class (1.8a–b), and let v be given by (2.3). Then,* *with reference to $\mathcal{E}_w(t)$ defined in (1.6a), the following inequality holds true:*

$$\int_0^T \|Kw\|^2_{L^2(\Omega)} dt \leq C\left[\int_0^T \|w|_{\Gamma_1}\|^2_{L^2(\Gamma_1)} dt + \left\|\frac{\partial w}{\partial \nu_{\mathcal{A}}}\right\|^2_{H^{-1}(\Sigma_{1T})} + \int_{-\infty}^{\infty} \|v\|^2_{L^2(\Omega)} dt + \mathcal{E}_w(T) + \mathcal{E}_w(0)\right], \qquad (2.6)$$

where the constant C does not depend on T.

Remark 2 In effect, the proof in [L-T.2] refers explicitly to the feedback problem (1.11) rather than to problem (1.1a–b) in the class (1.8a–b). But problem (1.11) does satisfy the property (1.8a) (see Thm. 2) as well as (1.8b): see [L-T.1] for $w|_{\Sigma_1}$ and see [L-T.2, Lemma 2.1] stating that

$$\left\|\frac{\partial w}{\partial \nu_{\mathcal{A}}}\right\|_{H^{-1}(\Sigma_\infty)} \leq C\|w|_\Sigma\|_{L_2(\Sigma_{1T})}. \qquad (2.7)$$

The proof of [L-T.2] takes advantage of estimate (2.7), see [L-T.2, below (5.27)], as well as of the dissipativity relation for (1.11): in the general case of Eqn. (1.1) (with $f \equiv 0$) with no boundary conditions imposed, the counterpart of the dissipativity relation for problem (1.11) is now

$$\mathcal{E}_w(t) + 2\int_0^t \left(w|_{\Gamma_1}, \frac{\partial \mathcal{A}_0^{-1} w_t}{\partial \nu_{\mathcal{A}}}\right)_{L_2(\Gamma_1)} d\tau = \mathcal{E}_w(0). \qquad (2.8)$$

Thus, in treating the more general case (1.1a–b) within the class (1.8a–b), we have that the estimates of Lemma 3.1 and of Corollary 3.3 in [L-T.2] must now be augmented with the term $\left\|\frac{\partial w}{\partial \nu_{\mathcal{A}}}\right\|^2_{H^{-1}(\Sigma_{1T})}$ to obtain (2.5) of Proposition 4, and (2.6) of Proposition 5, while this term was absorbed by (2.7) in [L-T.2] following Eqn. (5.27) there, p. 212. Moreover, in the present proof, (2.8) replaces the dissipative version specialized for (1.11) in [L-T.2], which was used following (4.24) there, p. 203.

2.2 Analysis of the v-problem (2.4) by Riemann geometric energy methods

The following result is the counterpart of [L-T.2, Theorem 4.1, p. 200], where it was obtained for $\mathcal{A} = -\Delta$ by using "classical" multipliers in the Euclidean metric. The present proof treats instead the variable coefficient operator $\mathcal{A}$ in (1.2) and uses energy methods in the Riemann metric $(\mathbb{R}^n, g)$.

Theorem 6 *Let w be a solution of (1.1a–b) with $f \equiv 0$ in the class (1.8a–b) subject to hypothesis (1.9). Assume (H.2) = (1.4). Then, for T large enough,* *the following inequality holds true*

$$\int_0^T E_v(t)dt \leq C_T \left[\|w|_{\Sigma_1}\|^2_{L^2(\Sigma_{1T})} + \left\| \frac{\partial w}{\partial \nu_{\mathcal{A}}} \right\|^2_{H^{-1}(\Sigma_{1T})} \right] + \|v\|^2_{L_2(Q_T)} + \|w\|^2_{H^{-1}((0,T)\times\Omega)} + \left| \int_0^T \left(w|_{\Gamma_1}, \frac{\partial \mathcal{A}_0^{-1} w_t}{\partial \nu_{\mathcal{A}}} \right)_{L_2(\Gamma_1)} dt \right| \Bigg], \tag{2.9}$$

where $\mathcal{E}_w(t)$ is defined in (1.6a), and where we have set

$$E_v(t) = \int_\Omega [v_t(t) + |\nabla_g v(t)|_g^2] dx. \quad \square \tag{2.10}$$

Proof. We will use the Riemann geometry energy method given in [Y.1], combined with the strategy of [L-T.2, Theorem 4.1, p. 200]. To do so, we need several multiplier identities.

(i) Multiplying (2.4) by $H(v) = \langle \nabla_g v, H \rangle_g$ and integrating by parts, we obtain e.g., [Y.1, Proposition 2.1(1)],

$$\begin{aligned} \int_{Q_T} DH(\nabla_g v, \nabla_g v) dQ &= \int_{\Sigma_T} \frac{\partial v}{\partial \nu_{\mathcal{A}}} H(v) d\Sigma + \frac{1}{2} \int_{\Sigma_T} (v_t^2 - |\nabla_g v|_g^2)\, H \cdot \nu \, d\Sigma \\ &\quad - \frac{1}{2} \int_{Q_T} (v_t^2 - |\nabla_g v|_g^2) \operatorname{div}_0 H \, dQ \\ &\quad + \int_{Q_T} (Kw) H(v) dQ - [(v_t, H(v))]_0^T \end{aligned} \tag{2.11}$$

(counterpart of [L-T.2, Eqn. (4.3)] in the constant coefficient case $\mathcal{A} = -\Delta$) where $DH(\,\cdot\,,\,\cdot\,)$ is the covariant differential of the vector field H.

(ii) Multiplying (2.4) by $v \operatorname{div}_0 H$ and integrating by parts, we obtain, by the boundary condition $v|_{\Sigma_0}$ in (2.4b), e.g., [Y.1, Proposition 2.1(2)],

$$\begin{aligned} \int_{Q_T} (v_t^2 - |\nabla_g v|_g^2) P \, dQ &= [(v_t, vP)_{L_2(\Omega)}]_0^T + \int_{Q_T} v \nabla_g P(v) dQ \\ &\quad - \int_{Q_T} (Kw) vP \, dQ - \int_{\Sigma_{1T}} \frac{\partial v}{\partial \nu_{\mathcal{A}}} vP \, d\Sigma, \end{aligned} \tag{2.12}$$

where $P = \operatorname{div}_0 H$, $\nabla_g P(v) = \langle \nabla_g P, \nabla_g v \rangle_g = \langle A \nabla_0 P, \nabla_g v \rangle_g = \nabla_0 P \cdot \nabla_g v$. (Counterpart of [L-T.2, Eqn. (4.4)]) in the constant coefficient case.

(iii) Multiplying (2.4) by v_t and integrating by parts, we obtain, by the boundary condition in (2.4b),

$$E_v(t) - E_v(0) = 2 \int_{Q_T} (Kw) v_t dQ + \int_{\Sigma_{1T}} \frac{\partial v}{\partial \nu_{\mathcal{A}}} v_t \, d\Sigma \tag{2.13}$$

as in [L-T.2, Eqn. (4.5)]. With the identities (2.11)–(2.13) at hand and using the inequality in (1.4) to take the place of that in [L-T.2, Eqn. (1.10)], we may get the inequality in (2.9) by following the proof of [L-T.2, Theorem 4.1, p. 200] using (2.8) in place of its dissipative version for (1.11) valid there. $\square$

2.3 Return from variable v to original variable w: Proof of Theorem 1

We shall establish the desired estimate (1.10) of Theorem 1 for w starting from inequality (2.9) of Theorem 6 for v. Our proof here follows closely [L-T.2, Theorem 4.2, p. 204–205]. Moreover, the norm of $v = \Lambda \bar{w}$ in $L_2(Q_T)$ is dominated by the norm of w in $H^{-1}(Q_T)$: Λ lifts in t and in the tangential direction; then, Eqn. (1.1a) lifts by one in the normal direction as well.

References

[B-L-R] C. Bardos, J. Lebeau, J. Rauch, Sharp sufficient conditions for the observation, control and stabilization of waves from the boundary, *SIAM J. Control & Optimiz.* 30, N.S. (1992), 1024–1065.

[B-H-L-R-Z] C. Bardos, L. Halperin, G. Lebeau, J. Rauch, E. Zuazua, Stabilization de l'equation des ondes au moyen d'un feedback portant sur la condition aux limites de Dirichlet, *Asymptotic Analysis.*

[L-T.1] I. Lasiecka and R. Triggiani, Uniform exponential energy decay of the wave equation in a bounded region with $L_2(0, \infty; L_2(\Gamma))$-feedback control in the Dirichlet boundary conditions, *J. Diff. Eqns.* 66 (1987), 340-390.

[L-T.2] I. Lasiecka and R. Triggiani, Uniform stabilization of the wave equation with Dirichlet or Neumann feedback control without geometrical conditions, *Appl. Math. & Optim.* 25 (1992), 189–224. Preliminary version in LNCIS vol. 147, Springer-Verlag, pp. 62–108, Proceedings of the Third Working Conference, Montpellier, France, January 1989.

[L-T-Y.1] I. Lasiecka, R. Triggiani, P. F. Yao, Exact controllability for second-order hyperbolic equations with variable coefficients–principal part and and first order terms, *Non-Linear Analysis Theory, Methods & Applications* 30 (1997), 111–122.

[T.1] D. Tataru, A-priori estimates of Carleman's type in domains with boundary, *J. Math. Pure & Appl.* 73 (1994), 355–387.

[Y.1] P. F. Yao, Exact controllability of wave equations with variable coefficients in a bounded region, to appear.

IDENTIFICATION PROBLEM FOR A WAVE EQUATION VIA OPTIMAL CONTROL

Suzanne Lenhart and Min Liang

University of Tennessee
Mathematics Department
Knoxville, TN 37996-1300

Vladimir Protopopescu

Oak Ridge National Laboratory
Computer Science and Mathematics Division
Oak Ridge, TN 37831-6364

Abstract We approximate an identification problem by applying optimal control techniques to a Tikhonov's regularization. We seek to identify the dispersive coefficient in a wave equation and allow for the case of error or uncertainty in the observations used for the identification.

1 INTRODUCTION

We apply optimal control techniques to find approximate solutions to an identification problem for a wave equation. Consider the wave equation in $Q = \Omega \times (0,T)$:

$$\begin{aligned} w_{tt} &= \Delta w + hw + f && \text{in } Q \\ w &= 0 && \text{on } \partial\Omega \times (0,T) \\ w &= w_0(x) && t=0, \quad x \in \Omega \\ w_t &= w_1(x) && t=0, \quad x \in \Omega \end{aligned} \tag{1.1}$$

where $f \in L^2(Q)$, $w_0 \in H_0^1(\Omega)$, $w_1 \in L^2(\Omega)$ and $\Omega \subset R^n$ with C^2 boundary. We seek to identify the dispersive coefficient h from observations of the solution on a set $Q' \subset Q$.

The identification problem is to find bounded h such that the corresponding solution $w = w(h)$ to (1.1) is close to the observations z on Q'. Thus, we want to minimize the functional

$$J(h) = \frac{1}{2}\int_{Q'}(w(h)-z)^2 dxdt \tag{1.2}$$

over the class

$$U = \{h \in L^\infty(Q) | -M \le h(x,t) \le M\}.$$

To approximate this problem, we introduce the following control problem for $\beta > 0$:

$$\min_{h\in U} J_\beta(h) \tag{1.3}$$

with

$$J_\beta(h) = \frac{1}{2}\left(\int_{Q'}(w(h)-z)^2 dxdt + \beta\int_Q h^2 dxdt\right).$$

The coefficient to be identified is viewed as a control, which is adjusted to get the corresponding solution, $w(h)$, close to the observations z. This type of approximation is called Tikhonov's regularization [1].

See Liang [6] for some further results on this type of control problem (without the connection to identification). See Lenhart, Protopopescu and Yong [4, 5] for similar approximation techniques for other wave equation problems. See [1, 2] for other approaches to using Tikhonov's regularization for identification problems.

In section 2, we prove the existence of an optimal control to the approximate problem (1.3) and characterize it through an optimality system. In section 3, we connect these approximations to the identification problem as $\beta \to 0$. Specifically, we show that as $\beta \to 0$, the optimal control h_β for the functional J_β, converges to a feasible coefficient. We allow for the possibility that there is some error or uncertainty in the observations z. Thus the solution, which is the limit of $w_\beta = w(h_\beta)$ as $\beta \to 0$, may not match the observation z.

2 OPTIMAL CONTROL PROBLEM

We assume $f,\ f_t \in L^2(Q)$ and define the bilinear form:

$$B[u,v;t] = \int_\Omega \nabla u \nabla v dx - \int_\Omega huv dx.$$

Definition: Given $h \in U$, a function $w \in L^2(0,T;H_0^1(\Omega))$, with $w_t \in L^2(0,T;L^2(\Omega))$ and $w_{tt} \in L^2(0,T;H^{-1}(\Omega))$ is a weak solution of the problem (1.1) if

$$\int_0^T < w_{tt}, \phi > dt + \int_0^T B[w,\phi;t]dt = \int_Q f\phi dxdt$$

for any $\phi \in H_0^1(\Omega)$, and $w(0) = w_0$; $w_t(0) = w_1$, Here $< \, , \, >$ denotes the duality between $H^{-1}(\Omega)$ and $H_0^1(\Omega)$.

Lemma 2.1 For $h \in U$, the problem (1.1) has a unique weak solution w and

$$\sup_{0 \le t \le T} \left(\|w(t)\|_{H_0^1(\Omega)} + \|w_t(t)\|_{L^2(\Omega)} \right) + \|w_{tt}\|_{L^2(0,T;H^{-1}(\Omega))}$$
$$\le C(\|f\|_{L^2(Q)} + \|w_0\|_{H_0^1(\Omega)} + \|w_1\|_{L^2(\Omega)}).$$

The proof uses standard estimates on Galerkin approximations starting from smooth initial data [3, 6].

Theorem 2.1: There exists an optimal control $h_\beta \in U$, which minimizes the objectve functional $J_\beta(h)$ over $h \in U$.

Proof: Let $\{h^n\} \in U$ be a minimizing sequence such that

$$\lim_{n \to \infty} J(h^n) = \inf_{h \in U} J(h).$$

Lemma 2.1 gives a priori estimates on $w^n = w(h^n)$ with bounds independent of n. On a subsequence, by weak compactness, there exists w_β in $L^2(0,T;H_0^1(\Omega))$ such that

$$\begin{aligned} w^n &\rightharpoonup w_\beta \text{ weakly in } L^2(0,T;H_0^1(\Omega)), \\ w_t^n &\rightharpoonup (w_\beta)_t \text{ weakly in } L^2(Q), \\ h^n &\rightharpoonup h_\beta \text{ weakly in } L^2(Q), \\ w_{tt}^n &\rightharpoonup (w_\beta)_{tt} \text{ weakly in } L^2(0,T;H^{-1}(\Omega)). \end{aligned}$$

By a compactness result [7], we have $w^n \to w^\beta$ strongly in $L^2(Q)$ and hence $w_\beta = w(h_\beta)$. Using the lower-semicontinuity of L^2 norm with respect to weak convergence, we conclude that h_β is an optimal control. □

To characterize the optimal control, we need to differentiate the maps

$$h \to J_\beta(h) \text{ and } h \to w(h).$$

Lemma 2.2: The mapping

$$h \in U \longrightarrow w(h)$$

is differentiable in the following sense;

$$\frac{w(h + \varepsilon\ell) - w(h)}{\varepsilon} \rightharpoonup \Psi \text{ weakly in } L^2(0,T;H_0^1(\Omega))$$

as $\varepsilon \to 0$, for any $h, h + \varepsilon\ell \in U$. Moreover Ψ is a weak solution of

$$\begin{aligned} \Psi_{tt} &= \Delta\Psi + h\Psi + \ell w, && \text{in } Q \\ \Psi &= 0 && \text{on } \partial\Omega \times (0,T) \\ \Psi &= 0 && t = 0, \ x \in \Omega \\ \Psi_t &= 0 && t = 0, \ x \in \Omega \end{aligned} \tag{2.1}$$

where $w = w(h)$.
Proof: Denoting $w^\varepsilon = w(h+\varepsilon\ell)$ and $w = w(h)$, we have

$$\begin{aligned} &\sup_{0\le t\le T} \left\| \frac{(w^\varepsilon - w)(t)}{\varepsilon} \right\|_{H_0^1(\Omega)} + \left\| \frac{w_{tt}^\varepsilon - w_{tt}}{\varepsilon} \right\|_{L^2([0,T];H^{-1}(\Omega))} \\ \le\ & C\|\ell\|_{L^\infty}. \end{aligned}$$

On a subsequence, we have the existence of Ψ in $L^2(0,T;H_0^1(\Omega))$ and Ψ solves (2.1). □

We derive the necessary conditions to characterize an optimal control.

Theorem 2.2: Given an optimal control, h_β, and corresponding state, $w_\beta = w(h_\beta)$, there exists a weak solution p in $L^2(0,T;H_0^1(\Omega))$ to the adjoint problem:

$$\begin{aligned} p_{tt} &= \Delta p + hp + (w_\beta - z)\chi_{Q'}, && \text{in } Q \\ p &= 0, && \text{on } \partial\Omega \times (0,T) \\ p &= p_t = 0, && t = T,\ x \in \Omega \end{aligned} \tag{2.2}$$

where $\chi_{Q'}$ is the characteristic function of the set Q'. Furthermore, h_β satisfies

$$h_\beta = \max(-M,\ \min(-\frac{wp}{\beta}, M)). \tag{2.3}$$

Proof: Let $h_\beta + \varepsilon\ell \in U$ for $\varepsilon > 0$ and $w^\varepsilon = w(h_\beta + \varepsilon\ell)$. Since the minimum of J_β is achieved at h_β, we have

$$\begin{aligned} 0 &\le \lim_{\varepsilon\to 0+} \frac{J_\beta(h_\beta + \varepsilon\ell) - J_\beta(h_\beta)}{\varepsilon} \\ &= \int_{Q'} \Psi(w_\beta - z)\,dx\,dt + \beta \int_Q h_\beta \ell\,dx\,dt. \end{aligned} \tag{2.4}$$

Let p be the weak solution of the adjoint problem (2.2). Using (2.1) and (2.2) in (2.4), we obtain

$$0 \le \int_Q \ell(pw_\beta + \beta h_\beta)\,dx\,dt,$$

which gives the desired characterization (2.3). □

Thus, for a fixed β, the optimal control can be expressed from (2.3) in terms of the solution of the optimality system

$$\begin{aligned} w_{tt} &= \Delta w + \max(-M,\ \min(-\frac{wp}{\beta}, M))w + f \quad \text{in } Q \\ p_{tt} &= \Delta p + \max(-M,\ \min(-\frac{wp}{\beta}, M))p + (w - z)\chi_{Q'} \quad \text{in } Q \\ w &= p = 0 \qquad \text{on } \partial\Omega \times (0,T) \\ w &= w_0, \quad w_t = w_1 \qquad t = 0,\ x \in \Omega \\ p &= p_t = 0 \qquad t = T,\ x \in \Omega. \end{aligned} \tag{2.5}$$

Note that bounded solutions of the optimality system are unique if T is sufficiently small. See [4], [6] for similar results. If one assumes h is a function of x only or t only, better uniqueness results can be obtained [6].

3 IDENTIFICATION PROBLEM

We now use a sequence of optimal controls (as $\beta \to 0$) to identify h from observations z. We do not assume that z is in the range of the map

$$h \in U \;\to\; w(h)|_{Q'}. \tag{3.1}$$

This situation could occur from inaccurate observations.

__Theorem 3.1:__ There exist: (i) a sequence $\beta_n \to 0$, (ii) corresponding optimal controls, h_{β_n}, for the functional $J_{\beta_n}(h)$, (iii) $h^* \in U$, and (iv) $w^* = w(h^*)$ (solution of (1.1)), such that

$$\begin{aligned} h_{\beta_n} &\rightharpoonup h^* \text{ weakly in } L^2(Q), \\ w_{\beta_n} = w(h_{\beta_n}) &\rightharpoonup w^* \text{ weakly in } L^2(0,T;H_0^1(\Omega)), \end{aligned}$$

and

$$\int_{Q'} (w^* - z)^2 dx dt = \inf_{h \in U} \int_{Q'} (w(h) - z)^2 dx dt. \tag{3.2}$$

__Note:__ The limit w^* can be interpreted as a (not necessarily unique) projection of z onto the range of the map (3.1).

Proof: Note that the a priori estimates from Lemma 1.1 are independent of β. Thus we can find a sequence $\beta_n \to 0$ and $h^* \in U, w^* \in L^2(0,T;H_0^1(\Omega))$ such that

$$J_{\beta_n}(h_{\beta_n}) = \inf_{h \in U} J_{\beta_n}(h)$$

and

$$\begin{aligned} h_{\beta_n} &\rightharpoonup h^* \text{ weakly in } L^2(Q), \\ w_{\beta_n} &\rightharpoonup w^* \text{ weakly in } L^2(0,T;H_0^1(\Omega)), \\ w_{\beta_n} &\to w^* \text{ strongly in } L^2(Q), \\ (w_{\beta_n})_{tt} &\rightharpoonup (w^*)_{tt} \text{ weakly in } L^2(0,T;H^{-1}(\Omega)). \end{aligned}$$

where $w^* = w(h^*)$, is the weak solution of (1.1). Now for any $\bar{h} \in U, \bar{w} = w(\bar{h})$, we have

$$J_\beta(h_\beta) \le J_\beta(\bar{h}). \tag{3.3}$$

Letting $\beta_n \to 0$ in (3.3) gives

$$\int_{Q'} (w^* - z)^2 dxdt \leq \int_{Q'} (\bar{w} - z)^2 dxdt$$

which yields the desired result (3.3). □

Acknowledgments

S.L and V.P. acknowledge partial support from the US Department of Energy, Office of Basic Energy Sciences under contract No. ACO5-96OR22464 with Lockheed Martin Energy Research Corporation.

References

[1] Banks, H.T. and Kunisch, K., *Estimation Techniques for Distributed Parameter Systems,* Birkhäuser, Boston, 1989

[2] Borggaard, J. and Burns, J., *A PDE Sensitivity Equation Method for Optimal Aerodynamic Design,* ICASE Report 96-44, NASA Langley Research Center

[3] Evans, L.C., *Partial Differential Equations,* AMS, Providence, 1998

[4] Lenhart, S., Protopopescu, V., and Yong J., "Optimal Control of a Reflection Boundary Coefficient in an Acoustic Wave Equation", Applicable Analysis, 68(1998), 179-194

[5] Lenhart, S., Protopopescu, V., and Yong J., "Solving Inverse Problems of Identification Type by Optimal Control Methods", in *Applied Nonlinear Dynamics and Stochastic Systems near the Millenium,* J.B. Kadtke and A. Bulsara, eds., American Institute of Physics, 1998, 87-94

[6] Liang, M., "Bilinear Optimal Control of a Wave Equation", University of Tennessee PhD Dissertation, 1998, and Mathematical Models and Methods in Applied Sciences, 9(1), 1999.

[7] Simon, J., "Compact Sets in The Space $L^p(0, T; B)$" , Annali di Matematica Pura ed Applicata CXLVI (1987), 65-96

OPTIMAL CONTROL THEORY: FROM FINITE DIMENSIONS TO INFINITE DIMENSIONS *

Xunjing Li

Laboratory of Mathematics for Nonlinear Sciences
and Department of Mathematics
Fudan University, Shanghai 200433, China

xjli@ms.fudan.edu.cn

1 INTRODUCTION

Optimal control theory has more than 40 years of history. For deterministic finite dimensional case, the works of Pontryagin *et al* (maximum principle [14]), Bellman (dynamic programming [1]) and Kalman (linear quadratic regulator problem [6]) have been regarded as three main milestones in the field.

The purpose of this paper is to give a brief survey on the works done by the people from Fudan Group related to optimal control theory for deterministic infinite dimensional systems.

2 MAXIMUM PRINCIPLE FOR EVOLUTION SYSTEMS

Maximum principle is the necessary conditions of optimal controls for finite dimensional systems. The infinite dimensional version of maximum principle was firstly proved by A. G. Butkovsky. Yu. V. Egorov [3] constructed an example showing that the maximum principle does not necessarily hold for infinite dimensional systems. In past 12 years, the people from Fudan Group makes an effort to derive the Pontryagin type maximum principle for infinite dimensional case under some assumptions.

Let us begin with the following hypothesis.

(H1) Let X be a Banach space, S be a closed and convex subset of $X \times X$, U be a separable metric space. Let e^{At} be a C_0-semigroup on X with generator $A : \mathcal{D}(A) \subset X \to X$.

*This work is supported in part by the NSFC, under grant 79790130, the National Key Project of China, and the Chinese State Education Commission Science Foundation.

(H2) Let $f : [0,T] \times X \times U \to X$ and $f^0 : [0,T] \times X \times U \to \mathbb{R}$ such that $f(t,y,u)$ and $f^0(t,y,u)$ are strongly measurable in $t \in [0,T]$, continuously Fréchet differentiable in $y \in X$ with $f(t,\cdot,\cdot)$, $f_y(t,\cdot,\cdot)$, $f^0(t,\cdot,\cdot)$ and $f^0_y(t,\cdot,\cdot)$ being continuous. Consider the following evolution equation

$$y(t) = e^{At}y(0) + \int_0^t e^{A(t-s)} f(s,y(s),u(s))ds, \qquad t \in [0,T], \tag{2.1}$$

where $u(\cdot) \in \mathcal{U}[0,T] \equiv \{u : [0,T] \to U \mid u(\cdot) \text{ measurable }\}$.

Any $(y(\cdot),u(\cdot)) \in C([0,T];X) \times \mathcal{U}[0,T]$ satisfying (1.1) is called a *feasible pair*. The set of all feasible pairs is denoted by $\mathcal{A}$. Next, we let $\mathcal{A}_{ad}$ be the set of all feasible pairs $(y(\cdot),u(\cdot)) \in \mathcal{A}$, such that

$$(y(0),y(T)) \in S, \tag{2.2}$$

For any $(y(\cdot),u(\cdot)) \in \mathcal{A}_{ad}$, we define

$$J(y(\cdot),u(\cdot)) = \int_0^T f^0(s,y(s),u(s))ds. \tag{2.3}$$

Optimal Control Problem (C). Find $(\bar{y}(\cdot),\bar{u}(\cdot)) \in \mathcal{A}_{ad}$, such that

$$J(\bar{y}(\cdot),\bar{u}(\cdot)) = \inf_{\mathcal{A}_{ad}} J(y(\cdot),u(\cdot)). \tag{2.4}$$

Definition. Let (H1)–(H2) hold and $(\bar{y}(\cdot),\bar{u}(\cdot)) \in \mathcal{A}_{ad}$ be an optimal pair. We say that $(\bar{y}(\cdot),\bar{u}(\cdot))$ satisfies the *Maximum Principle*, if there exists a nontrivial pair $(\psi^0,\psi(\cdot)) \in \mathbb{R} \times C([0,T];X^*)$, i.e., $(\psi^0,\psi(\cdot)) \neq 0$, such that

$$\psi^0 \leq 0, \tag{2.5}$$

$$\begin{aligned}\psi(t) = e^{A^*(T-t)}\psi(T) &+ \int_t^T e^{A^*(s-t)} f_y(s,\bar{y}(s),\bar{u}(s))^*\psi(s)ds \\ &+ \int_t^T e^{A^*(s-t)}\psi^0 f^0_y(s,\bar{y}(s),\bar{u}(s))ds, \qquad \text{a.e. } t \in [0,T],\end{aligned} \tag{2.6}$$

$$\langle \psi(0), x_0 - \bar{y}(0) \rangle - \langle \psi(T), x_1 - \bar{y}(T) \rangle \leq 0, \qquad \forall (x_0,x_1) \in S, \tag{2.7}$$

$$\begin{aligned}H(t,\bar{y}(t),\bar{u}(t),\psi^0,\psi(t)) = \max_{u\in U} H(t,\bar{y}(t),u,\psi^0,\psi(t)),& \\ \text{a.e. } t \in [0,T],&\end{aligned} \tag{2.8}$$

where

$$\begin{aligned}&H(t,y,u,\psi^0,\psi) = \psi^0 f^0(t,y,u) + \langle \psi, f(t,y,u) \rangle, \\ &\forall (t,y,u) \in [0,T] \times X \times U, \quad \forall (\psi^0,\psi) \in \mathbb{R} \times X^*.\end{aligned} \tag{2.9}$$

In 1985, Li and Yao [10] consider the problem with the endpoint constraints: $y(0) = x_0, \quad y(T) \in Q_1$. They proved that if Q_1 is finite codimensional in X, then, maximum principle holds. In 1987, H.O.Fattorini [4] consider the problem with the endpoint constraints: $y(0) = x_0, \quad y(T) = x_1$. He proved that if X is Hilbert space and the associate reachable set is finite codimension in X, then the maximum principle holds.

In 1991, Li and Yong [11] consider the general case. Let $\mathcal{R}$ be *the associate reachable set*:

$$\begin{aligned}\mathcal{R} =\{\xi(T) \in X \mid \xi(t) = &\int_0^t e^{A(t-s)} f_y(s, \bar{y}(s), \bar{u}(s))\xi(s)ds \\ &+ \int_0^t e^{A(t-s)}[f(s, \bar{y}(s), u(s)) - f(s, \bar{y}(s), \bar{u}(s))]ds, \\ &t \in [0,T],\ u(\cdot) \in \mathcal{U}[0,T]\ \},\end{aligned} \tag{2.10}$$

and Q be the *modified endpoint constraint set*:

$$\begin{aligned}Q = \{y_1 - \eta(T) \mid \eta(t) = e^{At}y_0 + \int_0^t e^{A(t-s)} f_y(s, \bar{y}(s), \bar{u}(s))\eta(s)ds, \\ t \in [0,T],\ (y_0, y_1) \in S\ \}.\end{aligned} \tag{2.11}$$

Theorem 2.1 *Let (H1)–(H2) hold. Let $(\bar{y}(\cdot), \bar{u}(\cdot))$ be an optimal pair of Problem (C). Assume*

(H3) $\mathcal{R} - Q \equiv \{r - q \mid r \in \mathcal{R}, q \in Q\}$ is finite codimensional in X.

(H4) The dual X^* of X is strictly convex.

Then, the pair $(\bar{y}(\cdot), \bar{u}(\cdot))$ satisfies the Maximum Principle.

For the optimal periodic [9](or anti-periodic) control problem, i.e.,

$$y(0) = y(T) \qquad (or \quad y(0) = -y(T)), \tag{2.12}$$

we have the following

Theorem 2.2 *Let e^{At} be a compact semigroup. Let $(\bar{y}, \bar{u})$ be an optimal pair for the optimal periodic (or anti-periodic) control problem. Then, there exists a pair $(\psi^0, \psi(\cdot)) \neq 0$, such that (2.6) and (2.8) hold, and*

$$\psi(0) = \psi(T) \qquad (or \qquad \psi(0) = -\psi(T)). \tag{2.13}$$

Moreover, if $\mathcal{N}(I - G^*(T,0)) = \{0\}$ (or $\quad \mathcal{N}(I + G^*(T,0)) = \{0\}$), then $\psi^0 = -1$.

In 1993, Yong [18] discussed the optimal control problem for infinite dimensional Volterra-Steiltjees evolution equations and derived a maximum principle.

The optimal control problem of a semilinear second order elliptic partial differential equation as follows: minimize the cost functional

$$J(y,u) = \int_\Omega f^0(x, y(x), u(x))dx. \tag{3.1}$$

subject to

$$\begin{aligned} &Ay(x) = f(x, y(x), u(x)), \qquad \text{in } \Omega, \\ &y\big|_{\partial\Omega} = 0. \end{aligned} \tag{3.2}$$

Assumptions:

(S1) $\Omega \subset \mathbb{R}^n$ is a bounded region with $C^{1,\gamma}$ boundary $\partial\Omega$, for some $\gamma > 0$ and U is a separable metric space.

(S2) Operator A is defined by

$$Ay(x) = -\sum_{i,j=1}^{n} \left(a_{ij}(x)y_{x_j}(x)\right)_{x_i}, \tag{3.3}$$

with $a_{ij} \in C(\overline{\Omega})$, $a_{ij} = a_{ji}$, $1 \le i,j \le n$ and for some $\lambda > 0$,

$$\sum_{i,j=1}^{n} a_{ij}(x)\xi_i\xi_j \ge \lambda \sum_{i=1}^{n} |\xi_i|^2, \qquad \forall x \in \Omega,\ (\xi_1, \xi_2, \cdots, \xi_n) \in \mathbb{R}^n. \tag{3.4}$$

(S3) $f : \Omega \times \mathbb{R} \times U \to \mathbb{R}$ satisfies

$$f_y(x, y, u) \le 0, \forall (x, y, u) \in \Omega \times \mathbb{R} \times U.$$

(S4) $\mathcal{Y}$ is a Banach space with strict convex dual $\mathcal{Y}^*$, $F : W_0^{1,p}(\Omega) \to \mathcal{Y}$ is continuously Fréchet differentiable and $Q \subset \mathcal{Y}$ is closed and convex.

Let $\mathcal{A}_{ad}$ be the set of all pair (y, u), such that $u(\cdot) \in \mathcal{U} = \{u : \Omega \to U \mid u \text{ is measurable }\}$ and the state constraint $F(y) \in Q$.

Problem (SD). Find an admissible pair $(\bar{y}, \bar{u}) \in \mathcal{A}_{ad}$, such that $J(\bar{y}, \bar{u}) = \inf_{\mathcal{A}_{ad}} J(y, u)$.

Let $z = z(\cdot\,; u) \in W_0^{1,p}(\Omega)$ be the unique solution of *variational system*:

$$\begin{aligned} Az(x) = f_y(x, \bar{y}(x), \bar{u}(x))z(x) &+ f(x, \bar{y}(x), u(x)) \\ &- f(x, \bar{y}(x), \bar{u}(x)), \qquad \text{in } \Omega, \\ z\big|_{\partial\Omega} = 0. & \end{aligned} \tag{3.5}$$

$\mathcal{R}$ be the *reachable set* and the *Hamiltonian* H as follows:

$$\mathcal{R} = \{z(\cdot\,; u) \bigm| u \in \mathcal{U}\}; \tag{3.6}$$

$$H(x, y, u, \psi^0, \psi) = \psi^0 f^0(x, y, u) + \psi f(x, y, u),$$
$$\forall (x, y, u) \in \Omega \times \mathbb{R} \times U, \ (\psi^0, \psi) \in \mathbb{R} \times \mathbb{R}. \tag{3.7}$$

In 1992, Yong [17] proved

Theorem (Maximum Principle) *Let (S1)–(S5) hold. Let $(\bar{y}, \bar{u}) \in \mathcal{A}_{ad}$ be an optimal pair of Problem (SD). Let $F'(\bar{y})\mathcal{R} - Q$ be finite codimensional in $\mathcal{Y}$. Then, there exists a triplet $(\psi^0, \psi, \varphi) \in [-1, 0] \times W_0^{1,p'}(\Omega) \times \mathcal{Y}^*$, such that $(\psi^0, \varphi) \neq 0$,*

$$\langle \varphi, \eta - F(\bar{y}) \rangle_{\mathcal{Y}^*, \mathcal{Y}} \leq 0, \qquad \forall \eta \in Q. \tag{3.8}$$

$$\begin{aligned} A\psi(x) &= f_y(x, \bar{y}(x), \bar{u}(x))\psi(x) \\ &\quad + \psi^0 f_y^0(x, \bar{y}(x), \bar{u}(x)) - F'(\bar{y}(\cdot))^* \varphi, \qquad \text{in } \Omega, \\ \psi\big|_{\partial\Omega} &= 0. \end{aligned} \tag{3.9}$$

$$H(x, \bar{y}(x), \bar{u}(x), \psi^0, \psi(x)) = \max_{u \in U} H(x, \bar{y}(x), u, \psi^0, \psi(x)),$$
$$\text{a.e. } x \in \Omega. \tag{3.10}$$

In addition, if $F'(\bar{y})^*$ is injective (i.e., $\mathcal{N}(F'(\bar{y})^*) = \{0\}$), then, $(\psi^0, \psi) \neq 0$.

In 1996, Gao [5] discussed the optimal control problem when the operator A dependent to the control variable u and derived a maximum principle.

4 RELATIONS BETWEEN MP AND DP

The another important approach to the optimal control problems was originated by R. Bellman for finite dimensional optimal control problems and is called the *dynamic programming method.* Recent works done by M. G. Crandall and P. L. Lions on the *viscosity solutions* for Hamilton-Jacobi-Bellman equations is a breakthrough in this direction.

Both maximum principle (MP) and dynamic programming (DP) are all necessary conditions for optimal controls. Thus, it is very natural to ask: Are there any relations between them? The objective of this section is to answer this question.

Let X be a Banach space. Suppose $\varphi : X \to \mathbb{R}$ and $x_0 \in X$. We define

$$D^+\varphi(x_0) = \{p \in X^* \mid \overline{\lim_{x \to x_0}} \frac{\varphi(x) - \varphi(x_0) - \langle p, x - x_0 \rangle}{|x - x_0|} \leq 0 \}, \tag{4.1}$$

and

$$D^-\varphi(x_0) = \{p \in X^* \mid \underline{\lim_{x \to x_0}} \frac{\varphi(x) - \varphi(x_0) - \langle p, x - x_0 \rangle}{|x - x_0|} \geq 0 \}. \tag{4.2}$$

We call $D^+\varphi(x_0)$ and $D^-\varphi(x_0)$ the *superdifferential* and *subdifferential* of φ at x_0, respectively. It should be pointed out that $D^+\varphi(x_0)$ and/or $D^-\varphi(x_0)$ could be empty.

For functions $\varphi : [0,T]\times X \to \mathbb{R}$, fix t_0, $\varphi(t_0,x)$ is a function of x. We may define its superdifferential and subdifferential in x, denoted by $D_x^+\varphi(t_0,x_0)$ and $D_x^-\varphi(t_0,x_0)$, respectively.

We consider the following family of optimal control problems: For any given $(t,x)\in[0,T)\times X$, let us consider the following state equation:

$$y_{t,x}(s) = e^{A(s-t)}x + \int_t^s e^{A(s-r)} f(r, y_{t,x}(r), u(r))dr, \quad s\in[t,T]. \tag{4.3}$$

with $u(\cdot)\in\mathcal{U}[t,T]$ and the cost functional

$$J_{t,x}(u(\cdot)) = \int_t^T f^0(r, y_{t,x}(r), u(r))dr + h(y_{t,x}(T)). \tag{4.4}$$

And define the *value function* $V : [0,T]\times X \to \mathbb{R}$ by the following:

$$V(t,x) = \inf_{u(\cdot)\in\mathcal{U}[t,T]} J_{t,x}(u(\cdot)), \qquad V(T,x) = h(x). \tag{4.5}$$

The following Theorem (see Li-Yong [12]) gives the result concerning the relation between MP and DP. That is an infinite version of the result of Zhou [20](which was for finite dimensional cases).

Theorem *Let (H1) and (H2) hold. In addition, we assume that $h(x)$ are Fréchet differentiable in x with $h_x(x)$ being continuous in (x,u). Let $V(t,x)$ be the value function of optimal control problem. Suppose $(\bar y(\cdot),\bar u(\cdot))$ is an optimal pair for the problem starting from $(0,x)$. Let $\psi(\cdot)$ be the solution of the following equation*

$$\begin{aligned}\psi(t) = &-e^{A^*(T-t)}h_x(\bar y(T)) + \int_t^T e^{A^*(r-t)} f_x(r,\bar y(r),\bar u(r))^*\psi(r)dr \\ &- \int_t^T e^{A^*(r-t)} f_x^0(r,\bar y(r),\bar u(r))dr, \qquad t\in[0,T].\end{aligned} \tag{4.6}$$

Then,

$$\begin{aligned}&\langle \psi(t), f(t,\bar y(t),\bar u(t))\rangle - f^0(t,\bar y(t),\bar u(t)) \\ &= \max_{v\in U}\left[\langle \psi(t), f(t,\bar y(t),v)\rangle - f^0(t,\bar y(t),v)\right] \\ &\equiv -H(t,\bar y(t),-\psi(t)), \qquad \text{a.e. } t\in[0,T].\end{aligned} \tag{4.7}$$

$$D_x^-V(t,\bar y(t)) \subset \{-\psi(t)\} \subset D_x^+V(t,\bar y(t)), \quad \forall t\in[0,T], \tag{4.8}$$

$$\begin{aligned}D^-V(t,\bar y(t)) &\subset \Big\{\Big(\langle\psi(t),A\bar y(t)\rangle - H(t,\bar y(t),-\psi(t)), -\psi(t)\Big)\Big\} \\ &\subset D^+V(t,\bar y(t)), \qquad \forall t\in\mathcal{T}(\bar y(\cdot),\bar u(\cdot)),\end{aligned} \tag{4.9}$$

where for any admissible pair $(y(\cdot),u(\cdot))$,

$$\begin{aligned}\mathcal{T}(y(\cdot),u(\cdot)) \equiv \Big\{&t\in[0,T] \mid y(t)\in\mathcal{D}(A), \\ &\lim_{\delta\downarrow 0}\frac{1}{\delta}\int_t^{t+\delta} |f(r,y(r),u(r)) - f(t,y(t),u(t))|dr = 0\Big\}.\end{aligned} \tag{4.10}$$

5 LQ PROBLEM WITH UNBOUNDED CONTROL AND INDEFINITE COST FUNCTIONAL

The linear quadratic optimal control (LQ, for short) problems in finite dimensional system was first studied by Bellman-Glisksberg-Gross. In 1960, Kalman [6] founded the relation between the LQ problems and Riccati equation.

The LQ problems in infinite dimensional systems discussed by Lions [13], Lukes-Russell, Curtain-Pritchard, Balakrishnan, and Lasiecka-Triggiani [7, 8] etc. Flandoli and Da Prato-Ichikawa as well as Bensoussan, Da Prato, Delfour and Mitter had done a direct study of the Riccati equation of infinite dimensional case. However, all the above works are deal with the nonnegative cost functional.

Li-Yong [12] developed the results of You [19] and Chen [2] into the LQ problem with indefinite cost functional and unbounded control in Hilbert space including Neumann boundary control or pointwise control (for space dimension $n = 1$) for a parabolic equation.

Recently, Wu-Li [15] generalized and improve the results of You [19], Chen [2], Li-Yong [12] and Lasiecka-Triggiani [7, 8] etc into the finite horizon LQ problems with indefinite cost functional and unbounded control in Hilbert space, which covers Dirichlet, Neumann boundary control and pointwise control (for $n \leq 3$) for a parabolic equation. Wu-Li [16] also established the necessary-sufficient frequency theorem or called positive real lemma, and derive the synthesis for optimal control of the coressponding infinite horizon LQ problem, which does not require the assumption for the exact controllability of the system.

Let X and U be real Hilbert spaces, $-A$ generate an analytic semigroup e^{-At} on X and A^{γ} be defined for all $\gamma \in \mathbb{R}$. Denote F^* be the adjoint operators of F. Consider the evolution system

$$y(t) = e^{-At}x + \int_0^t A^{\alpha} e^{-A(t-s)} Bu(s)\, ds, \tag{5.1}$$

where $\alpha \in [0, 1)$ is constant, $x \in X$ and $B \in \mathcal{L}(U, X)$. In the case of parabolic equation on a bounded domain $\Omega \subset R^n$, the relevant values of α are as follows: $\alpha = \frac{3}{4}+\varepsilon$, $\forall \varepsilon > 0$, for Dirichlet boundary control with $U = L^2(\partial\Omega)$, $X = L^2(\Omega)$; $\alpha = \frac{1}{4}+\varepsilon$, $\forall\varepsilon > 0$ for Neumann boundary control with $U = L^2(\partial\Omega)$, $X = L^2(\Omega)$; and $\frac{n}{4} < \alpha < 1$, for pointwise control with $U = H^{-2\alpha}(\Omega)$, $X = L^2(\Omega)$.

Let the cost functional be

$$J(x; u(\cdot)) = \int_0^{\infty} \{\langle Qy(t), y(t)\rangle + 2\,\mathrm{Re}\langle Sy(t), u(t)\rangle + \langle Ru(t), u(t)\rangle\}\, dt,$$

where $S \in \mathcal{L}(X, U)$. A control $u(\cdot)$ is called admissible at an initial state condition $x \in X$ if $u(\cdot) \in L^2(0, \infty; U)$ and $y(\cdot) \in L^2(0, \infty; X)$.

Problem LQ *For given $x \in X$, find an admissible $u(\cdot) \in L^2(0, \infty; U)$ so that the cost functional is minimized.*

Wu-Li [16] assume that e^{-At} is exponentially stable. Let

$$[Lu(\cdot)](t) \triangleq \int_0^t A^\alpha e^{-A(t-s)} Bu(s)\,ds, \quad \forall u(\cdot) \in L^2(0,\infty;U). \tag{5.2}$$

$$\begin{aligned}\widetilde{\Phi}(\omega) \triangleq\, & R + B^*(A^*)^\alpha(-i\omega + A^*)^{-1} Q A^\alpha (i\omega + A)^{-1} B \\ & + SA^\alpha(i\omega + A)^{-1}B + B^*(A^*)^\alpha(-i\omega + A^*)^{-1}S^*, \quad \forall \omega \in \mathbb{R},\end{aligned} \tag{5.3}$$

where $\widetilde{\Phi}(\cdot)$ is called the frequency characteristic assciated with Problem LQ.

[16] proved

Theorem *The following statements are equivalent:*

(I) *Problem LQ is uniquely solvable at any $x \in X$ with the optimal pair $(\bar{u}(\cdot), \bar{y}(\cdot))$;*

(II) *The algebraic Riccati equation*

$$\begin{aligned}& -\langle Ax, Px\rangle - \langle Px, Ax\rangle + \langle Qx, x\rangle \\ & \quad - \langle [S + B^*(A^*)^\alpha P]^* R^{-1} [S + B^*(A^*)^\alpha P]x, x\rangle = 0, \quad \forall\ x \in D(A),\end{aligned} \tag{5.4}$$

admits a solution P satisfying $(A^*)^\alpha P \in \mathcal{L}(X)$ and

$$\widetilde{A} \triangleq -A^\alpha\{A^{1-\alpha} + BR^{-1}[S + B^*(A^*)^\alpha P]\}$$

with

$$D(\widetilde{A}) = \{x \in D(A^{1-\alpha}) | \{A^{1-\alpha} + BR^{-1}[S + B^*(A^*)^\alpha P]\}x \in D(A^\alpha)\}$$

generates an exponentially stable and analytic semigroup. In this case, the state feedback optimal control of Problem LQ is given by

$$\bar{u}(t) = -R^{-1}[S + B^*(A^*)^\alpha P]\bar{y}(t), \qquad \text{for} \quad t \geq 0;$$

(III) *There exists some constant $\sigma > 0$ such that*

$$\langle (R + L^*QL + SL + L^*S^*)u(\cdot), u(\cdot)\rangle \geq \sigma\langle u(\cdot), u(\cdot)\rangle, \ \forall\, u(\cdot) \in L^2(0,\infty;U);$$

(IV) *There exists some constant $\sigma > 0$ such that*

$$\langle \widetilde{\Phi}(\omega)u, u\rangle \geq \sigma\langle u, u\rangle, \qquad \forall\ u \in U,\ \omega \in \mathbb{R}.$$

References

[1] R. Bellman, *Dynamic Programming,* Princeton Univ. Press, N.J., 1957.

[2] S. Chen, *Riccati equations arising in infinite dimensional optimal control problem, Jour Control Theory Appl.*, 2(1985) 64-72 (in Chinese).

[3] Yu. V. Egorov, *Optimal control in Banach spaces, Dokl. Acad. Nauk. USSR,* 150(1963), 241-244(in Russian).

[4] H. O. Fattorini, *A unified theory of necessary conditions for nonconvex control systems, Appl. Math. Optim.*, 15(1987), 141-185.

[5] H. Gao, *Ph D Thesis, Institute of Mathematics, Fudan University,* 1996.

[6] R. E. Kalman, *Contributions to the theory of optimal control, Bol. Soc. Math. Mexicana,* 5 (1960), 102–119.

[7] I. Lasiecka & R. Triggiani, *Differential and Algebraic Riccati Equations with Application to Boundary/Point Control Problems: Continuous Theory and Approximation Theory* LNCS, Vol.164, Springer-Verlag, New York, 1991.

[8] I. Lasiecka & R. Triggiani, *Riccati differential equations with unbounded coefficients and non-smooth terminal condition— The case of analytic semigroups, SIAM J. Math. Anal.*, 23(1992), 449-481.

[9] X. Li and S.-N. Chow, *Maximum principle of optimal control for functional differential systems, J.Optim. Theory & Appl.*, 54(1987), 335-360.

[10] X. Li and Y. Yao, *Maximum principle of distributed parameter systems with time lags, Lecture Notes in Control & Information Sciences,* 75(1985), 410-427.

[11] X. Li and J. Yong, *Necessary conditions of optimal controls for distributed parameter systems, SIAM J.Control & Optim.*, 29(1991), 895-908.

[12] X. Li and J. Yong, *Optimal Control Theory for Infinite Dimensional Systems,* Birkhäuser, Boston, 1995.

[13] J. L. Lions, *Optimal control of systems governed by partial differential equations,* Springer-Verlag, New York, 1971.

[14] L. S. Pontryagin, V. G. Boltyanskii, R. V. Gamkrelidze and E. F. Mischenko, *Mathematical Theory of Optimal Processes,* Wiley, New York, 1962.

[15] H. Wu & X. Li, *A linear quadratic problem with unbounded control in Hilbert spaces,* Preprint, 1997.

[16] H. Wu & X. Li, *An infinite horizon linear quadratic problem with unbounded control in Hilbert space,* Preprint, 1998.

[17] J. Yong, *Pontryagin's maximum principle for semilinear second order elliptic partial differential equations and variational inequalities with state constraints, Diff. Int. Eqs.*, 5(1992), 1307-1334.

[18] J. Yong, *Infinite dimensional Volterra-Stietjes evolution equations and related optimal control problem, SIAM J. Control & Optim.*, 31(1993), 530-568.

[19] Y. You, *Optimal control for linear systems with quadratic indefinite criterion on Hilbert spaces, Chin. Ann. Math. Ser.B,* vol.4(1983), 21-32.

[20] X. Y. Zhou, *Maximum principle, dynamic programming and their connection in deterministic control, J. Optim. Theory & Appl.*, 65(1990), 363-373.

BOUNDARY STABILIZATION OF A HYBRID SYSTEM

Kangsheng Liu

Department of Applied Mathematics
Zhejiang University, Hangzhou, 310027, China.
ksliu@public.hz.zj.cn

Zhuangyi Liu

Department of Mathematics & Statistics
University of Minnesota, Duluth, MN 55812, USA.
zliu@d.umn.edu

Abstract: In this paper, we consider the boundary stabilization of a degenerate hybrid system composed of an Euler-Bernoulli beam with a tip mass. It is proved that the system is exponentially stabilizable when the usual velocity feedback controls are applied at the end with the tip mass. We also establish time reversibility and spectral completeness of the closed-loop system.

Key Words: boundary stabilization, beam, hybrid system, backward well-posedness, spectral completeness, multiplier technique
AMS subject classification: 93D15, 35B37, 35B40

1 INTRODUCTION

We consider a hybrid system which consists of an Euler-Bernoulli beam linked to a rigid body. The system is governed by the following equations:

$$\begin{cases} \rho w_{tt} + p w'''' = 0, \quad (x,t) \in (0,L) \times (0,\infty), \\ w(0,t) = w'(0,t) = 0, \\ J w'_{tt}(L,t) + p w''(L,t) = g(t), \\ M w_{tt}(L,t) - p w'''(L,t) = h(t). \end{cases} \tag{1.1}$$

where prime represents the derivative with respect to the spacial variable x, $p, \rho > 0$ are the elasticity modulus and mass density respectvely; J is the rotatory inertia of the tip mass M; $g(t), h(t)$ are controls applied at the end $x = L$.

There are many papers in the literature on the stabilizability of the system (1.1) including [3], and [7]-[8]. It was shown in [3] that using the feedback law

$$\begin{cases} g(t) = -\alpha w_t'(L,t) \\ h(t) = -\beta w_t(L,t) \end{cases} \tag{1.2}$$

where $\alpha, \beta > 0$, (1.1) is strongly stabilizable, but not exponentially stabilizable. Moreover, Rao[7] proved the lack of exponential stabilizability for a general feedback law

$$\begin{cases} g(t) = -\alpha_1 w(L,t) - \alpha_2 w'(L,t) - \alpha_3 w_t(L,t) - \alpha_4 w_t'(L,t) \\ h(t) = -\beta_1 w(L,t) - \beta_2 w'(L,t) - \beta_3 w_t(L,t) - \beta_4 w_t'(L,t) \end{cases} \tag{1.3}$$

with $\alpha_i, \beta_i, i = 1, \cdots, 4$ being any real numbers. In the same paper, He also obtained the exponential stabilizability of (1.1) for the high-order feedback law

$$\begin{cases} g(t) = w_t'''(L,t) \\ h(t) = -w_t''(L,t) \end{cases} \tag{1.4}$$

Recently, Rao in [8] studied the plate version of the hybrid system (1.1). When reduced to the beam problem, his result implies that when $J = 0$ and $M, \alpha, \beta > 0$, (1.1) is exponentially stabilizable by the feedback law (1.2). However, the case of $J > 0$ and $M = 0$ was left as an open question. The main purpose of this paper is to give an affirmative answer to this question. Moreover, we will show that the C_0 semigroup associated with the closed-loop system is actually a C_0 group and its infinitesimal generator has a complete system of generalized eigenfunctions.

2 EXPONENTIAL STABILIZATION

With $M = 0$, the controlled system (1.1) with feedback law (1.2) takes the form

$$\begin{cases} \rho w_{tt} + p w'''' = 0, \quad (x,t) \in (0,L) \times (0,\infty), \\ w(0,t) = w'(0,t) = 0, \\ J w_{tt}'(L,t) + p w''(L,t) = -\alpha w_t'(L,t), \\ p w'''(L,t) = \beta w_t(L,t). \end{cases} \tag{2.1}$$

The physical interpretation of this system is that the tip mass is small enough to be neglected, but not the rotatory inertia. This situation occurs when the beam is linked with a large, but very light antena. Let

$$W = \{ w \in H^2(0,L) |\, w(0) = w'(0) = 0 \}, \qquad V = L^2_\rho(0,L).$$

Define the Hilbert space

$$\mathcal{H} = W \times V \times \mathbb{C}$$

equipped with the inner product

$$\langle (w_1, v_1, z_1),\ (w_2, v_2, z_2) \rangle_{\mathcal{H}} = \int_0^L (p w_1'' \bar{w}_2'' + \rho v_1 \bar{v}_2)\, dx + J z_1 \bar{z}_2. \tag{2.2}$$

Furthermore, we define an operator $\mathcal{A}$ in $\mathcal{H}$ by

$$\mathcal{D}(\mathcal{A}) = \left\{ (w, v, z) \;\middle|\; \begin{array}{l} w, v \in W,\ w \in H^4(0, L), \\ pw'''(L) = \beta v(L), z = v'(L) \end{array} \right\}, \tag{2.3}$$

$$\mathcal{A}(w, v, z) = (v\ ,\ -\frac{p}{\rho}w'''', -\frac{1}{J}(pw''(L) + \alpha z)). \tag{2.4}$$

Let $Y = (w, v, z)$. Then system (2.1) can be written as an abstract evolution equation in $\mathcal{H}$

$$\frac{dY}{dt} = \mathcal{A}Y. \tag{2.5}$$

Theorem 1 *$\mathcal{A}$ generates a C_0-semigroup, $e^{t\mathcal{A}}$, of contractions on $\mathcal{H}$, and $\mathcal{A}^{-1} \in \mathcal{L}(\mathcal{H})$ is a compact operator.*

Proof. Since for $z = (w, v, z) \in \mathcal{D}(\mathcal{A})$ we have

$$\begin{aligned} \mathrm{Re}\langle \mathcal{A}z,\ z\rangle_{\mathcal{H}} &= \mathrm{Re} \int_0^L [pv''\bar{w}'' - pw''''\bar{v}]dx - (pw''(L) + \alpha z)\bar{z} \\ &= -\beta|v(L)|^2 - \alpha|z|^2 \le 0, \end{aligned} \tag{2.6}$$

Hence, $\mathcal{A}$ is dissipative. It is easy to verify that for any $(f_1, f_2, f_3) \in \mathcal{H}$, equation

$$\mathcal{A}(w, v, z) = (f_1, f_2, f_3), \qquad (w, v, z) \in \mathcal{D}(\mathcal{A})$$

has unique solution such that

$$||(w, v, z)||_{\mathcal{H}} \le M||(f_1, f_2, f_3)||_{\mathcal{H}},$$

where the constant $M > 0$ is independent of (f_1, f_2, f_3). Therefore, $\mathcal{A}^{-1} \in \mathcal{L}(\mathcal{H})$, $0 \in \rho(\mathcal{A})$, and $\mathcal{A}$ is closed. It follows that the range $R(\lambda - \mathcal{A}) = \mathcal{H}$ for sufficiently small $\lambda > 0$. By Theorem 4.6 in [5], $\overline{\mathcal{D}(\mathcal{A})} = \mathcal{H}$. The generation of C_0-semigroup now follows from the Lumer-Phillips theorem. By the compactness of embedding $H^2(0, 1) \hookrightarrow C^1[0, 1]$, we know that $\mathcal{A}^{-1}$ is also a compact operator. □

It is clear that exponential stabilizability holds if $e^{t\mathcal{A}}$ is exponentially stable. We will employ the following frequency domain theorem for exponential stability of a C_0 -semigroup of contractions on a Hilbert space [[1],[2], [6]]:

Lemma 2.1 *A C_0-semigroup $e^{t\mathcal{A}}$ of contractions on a Hilbert space is exponentially stable if and only if*

$$\rho(\mathcal{A}) \supset \{i\lambda \mid \lambda \in \mathbb{R}\} \equiv i\mathbb{R}, \tag{2.7}$$

and

$$\overline{\lim_{|\lambda|\to\infty}} ||(i\lambda - \mathcal{A})^{-1}|| < \infty. \tag{2.8}$$

Now, we are ready to state the main result in this section.

Theorem 2 *The semigroup $e^{t\mathcal{A}}$ defined above is exponentially stable.*

Proof. We only need verify conditions (2.7) and (2.8).

(i) Suppose (2.7) is false. Then, there exsit a nonzero $\lambda \in \mathbb{R}$ and $Y \in \mathcal{D}(\mathcal{A})$ with $||Y||_{\mathcal{H}} = 1$ such that

$$(i\lambda - \mathcal{A})Y = 0. \tag{2.9}$$

Take real part of the inner product of (2.9) with z in $\mathcal{H}$, then apply (2.6). We obtain that

$$v(L) = z = 0. \tag{2.10}$$

Thus, (2.9)-(2.10) can be reduced to the following initial value problem

$$\begin{cases} -\lambda^2 \rho w + p w'''' = 0, \\ w(L) = w'(L) = w''(L) = w'''(L) = 0. \end{cases} \tag{2.11}$$

There is nothing but $w = 0$. Furthermore, $v = z = 0$ follows from the first and third equation in (2.9). This contradicts to $||Y||_{\mathcal{H}} = 1$.

(ii) Suppose (2.8) is false. Then by the Resonance Theorem, there exist a sequence of real numbers $\lambda_n \to \infty$ and a sequence of vectors $Y_n = (w_n, v_n, z_n) \in \mathcal{D}(\mathcal{A})$ with $||Y_n||_{\mathcal{H}} = 1$ such that

$$||(i\lambda_n I - \mathcal{A})Y_n||_{\mathcal{H}} \to 0 \quad as \quad n \to \infty. \tag{2.12}$$

i.e.,

$$i\lambda_n w_n - v_n \quad \equiv f_n \to 0 \quad in \;\; W, \tag{2.13}$$

$$i\lambda_n \rho v_n + \frac{p}{\rho} w_n'''' \quad \equiv g_n \to 0 \quad in \;\; V. \tag{2.14}$$

$$i\lambda_n z_n + \frac{1}{J}(p w''(L) + \alpha z_n) \quad \equiv h_n \to 0 \quad in \;\; \mathbb{C}. \tag{2.15}$$

In view of (2.6) and (2.12), we have

$$|v_n(L)|, \;\; |z_n| \to 0, \tag{2.16}$$

which further implies that

$$|w_n'''(L)|, \; |\lambda_n w_n'(L)|, \; |\lambda_n w_n(L)| \to 0. \tag{2.17}$$

In what follows we first show that

$$|w''(L)| \to 0. \tag{2.18}$$

Solve v_n from equation (2.13) and substituting it into (2.14), we have

$$-\lambda_n^2 w_n + \frac{p}{\rho} w_n'''' = g_n + i\lambda_n f_n. \tag{2.19}$$

Denote $\phi_n = \sqrt{|\lambda_n|}$ and $\gamma = (\rho/p)^{1/4}$. We take the inner product of (2.19) with $\frac{1}{\phi_n}e^{-\gamma\phi_n(L-x)}$ in $L^2(0,L)$ to get

$$\begin{aligned}&\langle -\phi_n^3 w_n \,,\, \rho e^{-\gamma\phi_n(L-x)}\rangle + \langle \frac{p}{\rho} w_n'''' \,,\, \frac{1}{\phi_n} e^{-\gamma\phi_n(L-x)}\rangle \\ = \;& \langle g_n + i\lambda_n f_n \,,\, \frac{1}{\phi_n} e^{-\gamma\phi_n(L-x)}\rangle. \end{aligned} \tag{2.20}$$

Clearly, the inner product on the right-hand side of (2.20) converges to zero. After integrating by parts four times to the second inner product on the left-hand side of (2.20), we can cancel the resulting inner product with the first inner product on the left-hand side of (2.20). Using the boundary conditions of w_n at $x = 0$ in (2.1) and at $x = L$ in (2.17), we rewrite (2.20) as

$$\frac{p}{\rho}\phi_n e^{-\gamma L\phi_n}\left(-\frac{w_n'''(0)}{\phi_n^2} + \gamma\frac{w_n''(0)}{\phi_n}\right) - \frac{p\gamma}{\rho} w_n''(L) \to 0. \tag{2.21}$$

From equation (2.14), we see that w_n/λ_n is bounded in $H^4(0,L)$. Applying the trace theorem, we obtain

$$\frac{p}{\rho}\phi_n e^{-\gamma L\phi_n}\left|-\frac{w_n'''(0)}{\phi_n^2} + \gamma\frac{w_n''(0)}{\phi_n}\right| \leq C\phi_n e^{-\gamma L\phi_n} \to 0. \tag{2.22}$$

The claim in (2.18) follows from (2.21)-(2.22). With these boundary conditions of w_n at hand, we now take the inner product of (2.19) with the standard multiplier xw_n' in $L^2(0,L)$. A straight forward calculation via integration by parts leads to

$$\frac{\rho}{2}||\lambda_n w_n||^2 + \frac{3p}{2}||w_n''||^2 \to 0. \tag{2.23}$$

This, together with equation (2.13), yields

$$||w_n''||, ||v_n|| \to 0. \tag{2.24}$$

In summary, we get that $||Y_n||_{\mathcal{H}}$ converges to zero. A contradiction. □

3 BACKWARD WELLPOSEDNESS AND SPECTRAL COMPLETENESS

In this section, by means of theory developed in Liu and Russell [4], we will prove the following:

Theorem 1 *$\mathcal{A}$ generates a C_0 group and has a complete system of generalized eigenfunctions.*

Proof. From discussions in [4],§3,4, we need prove that there exist $T, \delta > 0$ such that

$$\int_0^T ||e^{t\mathcal{A}}Y_0||_{\mathcal{H}}^2\, dt \geq \delta||Y_0||_{\mathcal{H}} \qquad \forall\; Y_0 \in \mathcal{D}(\mathcal{A}). \tag{3.1}$$

Set $(w(\cdot,t), v(\cdot,t), z(t)) = e^{t\mathcal{A}}Y_0$, $E(t) = \frac{1}{2}\|e^{t\mathcal{A}}Y_0\|^2_{\mathcal{H}}$. For $Y_0 = (\xi, \eta, z_0) \in \mathcal{D}(\mathcal{A})$, we know that

$$\begin{cases} w \in C^2([0,\infty); V) \cap C^1([0,\infty); W) \cap C([0,\infty); H^4(0,L)), \\ w_t = v, \; z(t) = w_t'(L,t) \end{cases} \tag{3.2}$$

and w satisfies (2.1). Thus,

$$E(t) = \frac{1}{2}\int_0^L (|pw''(x,t)|^2 + |\rho w_t(x,t)|^2)\, dx + \frac{1}{2}J|w_t'(L,t)|^2. \tag{3.3}$$

We may assume without loss of generality that $w(x,t)$ is real-valued. Multiplying the first equation of (2.1) by w_t and integrating by parts, we obtain

$$E(0) - E(t) = \int_0^t \left[\beta w_t(L,s)^2 + \alpha w_t'(L,s)^2\right] ds. \tag{3.4}$$

It follows that

$$TE(0) - \int_0^T E(t)\, dt = \int_0^T (T-t)\left[\beta w_t(L,t)^2 + \alpha w_t'(L,t)^2\right] dt. \tag{3.5}$$

We now multiply the first equation of (2.1) by $(T-t)xw'$. Then we integrate by parts to get

$$\begin{aligned} &\int_0^T (T-t)\left[\rho w_t(L,t)^2 + Lpw''(L,t)^2\right] dt + 2T\int_0^L x\xi'\eta\, dx \\ \le\, & 2\int_0^T (T-t)w'(L,t)[L\beta w_t(L,t) - pw''(L,t)]\, dt + C_1\int_0^T E(t)\, dt \end{aligned} \tag{3.6}$$

where C_1 is some positive number independent of (ξ,η). Combination of (3.5) and (3.6) yields

$$TE(0) \le -2T\int_0^L x\xi'\eta\, dx + C_T\int_0^T E(t)\, dt \tag{3.7}$$

for some $C_T > 0$ independent of (ξ,η). Applying a standard compactness argument, we can now obtain the desired estimate. □

References

[1] Gearhart, L. M. (1978). Spectral theory for contraction semigroups on Hilbert space. *Trans. Amer. Math. Soc.*, 236:385-394.

[2] Huang, F. (1985). Characteristic condition for exponential stability of linear dynamical systems in Hilbert spaces. *Chinese Ann. of Diff. Eqs*, 1:43-56.

[3] Littman, W. and Markus, L. (1988). Stabilization of a hybrid system of elasticity by feedback damping. *Ann. Mat. Pura Appl.*, 52:281-330.

[4] Liu, K. and Russell, D. L. (1998). Exact controllability, time reversibility and spectral completeness of linear systems in Hilbert spaces. preprint.

[5] Pazy, A. (1983). *Semigroups of Linear Operators and Applications to Partial Differential Equations.* Springer, New York.

[6] Prüss, J. (1984). On the spectrum of C_0-semigroups. *Trans. Amer. Math. Soc.*, 284:847-857.

[7] Rao, B. (1995). Uniform stabilization of a hybrid system of elasticity. *SIAM J. Cont. Optim.*, 33:440-454.

[8] Rao, B. (1998). Stabilization of a elastic plates with dynamical boundary control. *SIAM J. Cont. Optim.*, 36:148-163.

NEW MEANING OF EXACT CONTROLLABILITY OF LINEAR SYSTEMS IN HILBERT SPACES

Kangsheng Liu*

Department of Applied Mathematics
Zhejiang University, Hangzhou, 310027, China
Email: ksl@math.zju.edu.cn

David L. Russell

Department of Mathematics
Virginia Polytechnic Institute and State University
Blacksburg, Virginia, 24061, USA

Abstract: It is showed that under reasonable conditions, exact controllability of a linear system in a Hilbert space with bounded control implies time reversibility and spectral completeness of the system. The theory is applied to the beam and plate equations with boundary damping to establish their backward-time wellposedness and completeness of associated generalized eigenfunctions.

Key Words: exact controllability, C_0 group, generalized eigenfunctions, completeness, elastic system, damping

1 INTRODUCTION

There is an extensive literature on exact controllability for both abstract systems and concrete systems governed by partial differential equations (cf. Russell [18], Pritchard and Zabczyk [17] Lions [6, 7], Komornik [5], Liu [8] and references cited therein). So far, exact controllability has been established almost in the systems which are reversible in time. Consider the wave and plate equations with boundary damping and distributed controls:

$$\begin{cases} \ddot{w} - \Delta w = u & \text{in } \Omega \times \mathbb{R}^+ \\ \dfrac{\partial w}{\partial \nu} + \alpha(x)\dot{w} = 0 & \text{on } \partial\Omega \times \mathbb{R}^+, \end{cases} \tag{1.1}$$

*Supported partially by the National Key Project of China and by NSFC grant 19571073.

$$\begin{cases} \ddot{w} + \Delta^2 w = u & \text{in} \quad \Omega \times \mathbb{R}^+ \\ y = 0 \quad \text{on} \quad \partial\Omega \times \mathbb{R}^+, & \dfrac{\partial w}{\partial \nu} = 0 \quad \text{on} \quad \Gamma_0 \times \mathbb{R}^+ \\ \Delta w = -\dfrac{\partial \dot{w}}{\partial \nu} & \text{on} \quad \Gamma_1 \times \mathbb{R}^+ \end{cases} \tag{1.2}$$

where Ω is a bounded open subset in $\mathbb{R}^n$ with the boundary $\partial\Omega$ smooth enough, ν is the unit normal vector of $\partial\Omega$ pointing towards the exterior of Ω, $\alpha(x) \geq 0$ is a smooth function on $\partial\Omega$, $u \in L^2(\Omega \times \mathbb{R})$ is control, $\overline{\Gamma_0} \cup \overline{\Gamma_1} = \partial\Omega$, $\Gamma_0 \cap \Gamma_1 = \emptyset$. It has not been known whether the system (1.1) / (1.2) is exactly controllable in the finite energy state space, even in the one-dimensional case ($n = 1$). Majda [15] proved that the backward mixed problem of (1.1) is wellposed in the energy norm if and if $\alpha \equiv 0$. He [14] also showed that for certain $\alpha(x)$ the span of the generalized eigenfunctions associated with (1.1) has infinite codimension.

In this note, we announce the main results in Liu and Russell [12] (without the proofs) and apply them to establish backward-time wellposedness of (1.2), completeness of generalized eigenfunctions associated with (1.2), as well as lack of exact controllability of (1.1) when $\alpha(x_0) \neq 0$ for some $x_0 \in \partial\Omega$.

2 EXACT CONTROLLABILITY AND THE C_0 GROUP

Let H be a Hilbert space over the field $\mathbb{C}$ of complex numbers with the inner product $\langle \cdot, \cdot \rangle$ and the norm $\| \cdot \|$. Let A generate a C_0 semigroup e^{tA} on H. Consider the control system (A, B):

$$y(u,t) = e^{tA} y_0 + \int_0^t e^{(t-s)A} B u(s) ds \tag{2.1}$$

where $y_0 \in H$, $B \in \mathcal{L}(U; H)$, U is another Hilbert space.

Definition 2.1 *The system (A, B) is said to be exactly controllable if for every $y_0, y_1 \in H$ there exist $T > 0$, $u(\cdot) \in L^2(0,T;U)$ such that $y(u,0) = y_0$, $y(u,T) = y_1$. It is said to be reversible in time if A generates a C_0 group.*

Let us recall a fundamental result.

Proposition 2.1 (Zabczyk [20]) *The following conditions are equivalent:*

(1) *The system (A, B) is exactly controllable.*

(2) *There is a $T > 0$ such that for every $y_0, y_1 \in H$ there exists $u(\cdot) \in L^2(0,T;U)$ for which $y(u,0) = y_0$, $y(u,T) = y_1$.*

(3) *(observability inequality) There exist $T, \delta > 0$ such that*

$$\int_0^T \|B^* e^{tA^*} y\|_U^2 dt \geq \delta \|y\|^2 \qquad \forall \ y \in H. \tag{2.2}$$

In infinite-dimensional spaces, the class of systems (A, B) which are exactly controllable is quite restricted. Pazy [16] proved that the observability inequality (2.2) with $U = H$ and $B = I$ holds if and only if for any $T > 0$, there exist $\delta_T > 0$ such that

$$\|e^{tA^*}y\| \geq \delta_T \|y\| \qquad \forall\, t \in [0, T], y \in H. \tag{2.3}$$

As an immediate consequence of Pazy's result, the system (A, B) is never exactly controllable if e^{tA} is either compact or differentiable for $t \geq t_0 > 0$. Triggiani [19] proved that the system (A, B) is never exactly controllable if B is a compact operator. Louis and Wexler [13] proved that (A, I) is exactly controllable if and only if there exists a C_0 semigroup $\hat{S}(t)$ on H such that $e^{tA}\hat{S}(t) = I$ for all $t \geq 0$.

Let L be a densely defined linear operator in H. We denote its defined domain, range, null space, adjoint operator, resolvent set , point spectrum by $D(L)$, $R(L)$, $\mathrm{Ker}L$, L^*, $\rho(L)$, $\sigma_p(L)$, respectively. Let $\mathcal{L}(H)$ be the linear space of bounded linear operators on H . We define

$$\mathcal{S} = \{L \in \mathcal{L}(H) |\, L^* = L\}, \tag{2.4}$$

$$\mathcal{S}_d^+ = \{L \in \mathcal{S} |\, \exists\, a > 0 \text{ such that } \langle Ly, y\rangle \geq a\|y\|^2 \;\; \forall y \in H\}. \tag{2.5}$$

$$\sigma_p^*(A) = \{\lambda \in \sigma_p(A) \mid R(\lambda I - A) = H\}. \tag{2.6}$$

An operator $L \in \mathcal{L}(H)$ is said to be *isometric* if $\|Ly\| = \|y\|$ for all $y \in H$; it is said to be *unitary* if $L^* = L^{-1}$. Now, we are in a position to present our results.

Theorem 1 (Liu and Russell [12]) *The following statements are equivalent:*

(a) *The system (A,I) is exactly controllable.*

(b) *There exist $P \in \mathcal{S}_d^+$, $Q \in \mathcal{S}$ such that $Pe^{t(A^*+Q)}P^{-1}$ is a C_0 semigroup of isometric operators on H.*

(c) *There exist $P \in \mathcal{S}_d^+$, $Q \in \mathcal{S}$ such that*

$$e^{tA}Pe^{t(A^*+Q)}P^{-1} = I \qquad \forall\, t \geq 0. \tag{2.7}$$

Theorem 2 (Liu and Russell [12]) *The following statements are equivalent:*

(a) *A generates a C_0 group on H.*

(b) *There exist $P \in \mathcal{S}_d^+$, $Q \in \mathcal{S}$ such that $Pe^{t(A+Q)}P^{-1}$ is a C_0 group of unitary operators on H, i.e., $P(A+Q)P^{-1}$ is skew-adjoint.*

(c) *The system (A, I) is exactly controllable and there exists a sequence $\{\lambda_n\}$ satisfying $\lambda_n \notin \sigma_p^*(A)$ and $\mathrm{Re}\lambda_n \to -\infty$.*

Liu [8] also proved that e^{tA} can be embedded in a C_0 group on H if and only if

(i) There exists $\sigma > 0$ such that $\{\lambda \in \mathbb{C} \mid \mathrm{Re}\lambda \geq \sigma\} \subset \rho(-A)$.

(ii) There exists a nonincreasing function ψ: $[\sigma, \infty) \to (0, \infty)$ satisfying $\lim_{\tau\to+\infty} \psi(\tau) = 0$, such that

$$\|(\lambda + A)^{-1}\| \leq \psi(\mathrm{Re}\lambda) \qquad \forall\, \mathrm{Re}\lambda \geq \sigma. \tag{2.8}$$

Corollary 2.1 *If the system (A, B) with some bounded B is exactly controllable and A has compact resolvent, then A generates a C_0 group on H.*

Combination of Majda's result [15] and our corollary here yields the following:

Theorem 3 *The system (1.1) is never exactly controllable if $\alpha(x_0) \neq 0$ for some $x_0 \in \partial\Omega$.*

Remark 2.1 From Theorem 2.2, if (A, I) is exactly controllable but A does not generate a C_0 group, then there exists $\sigma_0 \in \mathbb{R}$ such that $\{\lambda \in \mathbb{C} \mid \mathrm{Re}\,\lambda \leq \sigma_0\} \subset \sigma_p^*(A)$. This case occurs when e^{tA} is the semigroup of left translations on $L^2(\mathbb{R}^+)$ (see Zabczyk [20]).

3 THE COMPLETENESS OF GENERALIZED EIGENVECTORS

Let A be a closed linear operator defined in H. A nonzero vector y is called a generalized eigenvector of A, corresponding to the eigenvalue $\lambda \in \sigma_p(A)$, if $(\lambda I - A)^n y = 0$ for some positive integer n. Denote by $\mathrm{Sp}(A)$ the set of all finite linear combinations of the generalized eigenvectors. A is said to have a complete system of generalized eigenvectors if $\mathrm{Sp}(A)$ is dense in H. The completeness of generalized eigenvectors is a desirable property in spectral analysis of linear operators. In the present section we will present this property for the infinitesimal generator of a C_0 group, by the theory in Gokhberg and Krein [3].

Let $\mathcal{M}_\infty$ be the set of all compact operators on H. For $L \in \mathcal{M}_\infty$, we have $0 \leq F \equiv (L^*L)^{\frac{1}{2}} \in \mathcal{M}_\infty \cap \mathcal{S}$. Following [3, §II.2], we enumerate the nonzero eigenvalues of F in decreasing order, taking account of their multiplicities,

$$s_j(L), \qquad j = 1, 2, \cdots, r(F) = \dim R(F).$$

If $r(F) < \infty$, we put $s_j(L) = 0$ for $j = r(F) + 1, \cdots$. For $1 \leq p < \infty$, we define

$$\mathcal{M}_p = \{L \in \mathcal{M}_\infty \mid \sum_{j=1}^{\infty} s_j^p(L) < +\infty\}.$$

By the discussions in [3, Chap.III], we know that $\mathcal{M}_p$ is a two-sided ideal of the ring $\mathcal{L}(H)$. A compact operator is said to be of finite order if it is in $\mathcal{M}_p$ for some $1 \leq p < \infty$.

Now, we are ready to present our completeness result.

Theorem 1 (Liu and Russell [12]) *If A is the infinitesimal generator of a C_0 group on a Hilbert space and $(\lambda_0 I + A)^{-1}$ is a compact operator of finite order for some $\lambda_0 \in \rho(A)$, then A has a complete system of generalized eigenvectors.*

We will apply our completeness theorem to the following variational evolution equation:

$$\mathbf{c}(\ddot{w}(t), v) + \mathbf{b}(\dot{w}(t), v) + \mathbf{a}(w(t), v) = 0, \quad \forall\, v \in V, \;\; t > 0. \tag{3.1}$$

A damped linear elastic system can be written in the form (3.1), where $\mathbf{a}(v, v)$ and $\mathbf{c}(v, v)$ are the strain energy and kinetic energy quadratic forms, respectively, $\operatorname{Re}\mathbf{b}(v, v)$ and $\operatorname{Im}\mathbf{b}(v, v)$ represent damping and gyroscopic forces, respectively. See K. Liu and Z. Liu [11], S. Chen, K. Liu and Z. Liu [2] for various properties of the C_0 semigroup associated with (3.1). We assume that

(H1) V and W are Hilbert spaces with inner products $\mathbf{a}(\cdot, \cdot)$ and $\mathbf{c}(\cdot, \cdot)$, respectively;

(H2) $V \hookrightarrow W$ is a compact and dense embedding;

Assumptions (H1) and (H2) imply that there exists a self-adjoint, positive definite operator A_0 in W with compact resolvent such that ([4])

$$D(A_0^{\frac{1}{2}}) = V, \quad \mathbf{a}(u, v) = \mathbf{c}(A_0^{\frac{1}{2}} u, A_0^{\frac{1}{2}} v), \quad \forall\, u, v \in V. \tag{3.2}$$

It is clear that for every $0 < \beta \leq 1$, $D(A_0^{\beta})$ is also a Hilbert space with the inner product $\mathbf{c}(A^{\beta}\cdot, A^{\beta}\cdot)$. We further assume that

(H3) $\mathbf{b}(\cdot, \cdot)$ is a continuous sesquilinear form on $D(A_0^{\theta})$ for some $0 \leq \theta < \frac{1}{2}$.

By the Riesz representation theorem, assumption (H3) implies that there exists a compact operator $B_1 \in \mathcal{L}(D(A_0^{\theta}))$ such that

$$\mathbf{b}(u, v) = \mathbf{c}(A_0^{\theta} B_1 u, A_0^{\theta} v), \quad \forall\;\; u, v \in D(A_0^{\theta}). \tag{3.3}$$

Denote by $B = A_0^{\theta} B_1$. Then $B \in \mathcal{L}(D(A_0^{\theta}); W)$. Let $\mathcal{H} = V \times W$ with the naturally induced inner product

$$\langle (u, v), (\hat{u}, \hat{v}) \rangle_{\mathcal{H}} = \mathbf{a}(u, \hat{u}) + \mathbf{c}(v, \hat{v}); \tag{3.4}$$

$\mathcal{H}$ may be regarded as the *finite energy state space* of the system (3.1). We define the operator $\mathcal{A}$ in $\mathcal{H}$ as follows:

$$\begin{cases} D(\mathcal{A}) = \{(u, v) \mid u \in D(A_0^{1-\theta}),\ v \in V,\ A_0^{1-\theta} u + Bv \in D(A_0^{\theta})\}, \\ \mathcal{A}(u, v) = (v\ ,\ -A_0^{\theta}[A_0^{1-\theta} u + Bv]). \end{cases} \tag{3.5}$$

It is easy to check that

$$\mathcal{A}^{-1}(u, v) = \begin{bmatrix} -A_0^{\theta-1} B & -A_0^{-1} \\ I & 0 \end{bmatrix} (u, v) \tag{3.6}$$

for $(u,v) := (u,v)^{\tau} \in \mathcal{H}$. Obviously, $\mathcal{A}^{-1}$ is a compact operator in $\mathcal{L}(\mathcal{H})$. We will show the connection between the finite-order property of $\mathcal{A}^{-1}$ and the asymptotic distribution of eigenvalues of A_0.

Theorem 2 *Let the assumptions* (H1)-(H3) *hold. Then the compact operator* $\mathcal{A}^{-1}$ *is of finite order if* A_0^{-1} *is so.*

Enumerating the eigenvalues of A_0 in increasing order, counting the multiple eigenvalues according to their multiplicity, we have

$$0 < \lambda_1 \leq \lambda_2 \cdots \lambda_n \leq \lambda_{n+1} \cdots. \tag{3.7}$$

We know that $\lambda_n \to +\infty$ (since $\dim W = \infty$). We define

$$N(\lambda) = \max\{j \mid \lambda_j \leq \lambda\} \quad \forall \lambda > 0. \tag{3.8}$$

It is easy to see that A_0^{-1} is finite order if

$$N(\lambda) = \mathcal{O}(\lambda^p), \quad \lambda \to +\infty \tag{3.9}$$

for some $p \geq 1$. We point out that the condition (3.9) always holds for a general self-adjoint elliptic operator A_0 on a bounded region (cf. Agmon [1] and the references therein).

4 BACKWARD WELLPOSEDNESS AND SPECTRAL COMPLETENESS

To write (1.2) with $u = 0$ in the form (3.1), we can introduce

$$W = L^2(\Omega), \quad V = \{v \in H^2(\Omega) |\ v = 0 \text{ on } \partial\Omega,\ \frac{\partial v}{\partial \nu} = 0 \text{ on } \Gamma_0\},$$

$$\mathbf{c}(v,v) = \int_\Omega |v|^2 dx, \quad \mathbf{a}(v,v) = \int_\Omega |\Delta v|^2 dx, \quad \mathbf{b}(v,v) = \int_{\Gamma_1} \left|\frac{\partial v}{\partial \nu}\right|^2 d\sigma.$$

Then, the assumptions (H1)-(H3) with $\theta = \frac{1-\epsilon}{2}$ for $0 < \epsilon < \frac{1}{4}$ and the condition (3.9) are valid.

Theorem 1 *If* $\overline{\Gamma_0} \cap \overline{\Gamma_1} = \emptyset$, *then* $\mathcal{A}$ *corresponding to (1.2) generates a* C_0 *group on* $V \times W$ *and has a complete system of generalized eigenvectors.*

Sketch of the Proof. We know that $\mathcal{A}$ generates a C_0 semigroup $S(t)$ of contractions on $\mathcal{H}$ and $\mathcal{A}(w,v) = (v, -\Delta^2 w)$,

$$\begin{aligned} D(\mathcal{A}) = \{(w,v) \mid\ & w \in H^4(\Omega),\ w, v \in V,\ w = 0 \text{ on } \partial\Omega, \\ & \frac{\partial w}{\partial \nu} = 0 \text{ on } \Gamma_0,\ \Delta w = -\frac{\partial v}{\partial \nu} \text{ on } \Gamma_1\}. \end{aligned} \tag{4.1}$$

For $(\xi, \eta) \in D(\mathcal{A})$, set $(w(t), w_1(t)) = S(t)(\xi, \eta)$. Then the function $w(\cdot)$ is in

$$C([0,\infty); H^4(\Omega)) \cap C^1([0,\infty); V) \cap C^2([0,\infty); W). \tag{4.2}$$

It satisfies $\dot{w}(\cdot) = w_1(\cdot)$ and (1.2) with $u = 0$. Let

$$E(t) = \frac{1}{2}||S(t)(\xi,\eta)||^2_{\mathcal{H}} = \frac{1}{2}\int_\Omega (|\Delta w|^2 + |\dot{w}|^2)\, dx. \tag{4.3}$$

We only need to prove the observability inequality

$$\int_0^T E(t)\, dt \geq \delta E(0) \tag{4.4}$$

for some $T, \delta > 0$ and all real-valued $(\xi, \eta) \in D(\mathcal{A})$. Multiplying the first equation of (1.2)($u = 0$) by $\dot{w}$ and integrating by parts, we obtain

$$TE(0) - \int_0^T E(t)\, dt = \int_0^T (T-t) \int_{\Gamma_1} \left(\frac{\partial \dot{w}}{\partial \nu}\right)^2 d\sigma\, dt. \tag{4.5}$$

We choose $h \in C^1(\overline{\Omega}; \mathbb{R}^n)$ such that $h = 0$ on Γ_0 and $h = \nu$ on Γ_1. Then we multiply the first equation of (1.2)($u = 0$) by $(T-t)h \cdot \nabla w$ to get

$$\begin{aligned} &\int_0^T (T-t) \int_{\Gamma_1} \left[\left(\frac{\partial \dot{w}}{\partial \nu}\right)^2 + \frac{1}{2}\text{div}\frac{\partial \dot{w}}{\partial \nu}\frac{\partial w}{\partial \nu}\right] d\sigma\, dt \\ \leq\ & -\int_\Omega \eta h \cdot \nabla \xi\, dx + C_T \int_0^T E(t)\, dt. \end{aligned} \tag{4.6}$$

where C_T is some positive number independent of (ξ, η). By a compactness argument, we can now conclude that the observability inequality (4.4) is valid. □

See Liu [10] for the details of proof and more examples dealing with backward wellposedness and spectral completeness of PDEs with boundary dissipation.

References

[1] S. Agmon, Asymptotic formulas with remainder estimates for eigenvalues of elliptic operators, *Arch. Rational Mech. Anal.*, 28(1968), pp. 165-183.

[2] S. Chen, K. Liu and Z. Liu, Spectrum and stability of elastic systems with global or local Kelvin-Voigt damping, *SIAM J. Appl. Math.*, to appear.

[3] I. C. Gokhberg and M. G. Krein, *Introduction to the Theory of Linear Nonselfadjoint Operators*, AMS, Providence, 1969.

[4] T. Kato, *Perturbation Theory for Linear Operators*, Springer-Verlag, New York, 1980.

[5] V. Komornik, *Exact Controllability and Stabilization* – The Multiplier Method, RAM, Masson, Paris, 1994.

[6] J.L. Lions, Exact controllability, stabilization and perturbations for distributed systems, *SIAM Rev.*, 30(1988), pp.1-68.

[7] —, *Controlabilitè Exacte, Perturbations et Stabilisation de Systèmes Distribues, Tome 1*, Collection RMA 8, Masson, Paris, 1988.

[8] K. Liu, Locally distributed control and damping for the conservative systems, *SIAM J. Cont. Optim.*, 35(1997)5, pp. 1574-1590.

[9] K. Liu, A characterization of strongly continuous groups of linear operators on a Hilbert space, in submission.

[10] K. Liu, Backward wellposedness and spectral completeness of PDEs with boundary dissipation via exact observability, preprint.

[11] K. Liu and Z. Liu, Analyticity and differentiability of semigroups associated with elastic systems with damping and gyroscopic forces, *J. Diff. Eqs.*, 141(1997)2, pp. 340-355.

[12] K. Liu and D. L. Russell, Exact controllability, time reversibility and spectral completeness of linear systems in Hilbert spaces, in submission.

[13] J. C. Louis and D. Wexler, On exact controllability in Hilbert spaces *J. Diff. Eqns.*, 49(1983), pp. 258-269.

[14] A. Majda, Disappearing solutions for the dissipative wave equation, *Indiana Univ. Math. J.*, 24(1975), pp. 1119-1133.

[15] A. Majda, The location of the spectrum for the dissipative açoustic operator, *Indiana Univ. Math. J.*, 25(1976), pp. 973-987.

[16] A. Pazy, On the applicability of Lyapunov's theorem in Hilbert space, *SIAM J. Math. Anal.*, 3(1972), pp. 291-294.

[17] A. J. Pritchard and J. Zabczyk, Stability and Stabilizability of infinite dimensional systems, *SIAM Rev.*, 23(1981), pp.25-52.

[18] D. L. Russell, Controllability and stabilizability theory for linear PDE's: recent progress and open questions, *SIAM Rev.*, 20(1978), pp.639-739.

[19] R. Triggiani, On the lack of exact controllability for mild solutions in Banach spaces, *J. Math. Anal. Appl.*, 50(1975), pp. 438-446.

[20] J. Zabczyk, Remarks on the algebraic Riccati equation in Hilbert space, *Appl. Math. Optim.*, 2(1976), pp. 251-258.

MINIMAX DESIGN OF CONSTRAINED PARABOLIC SYSTEMS*

Boris S. Mordukhovich

Department of Mathematics
Wayne State University
Detroit, MI 48202

boris@math.wayne.edu

Abstract: This paper relates to minimax control design problems for a class of parabolic systems with nonregular boundary conditions and uncertain distributed perturbations under pointwise control and state constraints. The main attention is paid to the Dirichlet boundary control that offers the lowest regularity properties. Our variational analysis is based on well-posed multistep approximations and involves the solving of constrained optimal control problems for ODE and PDE systems. The design procedure essentially employs monotonicity properties of the parabolic dynamics and its asymptotics on the infinite horizon. Finally we justify a suboptimal three-positional structure of feedback boundary controllers and provide calculations of their optimal parameters that ensure the required system performance and robust stability under any admissible perturbations.

1 INTRODUCTION

In this paper we formulate and study a minimax feedback control problem for linear parabolic systems with uncertain disturbances and pointwise constraints on state and control variables. We deal with boundary controllers acting through the Dirichlet boundary conditions that are the most challenging for the parabolic dynamics. The original motivation comes from applications to some environmental problems; see [6] where a groundwater control problem of this kind has been considered for the case of one-dimensional heat-diffusion equations. Here we study a general class of multidimensional parabolic control systems that cover a fairly broad range of practical applications.

Dynamical processes in these systems are subject to pointwise (hard) state and control constraints that are typical in applied problems. Moreover, the

*This research was partly supported by the National Science Foundation under grant DMS-9704751 and by the USA-Israel Binational Science Foundation under grant 94-00237.

only information available for uncertain disturbances/perturbations is an admissible region of their variations. A natural approach to control design of such uncertain systems is *minimax synthesis* which guarantees the best system performance under the worst perturbations and ensures an acceptable behavior for any admissible perturbations. This approach is related to H_∞-control and differential games; see, e.g., [1], [2], [3]. However, we are not familiar with any results in these theories that may be applied to parabolic systems with hard control and state constraints under consideration. In [4] the reader can find a number of feedback boundary control results for unconstrained parabolic systems based on Riccati equations.

In this paper we develop an efficient design procedure to solve minimax control problems for hard-constrained parabolic systems. This procedure takes into account monotonicity properties of the parabolic dynamics and asymptotic characteristics of transients on the infinite horizon. It was initiated in [6] for the case of one-dimensional heat-diffusion equations and then developed in [7] where partial results for multidimensional systems were reported. The results presented below include a justification of a suboptimal discontinuous feedback control structure and optimization of its parameters along the parabolic dynamics. In this way we minimize an energy-type cost functional in the case of maximal perturbations and ensure the desired state performance within the required constraints for all admissible disturbances. Based on a variational approach, we obtain verifiable conditions for "stability in the large" of the nonlinear closed-loop control system that excludes unacceptable self-vibrating regimes.

Our design and justification procedures involve multistep approximations and results from the optimal control theory for ordinary differential equations. As a by-product of this approach, we obtain a complete measure-free solution for a class of state-constrained optimal control problems related to approximations of the parabolic dynamics.

2 PROBLEM FORMULATION AND BASIC PROPERTIES

Consider a self-adjoint and uniformly strongly elliptic operator defined by

$$A := -\sum_{i,j=1}^{n} a_{ij}(x)\frac{\partial^2}{\partial x_i \partial x_j} - c \tag{1.1}$$

where $c \in \mathbb{R}$, $a_{ij} \in C^\infty(\text{cl}\Omega)$,

$$\sum_{i,j=1}^{n} a_{ij}(x)\xi_i\xi_j \geq \nu \sum_{i=1}^{n} \xi_i^2, \quad \nu > 0 \;\; \forall x \in \Omega, \; \xi = (\xi_1, \xi_2, \ldots, \xi_n) \in \mathbb{R}^n,$$

Ω is a bounded open region in $\mathbb{R}^n$ with the sufficiently smooth boundary Γ. Given positive numbers $\underline{\alpha}$, $\bar{\alpha}$, $\underline{\beta}$, and $\bar{\beta}$, we define the sets of *admissible controls* $u(t)$ and admissible *uncertain disturbances* $w(t)$ by

$$U_{ad} = \{u \in L^2(0,T) \mid u(t) \in [-\bar{\alpha}, \underline{\alpha}] \;\text{ a.e. }\; t \in [0,T]\},$$

$$W_{ad} = \{w \in L^2(0,T) \mid w(t) \in [-\underline{\beta}, \bar{\beta}] \text{ a.e. } t \in [0,T]\}.$$

Suppose that x_0 is a given point in Ω at which we are able to collect information about the system performances, and let $\eta > 0$. Consider the following *minimax feedback control problem* (P):

$$\text{minimize } J(u) = \max_{w(\cdot)\in W_{ad}} \int_0^T |u(y(t,x_0))|dt$$

over $u(\cdot) \in U_{ad}$ subject to the system

$$\begin{cases} \dfrac{\partial y}{\partial t} + Ay = w(t) \text{ a.e. in } Q := (0,T) \times \Omega \\ y(0,x) = 0, \ x \in \Omega \\ y(t,x) = u(t), \ (t,x) \in \Sigma := (0,T] \times \Gamma, \end{cases} \tag{1.2}$$

the pointwise state constraints

$$|y(t,x_0)| \leq \eta \ \forall t \in [0,T], \tag{1.3}$$

and the feedback control law

$$u(t) = u(y(t,x_0)) \tag{1.4}$$

acting through the Dirichlet boundary conditions in (2).

Problem (P) formulated above is one of the most difficult control problems unsolved in the general theory. Our purpose is to develop an approach that takes into account specific features of parabolic systems and allows us to find a feasible *suboptimal* feedback control. To furnish this, we employ the spectral representation of solutions to the parabolic system (2) with the Dirichlet boundary conditions.

Let $\lambda \in \mathbb{R}$ be an eigenvalue of the operator A in (1) and let $\phi \in L^2(\Omega)$ be the corresponding eigenfunction satisfying the condition $\phi|_\Gamma = 0$. It is well known that, under the assumptions made, one has the properties:

(a) All the eigenvalues λ_i, $i = 1, 2, \ldots$, of A form a nondecreasing sequence that accumulates only at ∞, and the first eigenvalue λ_1 is simple.

(b) The corresponding orthonormal system of eigenfunctions is complete in $L^2(\Omega)$.

Let $y \in L^2(Q)$ be a generalized solution to (2) which uniquely exists for each $(u,w) \in U_{ad} \times W_{ad}$; see [5]. Based on the properties (a) and (b) and taking into account that both admissible controls and perturbations in (2) depend only on t, we conclude that the generalized solution $y(\cdot)$ admits the representation

$$y(t,x) = \sum_{i=1}^{\infty} \mu_i \Big(\int_0^t w(\theta) e^{\lambda_i \theta} d\theta + (c + \lambda_i) \int_0^t u(\theta) e^{\lambda_i \theta} d\theta \Big) e^{-\lambda_i t} \phi_i(x) \tag{1.5}$$

where $\mu_i = \int_\Omega \phi_i(x)dx$ and series (5) is strongly convergent in $L^2(Q)$. This allows us to deduce, involving the maximum principle for parabolic equations, the following *monotonicity property* of solutions to (2) with respect to both controls and perturbations.

Theorem 1. *Let $(u_i, w_i) \in L^2(0,T) \times L^2(0,T)$ and let $y_i(\cdot)$, $i = 1, 2$, be the corresponding generalized solutions to* (2). *Then*

$$y_1(t,x) \geq y_2(t,x) \text{ a.e. in } Q$$

if $u_1(t) \geq u_2(t)$ and $w_1(t) \geq w_2(t)$ a.e. in $[0,T]$.

One can see from Theorem 1 that *the bigger magnitude of a perturbation is, the more control of the opposite sign should be applied* to neutralize this perturbation and to keep the corresponding transient within the state constraint (3). This leads us to consider feedback control laws (4) satisfying the *compensation property*

$$u(y) \leq u(\tilde{y}) \text{ if } y \geq \tilde{y} \text{ and } y \cdot u(y) \leq 0 \;\; \forall y, \tilde{y} \in \mathbb{R}. \tag{1.6}$$

The latter property implies that

$$\int_0^T |u(y(t))|dt \geq \int_0^T |u(\tilde{y}(t))|dt \text{ if } y(t) \geq \tilde{y}(t) \geq 0 \text{ or } y(t) \leq \tilde{y}(t) \leq 0$$

for all $t \in [0,T]$, i.e., the compensation of bigger (by magnitude) perturbations reguires more cost with respect to the maximized cost functional in (P). This allows us to seek a *suboptimal control structure* in (P) by examining the control response to feasible perturbations of the *maximal magnitudes* $w(t) = \bar{\beta}$ and $w(t) = -\underline{\beta}$ for all $t \in [0,T]$.

3 SUBOPTIMAL CONTROL UNDER MAXIMAL PERTURBATIONS

Taking into account the symmetry of (P) relative to the origin, we consider the case of upper level maximal perturbations $w(\cdot) = \bar{\beta}$ and the corresponding set of admissible controls

$$\overline{U}_{ad} := \{u(\cdot) \in U_{ad} \mid -\bar{\alpha} \leq u(t) \leq 0 \text{ a.e. } t \in [0,T]\}.$$

To find an optimal control $\bar{u}(t)$ in respose to the maximal perturbations, we have the following *open-loop* control problem $(\overline{\mathrm{P}})$:

$$\text{minimize } \bar{J}(u) = -\int_0^T u(t)dt \tag{1.7}$$

over $u(\cdot) \in \overline{U}_{ad}$ subject to system (2) with $w(\cdot) = \bar{\beta}$ and the constraint

$$y(t) \leq \eta \;\; \forall t \in [0,T]. \tag{1.8}$$

This is a state-constrained Dirichlet boundary control problem which was considered in [8] in more generality. In [8] we obtained necessary optimality conditions for $(\overline{\mathrm{P}})$ that involve the adjoint operator to the so-called Dirichlet map and Borel measures. Those conditions are rather complicated and do not allow us to determine an optimal control.

Following [6], let us explore another approach to solve problem $(\overline{\mathrm{P}})$. It leads to suboptimal feasible solutions of a simple structure that can be used to design and justify a required feedback law in the original minimax control problem (P). To furnish this, we approximate $(\overline{\mathrm{P}})$ by optimal control problems for ODE systems directly obtained from the spectral representation (5) as $x = x_0$ and $w(\cdot) = \bar{\beta}$. In what follows we suppose, additionally to the basic assumptions in Section 2, that the first eigenvalue λ_1 in (a) is *positive.* Together with the other properties in (a) this gives

$$0 < \lambda_1 < \lambda_i, \; i = 2, 3, \ldots,$$

which ensures that the first term in (5) *dominates* as $t \to \infty$. This allows us to employ the *first term rule* [6] when the time interval is sufficiently large and thus to confine our treatment of suboptimality to the first order ODE approximation. In this way we arrive at the following constrained ODE optimal control problem $(\overline{\mathrm{P}}_1)$: minimize the cost functional (7) along the controlled differential equation

$$\dot{y} = -\lambda_1 y + \mu_1 \phi_1(x_0)(\bar{\beta} + (c + \lambda_1)u(t)) \;\text{ a.e. } t \in [0, T], \;\; y(0) = 0 \tag{1.9}$$

subject to $u(\cdot) \in \overline{U}_{ad}$ and the state constraint (8).

The next theorem provides the complete *exact solution* of the state-constrained problem $(\overline{\mathrm{P}}_1)$ with no measure involved.

Theorem 2. *Let $\mu_1 \phi_1(x_0)\bar{\beta} > \lambda_1 \eta$. Assume in addition that either*

$$\mu_1 \phi_1(x_0)(\bar{\beta} - \bar{\alpha}(c + \lambda_1)) \le \lambda_1 \eta \;\text{ or }\; \tau_1 := \frac{1}{\lambda_1} \ln \frac{\mu_1 \phi_1(x_0)\bar{\beta}}{\mu_1 \phi_1(x_0)\bar{\beta} - \lambda_1 \eta} \ge T. \tag{1.10}$$

Then system (8), (9) *is controllable, i.e., there is $u(\cdot) \in \bar{U}_{ad}$ such that the corresponding trajectory of* (9) *satisfies the state constraint* (8). *Moreover, problem $(\overline{\mathrm{P}}_1)$ admits an optimal control of the form*

$$\bar{u}_1(t) = \begin{cases} 0 & \text{if } t \in [0, \bar{\tau}_1) \\ \dfrac{\lambda_1 \eta - \mu_1 \phi_1(x_0)\bar{\beta}}{\mu_1 \phi_1(x_0)(c + \lambda_1)} & \text{if } t \in [\bar{\tau}_1, T] \end{cases} \tag{1.11}$$

where $\bar{\tau}_1 = \min\{\tau_1, T\}$ with τ_1 computed in (10).

To prove this theorem, we first approximate $(\overline{\mathrm{P}}_1)$ by a parametric family of optimal control problems without state constraints. The latter problems can be completely solved by using the Pontryagin maximum principle [9] which provides necessary and sufficient conditions for optimality in this case. We prove that optimal controls to approximating problems are piecewise constant and contain both *bang-bang and singular modes.* Passing to the limit, we establish all the results of Theorem 2 and come to a surprising conclusion that the optimal control (11) for the state-constrained problem occurs to be *simpler* than the ones for the unconstrained approximations.

In the way we justify that a two-positional form in (11) can be accepted as a reasonable suboptimal control structure for problem $(\overline{\mathrm{P}})$ with the parabolic dynamics. Then we optimize this structure subject to (2), (8), and $u(\cdot) \in \overline{U}_{ad}$. This gives

$$u(t) = \begin{cases} 0 & \text{for } 0 \le t < \tau \\ -\bar{u} & \text{for } \tau \le t \le T \end{cases} \tag{1.12}$$

with the optimal parameters

$$\bar{u} := \frac{\gamma\bar{\beta} - \eta}{1 + c\gamma}, \quad \gamma := \sum_{i=1}^{\infty} \frac{\mu_i \phi_i(x_0)}{\lambda_i}, \tag{1.13}$$

and τ satisfying the equation

$$\sum_{i=1}^{\infty} \frac{\mu_i \phi_i(x_0)}{\lambda_i} e^{-\lambda_i T}[(c + \lambda_i)(\gamma\bar{\beta} - \eta)e^{\lambda_i \tau} - \bar{\beta}(1 + c\gamma)] = 0. \tag{1.14}$$

We establish that (14) has a unique solution $\tau = \bar{\tau}(T) \in (0, T)$ for all T sufficiently large and that any control (12), (13) is *feasible* to $(\overline{\mathrm{P}})$ for all positive $\tau \le \bar{\tau}(T)$. Moreover, the switching time $\tau = \bar{\tau}(T)$ is *optimal* in $(\overline{\mathrm{P}})$ and $\bar{\tau}(T) \downarrow \bar{\tau}$ as $T \to \infty$ where the *asymptotically optimal* switching time $\bar{\tau}$ is computed by

$$\bar{\tau} = \frac{1}{\lambda_1} \ln \frac{\bar{\beta}(1 + c\gamma)}{(c + \lambda_1)(\gamma\bar{\beta} - \eta)}.$$

4 FEEDBACK CONTROL DESIGN

The obtained results allow us to justify the *three-positional control law*

$$u(y) = \begin{cases} -\bar{u} & \text{if } y \ge \bar{\sigma} \\ 0 & \text{if } -\underline{\sigma} < y < \bar{\sigma} \\ \underline{u} & \text{if } y \le -\underline{\sigma} \end{cases} \tag{1.15}$$

as a suboptimal feedback structure in (P) with the compensation property (6). Now using the monotonicity of transients with respect to both controls and perturbations as well as their asymptotic properties as $t \to \infty$, we arrive at the following theorem.

Theorem 3. *Let the feedback control parameters* $(\bar{u}, \overline{u}, \bar{\sigma}, \overline{\sigma})$ *in* (15) *are computed by the formulas* (13),

$$\bar{\sigma}(T) := \bar{\beta}(\gamma - \sum_{i=1}^{\infty} \frac{\mu_i \phi_i(x_0)}{\lambda_i} e^{-\lambda_i \bar{\tau}(T)}),$$

and their counterparts for $\underline{\beta}$. Then the control law (15) *is feasible for any perturbations $w(\cdot) \in W_{ad}$ being optimal in the case of maximal perturbations when T is sufficiently large. Moreover, $\bar{\sigma}(T) \downarrow \bar{\sigma}$ and $\underline{\sigma}(T) \downarrow \underline{\sigma}$ as $T \to \infty$ where the positive numbers*

$$\bar{\sigma} := \bar{\beta}(\gamma - \sum_{i=1}^{\infty} \frac{\mu_i \phi_i(x_0)}{\lambda_i} [\frac{(c+\lambda_1)(\gamma\bar{\beta} - \eta)}{\bar{\beta}(1+c\gamma)}]^{\frac{\lambda_i}{\lambda_1}}), \tag{1.16}$$

$$\underline{\sigma} := \underline{\beta}(\gamma - \sum_{i=1}^{\infty} \frac{\mu_i \phi_i(x_0)}{\lambda_i} [\frac{(c+\lambda_1)(\gamma\underline{\beta} - \eta)}{\underline{\beta}(1+c\gamma)}]^{\frac{\lambda_i}{\lambda_1}}) \tag{1.17}$$

form the maximal dead region $[-\underline{\sigma}, \bar{\sigma}]$ under which feedback (15) *keeps the state constraints* (3) *on the infinite horizon* $[0, \infty)$ *for any admissible perturbations.*

Finally we observe that the feedback control (15) with the parameters calculated in Theorem 3 does not guarantee the *robust stability* of the highly nonlinear (discontinuous) closed-loop system (2), (4), (15) under any admissible perturbations. Indeed, this system may have a *self-vibrating regime* (i.e., its zero-equilibrium is not stable in the large) if the dead region $[-\underline{\sigma}, \bar{\sigma}]$ is not sufficiently wide. The next theorem excludes such a possibility and ensures the required robust stability of the closed-loop control systems. Its proof is based on a variational approach and turnpike asymptotic properties of the parabolic dynamics.

Theorem 4. *The closed-loop control system* (2), (4), (15) *with arbitrary parameters $(\bar{u}, \underline{u}, \bar{\sigma}, \underline{\sigma})$ is stable in the large if*

$$\bar{\sigma} + \underline{\sigma} \geq \min\{\bar{u}, \underline{u}\}[\frac{\mu_1 \phi_1(x_0)(c+\lambda_1)}{\lambda_1} - (1+c\gamma)] > 0. \tag{1.18}$$

When $\bar{\beta} \leq \underline{\beta}$, the stability condition (18) can be written in the simplified form

$$2\bar{\sigma}_1 + \underline{\sigma}_1 \geq \eta$$

through the suboptimal dead region bounds

$$\bar{\sigma}_1 := \bar{\beta}(\gamma - \frac{\mu_1 \phi_1(x_0)(c+\lambda_1)(\gamma\bar{\beta} - \eta)}{\lambda_1 \bar{\beta}(1+c\gamma)}),$$

$$\underline{\sigma}_1 := \underline{\beta}(\gamma - \frac{\mu_1 \phi_1(x_0)(c+\lambda_1)(\gamma\underline{\beta} - \eta)}{\lambda_1 \underline{\beta}(1+c\gamma)})$$

which correspond to the first terms in (16) and (17).

References

[1] Basar, T. and Bernard, P. (1995), *H_∞-Optimal Control and Related Minimax Design Problems*, Birkhäuser, Boston.

[2] van Keulen, B. (1993), *H_∞-Control for Distributed Parameter Systems: A State-Space Approach*, Birkhäuser, Boston.

[3] Krasovskii, N.N. and Subbotin, A.I. (1988), *Game-Theoretical Control Problems*, Springer, New York, 1988.

[4] Lasiecka, I. and Triggiani, R. (1991) *Differential and Algebraic Riccati Equations with Applications to Boundary/Point Control Problems*, Springer, Berlin.

[5] Lions, J.L. (1971), *Optimal Control of Systems Governed by Partial Differential Equations*, Springer, Berlin.

[6] Mordukhovich, B.S. (1989) "Minimax design for a class of distributed parameter systems," *Autom. Remote Control, Vol. 50*, 262–283.

[7] Mordukhovich, B.S. and Zhang, K. (1996), "Feedback control design of constrained parabolic systems in uncertainty conditions," *Prep. 13th IFAC World Congress, Vol. D*, 225–230.

[8] Mordukhovich, B.S. and Zhang, K. (1997), "Minimax control of parabolic systems with Dirichlet boundary conditions and state constraints," *Appl. Math. Optim., Vol. 36*, 323–360.

[9] Pontryagin, L.S., Boltyanskii, V.G., Gamkrelidze, R.V. and Michshenko, E.F. (1962), *The Mathematical Theory of Optimal Processes*, Wiley-Interscience, New York.

STABILIZATION OF LINEAR BOUNDARY CONTROL SYSTEMS OF PARABOLIC TYPE: AN ALGEBRAIC APPROACH

Takao Nambu

Department of Applied Mathematics
Kobe University
Nada, Kobe 657-8501, Japan
nambu@godel.seg.kobe-u.ac.jp

1 INTRODUCTION

We consider in this paper the stabilization problem of a class of linear boundary control systems of parabolic type by means of feedback control. Our boundary control system with state $u = u(t, \cdot)$ is described by the differential equation:

$$\frac{\partial u}{\partial t} + \mathcal{L}u = 0, \qquad \tau u = \sum_{k=1}^{M} f_k(t)\tilde{h}_k(\xi), \qquad u(0,x) = u_0(x). \tag{1.1}$$

Here, $f_k(t)$ denote control inputs; $\tilde{h}_k(\xi)$ actuators on the boundary; and $(\mathcal{L}, \tau)$ a system of linear differential operators in a bounded domain Ω of $\mathbb{R}^m$ with the boundary Γ consisting of a finite number of smooth components of $(m-1)$-dimension. Actually, let $\mathcal{L}$ denote a uniformly elliptic differential operator of order 2 in Ω defined by

$$\mathcal{L}u = -\sum_{i,j=1}^{m} \frac{\partial}{\partial x_i}\Big(a_{ij}(x)\frac{\partial u}{\partial x_j}\Big) + c(x)u,$$

where $a_{ij}(x) = a_{ji}(x)$ for $1 \leqslant i,\, j \leqslant m$, $x \in \overline{\Omega}$. Associated with $\mathcal{L}$ is the boundary operator τ defined by

$$\tau u = \alpha(\xi)u(\xi) + (1-\alpha(\xi))\frac{\partial u(\xi)}{\partial \nu} = \alpha(\xi)u(\xi) + (1-\alpha(\xi))\sum_{i,j=1}^{m} a_{ij}(\xi)\nu_i(\xi)\frac{\partial u}{\partial x_j},$$

where $\alpha(\xi)$ denotes a smooth function on Γ, satisfying $0 \leqslant \alpha(\xi) \leqslant 1$, and $(\nu_1(\xi), \dots, \nu_m(\xi))$ the unit outer normal at $\xi \in \Gamma$. The inner products in $L^2(\Omega)$ and $L^2(\Gamma)$ are denoted by $\langle \cdot, \cdot \rangle_\Omega$ and $\langle \cdot, \cdot \rangle_\Gamma$, respectively.

The output of the system (1.1) is given by

$$\langle u, w_k \rangle_\Omega, \quad 1 \leqslant k \leqslant N, \quad w_k \in L^2(\Omega). \tag{1.2}$$

By feeding the output back into $f_k(t)$'s, our task is to choose suitable feedback parameters so that the state $u(t, \cdot)$ decays exponentially as $t \to \infty$ for every initial state u_0. Stabilization results for (1.1) in the literature are limited to the case where $\alpha(\xi) \equiv 1$ (the Dirichlet boundary) or the case where $0 \leqslant \alpha(\xi) < 1$ (the generalized Neumann boundary), e.g., [1, 3, 7]: The stabilization has been achieved, mainly due to the fact that the structure of the fractional powers of the associated elliptic operator L is entirely known [2, 5, 8]. In the generalized Neumann case, for example, set $x(t) = L_c^{-1/4-\epsilon} u(t, \cdot)$, $0 < \epsilon < 1/4$ for a large $c > 0$, where $L_c = L + c$. Then $x(t)$ satisfies the differential equation with the homogeneous boundary condition [7]:

$$\frac{dx}{dt} + Lx = \sum_{k=1}^{M} f_k(t) L_c^{3/4-\epsilon} g_k, \quad x(0) = L_c^{-1/4-\epsilon} u_0,$$

where $g_k \in H^2(\Omega)$ denote the unique solutions to the boundary value problems: $(\mathcal{L} + c) g_k = 0$ in Ω, $\tau g_k = \tilde{h}_k$ on Γ, $1 \leqslant k \leqslant M$ [4]. Thus the control enters the equation as a distributed input in this transformed equation. This has made the problem considerably easy. The above transform works just like an *integral transform* which makes the state u smoother in space variables.

Our boundary condition is partly of the Dirichlet type on the set $\Gamma_1 = \{\xi \in \Gamma;\ \alpha(\xi) = 1\} \neq \emptyset$ and partly of the generalized Neumann type on $\Gamma \setminus \Gamma_1$. Unfortunately the structure of L_c^ω with this boundary condition is not well known at present in the context of the fractional Sobolev spaces. Thus the above approach seems no more available in our problem. The objective of this paper is to develop an alternative approach to the stabilization. By introducing an algebraic transform T, the whole stabilization procedure is of an algebraic nature as the title shows.

A *compensator* is a differential equation in $\mathbb{R}^\ell$ written by

$$\frac{dv}{dt} + B_1 v = \sum_{k=1}^{N} \langle u, w_k \rangle_\Omega \xi_k + \sum_{k=1}^{N} \langle v, \zeta_k \rangle_{\mathbb{R}^\ell} \eta_k, \quad v(0) = v_0. \tag{1.3}$$

Here, the matrix B_1; the vectors ξ_k, ζ_k, η_k; ρ_k appearing just below; and the dimension ℓ are the parameters to be determined. By setting $f_k(t) = \langle v, \rho_k \rangle_{\mathbb{R}^\ell}$, $1 \leqslant k \leqslant N$, equations (1.1) and (1.3) become a closed loop system.

2 PRELIMINARY RESULTS

Let us begin with reviewing the well known spectral properties of $(\mathcal{L}, \tau)$. Set

$$\hat{L}u = \mathcal{L}u, \quad \mathcal{D}(\hat{L}) = \{u \in C^2(\Omega) \cap C^1(\overline{\Omega});\ \mathcal{L}u \in L^2(\Omega),\ \tau u = 0\}. \tag{2.1}$$

The closure of $\hat{L}$ in $L^2(\Omega)$, denoted by L, is self adjoint and has a compact resolvent $(\lambda - L)^{-1}$. Thus there is a set of eigenpairs $\{\lambda_i, \varphi_{ij}\}$ such that [4]

(i) $\sigma(L) = \{\lambda_1, \lambda_2, \dots\}$, $\quad -\infty < \lambda_1 < \lambda_2 < \cdots < \lambda_i < \cdots \to \infty$;

(ii) $(\lambda_i - L)\varphi_{ij} = 0$, $\quad i \geqslant 1$, $1 \leqslant j \leqslant m_i \,(< \infty)$; and

(iii) the set $\{\varphi_{ij}\}$ forms a complete orthonormal system for $L^2(\Omega)$.

For λ in $\rho(L)$ and h in $C^{2+\omega}(\Gamma)$, the boundary value problem

$$(\lambda - \mathcal{L})u = 0 \quad \text{in } \Omega, \qquad \tau u = h \quad \text{on } \Gamma \tag{2.2}$$

admits a unique solution $u \in C^2(\Omega) \cap C^1(\overline{\Omega})$ such that $\mathcal{L}u$ is in $L^2(\Omega)$. The solution u is denoted by $N_\lambda h$. Set $C = -\sum_{k=1}^{M} \langle \cdot, \hat{\rho}_k \rangle_\Omega N_{-c} h_k$, where $\hat{\rho}_k \in L^2(\Omega)$ are to be given later in (3.15). If λ is in $\rho(L+C) \cap \rho(L)$, the problem

$$(\lambda - \mathcal{L} - C)u = 0 \quad \text{in } \Omega, \qquad \tau u = h \quad \text{on } \Gamma \tag{2.3}$$

also admits a unique solution $u \in C^2(\Omega) \cap C^1(\overline{\Omega})$ such that $\mathcal{L}u$ is in $L^2(\Omega)$. The solution is denoted by $N(\lambda)h$.

3 MAIN RESULT

Let us consider the feedback control system

$$\begin{aligned}
&\frac{\partial u}{\partial t} + \mathcal{L}u = 0, \quad \tau u = \sum_{k=1}^{M} \langle v, \tilde{\rho}_k \rangle_{\mathbb{R}^\ell} \tilde{h}_k(\xi), \quad u(0,x) = u_0(x) \in L^2(\Omega), \\
&\frac{dv}{dt} + B_1 v = \sum_{k=1}^{N} \langle u, w_k \rangle_\Omega \tilde{\xi}_k + \sum_{k=1}^{N} \langle v, \tilde{\zeta}_k \rangle_{\mathbb{R}^\ell} \tilde{\eta}_k, \quad v(0) = v_0 \in \mathbb{R}^\ell.
\end{aligned} \tag{3.1}$$

Associated with (3.1) is an auxiliary feedback control system described by

$$\begin{aligned}
&\frac{\partial u}{\partial t} + \mathcal{L}u = 0, \quad \tau u = \sum_{k=1}^{M} \langle v, \rho_k \rangle_\Omega \tilde{h}_k(\xi), \quad u(0,x) = u_0(x) \in L^2(\Omega), \\
&\frac{dv}{dt} + Bv = \sum_{k=1}^{N} \langle u, w_k \rangle_\Omega \xi_k + \sum_{k=1}^{N} \langle v, \zeta_k \rangle_\Omega \eta_k, \quad v(0) = v_0 \in L^2(\Omega).
\end{aligned} \tag{3.2}$$

Here, the differential equation for v is the one in $L^2(\Omega)$, and

$$B = L - \sum_{k=1}^{M} \langle \cdot, L_c \rho_k \rangle_\Omega N_{-c} h_k, \quad \mathcal{D}(B) = \mathcal{D}(L), \tag{3.3}$$

where $L_c = L + c$, $c > -\lambda_1$; $h_k \in C^{2+\omega}(\Gamma)$; and $\rho_k \in \mathcal{D}(\hat{L})$, $1 \leqslant k \leqslant M$. Set

$$\Phi_c = \left(\langle N_{-c} h_k, \rho_j \rangle_\Omega \; ; \; \begin{matrix} k \to 1, \dots, M \\ j \downarrow 1, \dots, M \end{matrix} \right). \tag{3.4}$$

Here, the vectors ρ_j's will be determined by the stabilization problem for B. We may assume with no loss of generality that $(1-\Phi_c)^{-1}$ exists. The actuators $\tilde{h}_k$'s in (3.1) are then defined by

$$[\tilde{h}_1 \dots \tilde{h}_M] = [h_1 \dots h_M](1-\Phi_c)^{-1}. \tag{3.5}$$

Define the matrices H_i of size $m_i \times M$ and W_i of size $m_i \times N$ by

$$H_i = \left(h_{ij}^k \, ; \begin{matrix} k \to 1, \dots, M \\ j \downarrow 1, \dots, m_i \end{matrix} \right), \quad \text{and} \quad W_i = \left(w_{ij}^k \, ; \begin{matrix} k \to 1, \dots, N \\ j \downarrow 1, \dots, m_i \end{matrix} \right), \tag{3.6}$$

respectively, where

$$h_{ij}^k = \left\langle h_k, \varphi_{ij} - \frac{\partial \varphi_{ij}}{\partial \nu} \right\rangle_\Gamma, \quad \text{and} \quad w_{ij}^k = \langle w_k, \varphi_{ij} \rangle_\Omega .$$

Our main result is stated as follows:

Theorem 3.1. *Suppose that* $\lambda_1 < \dots < \lambda_I \leqslant 0 < \lambda_{I+1}$, *and that*

$$\operatorname{rank} H_i = \operatorname{rank} W_i = m_i, \quad 1 \leqslant i \leqslant I. \tag{3.7}$$

Then, for any $0 < \lambda < \lambda_{I+1}$, *there exist parameters* ρ_k*'s,* ξ_k*'s,* ζ_k*'s and* η_k*'s such that every solution to* (3.2) *with* u_0, $v_0 \in L^2(\Omega)$ *satisfies the estimate*

$$\|u(t,\cdot)\| + \|v(t,\cdot)\| \leqslant \text{const}\, e^{-\lambda t}\{\|u_0\| + \|v_0\|\}, \quad t \geqslant 0. \tag{3.8}$$

Eqn. (3.2) *is reduced to* (3.1) *with some integer* ℓ, *which admits a unique genuine solution* $(u(t,x), v(t))$ *for every* $(u_0, v_0) \in L^2(\Omega) \times \mathbb{R}^\ell$ *such that* $\mathcal{L}u$ *is bounded in* $(t_1, t_2) \times \Omega$, $0 < {}^\forall t_1 < {}^\forall t_2$. *Every solution to* (3.1) *satisfies the estimate*

$$\|u(t,\cdot)\| + |v(t)|_\ell \leqslant \text{const}\, e^{-\lambda t}\{\|u_0\| + |v_0|_\ell\}, \quad t \geqslant 0. \tag{3.9}$$

Outline of the proof: *First Step* (*Operator* T). Let us consider the operator B in (3.3) and determine ρ_k's in it. Assumption (3.7) implies that

$$\operatorname{rank} \left(\langle N_{-c} h_k, \varphi_{ij} \rangle_\Omega \, ; \begin{matrix} k \to 1, \dots, M \\ j \downarrow 1, \dots, m_i \end{matrix} \right) = \operatorname{rank} H_i = m_i, \quad 1 \leqslant i \leqslant I.$$

Let P_I denote the projection operator mapping $L^2(\Omega)$ onto span $\{\varphi_{ij};\ 1 \leqslant i \leqslant I,\ 1 \leqslant j \leqslant m_i\}$. According to the pole assignment theory of finite dimension [10], (see also [9]), there exist ρ_k's in $P_I L^2(\Omega)$ such that $\sigma(B)$ is contained in $\{\lambda \in \mathbb{C};\ \operatorname{Re}\lambda \geqslant \lambda_{I+1}\}$. The analytic semigroup e^{-tB} thus satisfies the estimate

$$\|e^{-tB}\| \leqslant \text{const}\, e^{-\lambda_{I+1} t}, \quad t \geqslant 0. \tag{3.10}$$

Let us introduce the algebraic transform T in $\mathcal{L}(L^2(\Omega))$ by

$$p = Tu = u - \sum_{k=1}^{M} \langle u, \rho_k \rangle_\Omega \, N_{-c} h_k. \tag{3.11}$$

The operator T is injective, and $T^{-1} \in \mathcal{L}(L^2(\Omega))$ is expressed by

$$u = T^{-1}p = p + [N_{-c}h_1 \dots N_{-c}h_M](1 - \Phi_c)^{-1}\langle p, \boldsymbol{\rho}\rangle_\Omega, \tag{3.12}$$

where $\boldsymbol{\rho}$ denotes the transpose of $(\rho_1, \dots, \rho_M)$. Set

$$\zeta_k = (T^{-1})^* w_k = \hat{w}_k, \quad \eta_k = -\xi_k, \quad 1 \leqslant k \leqslant N. \tag{3.13}$$

Here, ξ_k's are to be determined in the Second Step. Then, $p - v$ satisfies

$$\begin{aligned} \frac{\partial(p-v)}{\partial t} + \mathcal{B}(p-v) &= \sum_{j,k=1}^{M} \langle p - v, \rho_j\rangle_\Omega \left\langle \tilde{h}_j, \rho_k - \frac{\partial \rho_k}{\partial \nu}\right\rangle_\Gamma N_{-c}h_k \\ &\quad - \sum_{k=1}^{N} \langle p - v, \hat{w}_k\rangle_\Omega \, \xi_k, \\ \tau(p-v) &= -\sum_{k=1}^{M} \langle p - v, \rho_k\rangle_\Omega \, \tilde{h}_k, \quad (p-v)(0,\cdot) = p_0 - v_0, \end{aligned} \tag{3.14}$$

where we have formally set $\mathcal{B} = \mathcal{L} - \sum_{k=1}^{M}\langle \cdot, L_c\rho_k\rangle_\Omega N_{-c}h_k$.

Second Step (Operator K). We introduce the operator $\hat{K}$ by

$$\begin{aligned} \hat{K}y &= \mathcal{B}y - \sum_{j,k=1}^{M} \langle y, \rho_j\rangle_\Omega \left\langle \tilde{h}_j, \rho_k - \frac{\partial \rho_k}{\partial \nu}\right\rangle_\Gamma N_{-c}h_k \\ &= \mathcal{L}y - \sum_{k=1}^{M} \langle y, \hat{\rho}_k\rangle_\Omega \, N_{-c}h_k, \quad \text{where} \end{aligned} \tag{3.15}$$

$$\mathcal{D}(\hat{K}) = \left\{ y \in C^2(\Omega) \cap C^1(\overline{\Omega});\ \mathcal{L}y \in L^2(\Omega),\ \tau y = -\sum_{k=1}^{M} \langle y, \rho_k\rangle_\Omega \, \tilde{h}_k \text{ on } \Gamma \right\}.$$

Proposition 3.2. *The operator $\hat{K}$ admits the closure K which is densely defined in $L^2(\Omega)$. There exists a sector $\overline{\sum}_{-a} = \{\lambda - a \in \mathbb{C};\ \theta_0 \leqslant |\arg\lambda| \leqslant \pi\}$, $0 < \theta_0 < \pi/2$, $a \in \mathbb{R}^1$, such that the resolvent of K is expressed in $\overline{\sum}_{-a}$ as*

$$\begin{aligned} (\lambda - K)^{-1} &= (\lambda - K^0)^{-1} \\ &\quad - [N(\lambda)\tilde{h}_1 \dots N(\lambda)\tilde{h}_M](1 + \Phi(\lambda))^{-1}\langle (\lambda - K^0)^{-1}\cdot, \boldsymbol{\rho}\rangle_\Omega, \end{aligned}$$

where $K^0 = L - \sum_{k=1}^{M}\langle \cdot, \hat{\rho}_k\rangle_\Omega N_{-c}h_k$, $\mathcal{D}(K^0) = \mathcal{D}(L)$, and

$$\Phi(\lambda) = \left(\langle N(\lambda)\tilde{h}_k, \rho_j\rangle_\Omega \, ; \begin{array}{l} k \to 1, \dots, M \\ j \downarrow 1, \dots, M \end{array} \right).$$

Proposition 3.3. *The operator T is an isomorphism which maps $\mathcal{D}(L)$ onto $\mathcal{D}(K)$. Furthermore, K is similar to L, that is,*

$$T^{-1}KT = L, \quad T\mathcal{D}(L) = \mathcal{D}(K). \tag{3.16}$$

Associated with (3.14) is the differential equation for $y = p - v$ in $L^2(\Omega)$:

$$\frac{dy}{dt} + Ky + \sum_{k=1}^{N} \langle y, \hat{w}_k \rangle_\Omega \, \xi_k = 0, \quad y(0) = y_0 = p_0 - v_0. \tag{3.17}$$

Set $z = T^{-1}y$. Then, by Proposition 3.3, (3.17) is equivalent to

$$\frac{dz}{dt} + Lz + \sum_{k=1}^{N} \langle z, w_k \rangle_\Omega \, T^{-1}\xi_k = 0, \quad z(0) = z_0.$$

(*Determination of* ξ_k*'s*) In view of the assumption (3.7) and the above equation for z, there exist $T^{-1}\xi_k$'s in $P_I L^2(\Omega)$ and thus ξ_k's in $C^2(\Omega) \cap C^1(\overline{\Omega})$ such that the following estimate holds:

$$\left\| \exp t \left(-K - \sum_{k=1}^{N} \langle \cdot, \hat{w}_k \rangle_\Omega \, \xi_k \right) \right\| \leqslant \text{const}\, e^{-\lambda_{I+1} t}, \quad t \geqslant 0. \tag{3.18}$$

Third Step (*Equation for* p *and* v). The differential equation for y and v is the one in the product space $L^2(\Omega) \times L^2(\Omega)$:

$$\frac{d}{dt}\begin{pmatrix} y \\ v \end{pmatrix} + (F+G)\begin{pmatrix} y \\ v \end{pmatrix} = \begin{pmatrix} 0 \\ 0 \end{pmatrix}, \quad \begin{pmatrix} y(0) \\ v(0) \end{pmatrix} = \begin{pmatrix} y_0 \\ v_0 \end{pmatrix}, \quad \text{where} \tag{3.19}$$

$$F\begin{pmatrix} y \\ v \end{pmatrix} = \begin{pmatrix} K & 0 \\ 0 & B \end{pmatrix}\begin{pmatrix} y \\ v \end{pmatrix}, \quad \text{and} \quad G\begin{pmatrix} y \\ v \end{pmatrix} = \sum_{k=1}^{N} \langle y, \hat{w}_k \rangle_\Omega \begin{pmatrix} \xi_k \\ -\xi_k \end{pmatrix}.$$

The operator $-F - G$ generates an analytic semigroup. The equation for $p\,(= y + v)$ and v is written as, by setting $\Lambda = S(F+G)S^{-1}$ with $S = \left(\begin{smallmatrix} 1 & 1 \\ 0 & 1 \end{smallmatrix}\right)$,

$$\frac{d}{dt}\begin{pmatrix} p \\ v \end{pmatrix} + \Lambda\begin{pmatrix} p \\ v \end{pmatrix} = \begin{pmatrix} 0 \\ 0 \end{pmatrix}, \quad \begin{pmatrix} p(0) \\ v(0) \end{pmatrix} = \begin{pmatrix} p_0 \\ v_0 \end{pmatrix}. \tag{3.20}$$

Proposition 3.4. *The operator* Λ *is expressed by*

$$\Lambda\begin{pmatrix} p \\ v \end{pmatrix} = \begin{pmatrix} B\left(p + \sum_{k=1}^{M} \langle p - v, \rho_k \rangle_\Omega \, N_{-c}\tilde{h}_k\right) - \sum_{k=1}^{M} \langle p - v, \rho_k \rangle_\Omega \, BN_{-c}\tilde{h}_k \\ - \sum_{j,k=1}^{M} \langle p - v, \rho_j \rangle_\Omega \left\langle \tilde{h}_j, \rho_k - \frac{\partial \rho_k}{\partial \nu} \right\rangle_\Gamma N_{-c}h_k \\ \\ Bv - \sum_{k=1}^{M} \langle p - v, \hat{w}_k \rangle_\Omega \, \xi_k \end{pmatrix},$$

$$\mathcal{D}(\Lambda) = \left\{ \begin{pmatrix} p \\ v \end{pmatrix};\ p - v \in \mathcal{D}(K), \quad v \in \mathcal{D}(B) \right\}.$$

Fourth Step (*Stabilization*). The decay estimates (3.10) and (3.18) imply that, for any λ'; $\lambda < \lambda' < \lambda_{I+1}$,

$$\|e^{-tA}\|_{\mathcal{L}(L^2(\Omega)\times L^2(\Omega))} \leqslant \text{const}\, e^{-\lambda' t}, \quad t \geqslant 0. \tag{3.21}$$

This establishes the decay estimate (3.8), since $u = T^{-1}p$.

In order to obtain a finite-dimensional compensator, let us add a small perturbation Π_n to (3.20) to obtain the equation:

$$\frac{d}{dt}\begin{pmatrix} p \\ v \end{pmatrix} + (\Lambda + \Pi_n)\begin{pmatrix} p \\ v \end{pmatrix} = \begin{pmatrix} 0 \\ 0 \end{pmatrix}, \quad \begin{pmatrix} p(0) \\ v(0) \end{pmatrix} = \begin{pmatrix} p_0 \\ v_0 \end{pmatrix}, \quad \text{where} \tag{3.22}$$

$$\Pi_n \begin{pmatrix} p \\ v \end{pmatrix} = \begin{pmatrix} 0 \\ \displaystyle\sum_{k=1}^{N}\Big(\langle p - v, \hat{w}_k\rangle_\Omega (1 - P_n)\xi_k - \langle v, (1-P_n)\hat{w}_k\rangle_\Omega P_n\xi_k\Big) \\ \displaystyle + \sum_{k=1}^{M}\langle v, L_c\rho_k\rangle_\Omega (1-P_n)N_{-c}h_k \end{pmatrix}$$

Since $\alpha_n = \|\Pi_n\|_{\mathcal{L}(L^2(\Omega)\times L^2(\Omega))} \to 0$ as $n \to \infty$, we see that, for a large $n \geqslant I$,

$$\|e^{-t(\Lambda+\Pi_n)}\|_{\mathcal{L}(L^2(\Omega)\times L^2(\Omega))} \leqslant \text{const}\, e^{(-\lambda'+\alpha_n)t} \leqslant \text{const}\, e^{-\lambda t}, \quad t \geqslant 0. \tag{3.23}$$

Set $v_1(t) = P_n v(t)$ for such an n. If v_0 belongs to $P_n L^2(\Omega)$, then $(p(t), v_1(t))$ is a solution to (3.22). In other words, *$v(t)$ remains in the finite-dimensional subspace $P_n L^2(\Omega)$ as long as the initial state v_0 belongs to $P_n L^2(\Omega)$.*

Last Step (*Regularity of u and v*). Henceforth we assume that v_0 belongs to $P_n L^2(\Omega)$ in (3.22) so that $v(t)$ remains in $P_n L^2(\Omega)$. It is shown -via a result in [6]- that the solution $p(t,x)$ in (3.22) is in $C^2(\Omega) \cap C^1(\overline{\Omega})$ for $^\forall t > 0$ and $\mathcal{L}p(t,x)$ is bounded in $(t_1, t_2) \times \Omega$, $0 < {}^\forall t_1 < {}^\forall t_2$, and is a genuine solution to the initial-boundary value problem:

$$\begin{aligned} &\frac{\partial p}{\partial t} + \mathcal{B}p - \sum_{j,k=1}^{M} \langle p - v, \rho_j\rangle_\Omega \left\langle \tilde{h}_j, \rho_k - \frac{\partial \rho_k}{\partial \nu}\right\rangle_\Gamma N_{-c}h_k = 0, \\ &\tau p = -\sum_{k=1}^{M}\langle p - v, \rho_k\rangle_\Omega \tilde{h}_k, \quad p(0,x) = p_0(x). \end{aligned} \tag{3.24}$$

Let us derive the equation for $u = T^{-1}p$ and v. By applying Green's formula to a term of $\mathcal{B}$, the equation for p is rewritten as

$$\frac{\partial p}{\partial t} + \mathcal{L}_c p - \sum_{k=1}^{M}\langle \mathcal{L}_c p, \rho_k\rangle_\Omega N_{-c}h_k - cp = \frac{\partial p}{\partial t} + T\mathcal{L}_c p - cp = 0.$$

Note that $\mathcal{L}_c p = \mathcal{L}_c u$ by (3.11). As for the boundary condition, we see that

$$\tau u = [h_1 \dots h_M](1 - \Phi_c)^{-1}\langle p, \boldsymbol{\rho}\rangle_\Omega - [\tilde{h}_1 \dots \tilde{h}_M]\langle p - v, \boldsymbol{\rho}\rangle_\Omega = \sum_{k=1}^{M}\langle v, \rho_k\rangle_\Omega \tilde{h}_k.$$

Thus (u, v) satisfies the differential equation

$$\frac{\partial u}{\partial t} + \mathcal{L}u = 0, \qquad \tau u = \sum_{k=1}^{M} \langle v, \rho_k \rangle_\Omega \, \tilde{h}_k, \qquad u(0,x) = u_0 \in L^2(\Omega)$$

$$\frac{dv}{dt} + Lv - \sum_{k=1}^{M} \langle v, L_c \rho_k \rangle_\Omega \, P_n N_{-c} h_k \tag{3.25}$$
$$- \sum_{k=1}^{N} \langle u, w_k \rangle_\Omega \, P_n \xi_k + \sum_{k=1}^{N} \langle v, P_n \hat{w}_k \rangle_\Omega \, P_n \xi_k = 0, \quad v(0) = v_0 \in P_n L^2(\Omega).$$

Recall that the state of the compensator $v(t)$, $t \geqslant 0$ in (3.25) remains in $P_n L^2(\Omega)$. Thus eqn. (3.25) is equivalent to (3.1) with $\ell = \dim P_n L^2(\Omega) = m_1 + \cdots + m_n$. The decay estimate (3.9) will be now clear. The uniqueness of solutions to (3.25) is also clear, since (3.25) is finally transformed to (3.22). This finishes the proof of Theorem 3.1. Q.E.D.

References

[1] R. F. Curtain, Finite dimensional compensators for parabolic distributed systems with unbounded control and observation, *SIAM J. Control Optim.*, **22** (1984), 255–276.

[2] D. Fujiwara, Concrete characterization of the domain of fractional powers of some elliptic differential operators of the second order, *Proc. Japan Acad. Ser. A Math Sci.* **43** (1967), 82–86 .

[3] J. S. Gibson and A. Adamian, Approximation theory for linear quadratic-Gaussian optimal control of flexible structures, *SIAM J. Control Optim.*, **29** (1991), 1–37.

[4] D. Gilbarg and N. S. Trudinger, "Elliptic Partial Differential Equations of Second Order," 2nd ed., Springer-Verlag, New York, 1983.

[5] P. Grisvard, Caractérisation de quelques espaces d'interpolation, *Arch. Rational Mech. Anal.* **25** (1967), 40–63.

[6] S. Itô, "Diffusion Equations," Amer. Math. Soc., Providence, 1992.

[7] T. Nambu, On stabilization of partial differential equations of parabolic type: Boundary observation and feedback, *Funkcial. Ekvac.* **28** (1985), 267–298.

[8] T. Nambu, Characterization of the domain of fractional powers of a class of elliptic differential operators with feedback boundary conditions, *J. Differential Equations* **136** (1997), 294–324.

[9] Y. Sakawa, Feedback stabilization of linear diffusion systems, *SIAM J. Control Optim.*, **21** (1983), 667–676.

[10] W. M. Wonham, On pole assignment in multi-input controllable linear systems, *IEEE Trans. Automat. Control*, **AC-12** (1967), 660–665.

A DISTRIBUTED BIOREMEDIATION PROBLEM WITH MODAL SWITCHING

Thomas I. Seidman

Department of Mathematics and Statistics
University of Maryland Baltimore County
1000 Hilltop Circle
Baltimore, MD 21250
USA

seidman@math.umbc.edu

Abstract: We consider a model of bioremediation characterized by discontinuous transitions of the bacteria between 'dormant' and 'active' states. For a distributed problem with nondispersive spatial coupling (pure convection) it is shown both that the direct problem is well-posed and that there exists an optimal control for the natural cost functional.

1 INTRODUCTION

We consider the use of bacteria for removal of some undesirable pollutant (or conversion to an innocuous form: cometabolism) — i.e., bioremediation. For some further exposition of the nature and significance of bioremediation as a practical application we refer the reader, e.g., to [2], [6] or the talk [4] of the present conference. Our present concerns, however, will be purely mathematical — an analysis dominated by the dynamical discontinuities implied by the characteristic feature of this model: at any moment t the bacteria at a given point x may be either in an *active* or a *dormant* mode.

The state of the system is thus a *hybrid:* partly continuous, partly discrete. The concentrations, β, π, of bacteria and pollutant are continuous-valued components of the state, as is the concentration α of some 'critical nutrient'. On the other hand, there is a state component χ, taking discrete values $\{0, 1\}$, which indicates the mode: 0 for 'dormant', 1 for 'active'.

An idealized version of the modal transitions (compare [1]) is given by the *switching rules*:

- If the bacteria are active ($\chi = 1$) and the concentration α drops below some minimum α_* (with $\alpha_* > 0$), then they become dormant.
- If the bacteria are dormant ($\chi = 0$), they will be re-activated when α rises

above α^* (with $\alpha^* > \alpha_*$).

It is clear that χ can take either value (depending on the preceding history) when $\alpha_* < \alpha < \alpha^*$, so χ is not simply a function of α and must be introduced as an additional (discrete) state component, along with the more traditional continuous components α, β, π.

We note that the rules above define the 'elementary hysteron operator' ($W : \alpha(\cdot) \mapsto \chi$) as treated, e.g., by Krasnosel'skiĭ and Pokrovskiĭ [3]; see also [8]. It should be noted that $W[\cdot]$ is generally well-behaved and isotone as a map from, e.g., Lipschitzian inputs to piecewise constant outputs — but is inherently multi-valued in certain anomalous cases (e.g., if $\chi = 0$ and α rises to α^* but not above; cf., e.g., [7]) which will not affect our present analysis..

We consider a fixed interval $[0, T]$ in time. The one-dimensional spatial region $[0, \ell]$ initially contains some pollutant (distribution π_0) and also some bacteria (β_0), which are present but dormant. We are assuming that the bacteria are fixed in position and, for simplicity, will assume that the activity of the biomass is controlled by availability of a single critical nutrient (concentration α, initially 0). This nutrient is supplied (as a control, at rate $u(\cdot)$) first to re-activate the bacteria and then keeping them active to reduce the pollutant. Our principal technical concern will be the existence and characterization of solutions for specified $u(\cdot)$ with the existence of optimal controls following easily from that for, e.g., the problem of minimizing $\mathcal{J} =$ *[cost of nutrient] – [value of pollutant removal].*

We assume that the nutrient is soluble in, e.g., an underwater groundflow of given velocity $v = v(t) > 0$, so we are injecting nutrient into this flow at $x = 0$, the left boundary. The pollutant may or may not also be carried by the flow (with mobility $0 \leq \mu \leq 1$). Thus, we have the system:

$$\begin{aligned}
\alpha_t + v\alpha_x &= D\alpha_{xx} - \varphi\chi\beta \\
&\begin{cases} -D\alpha_x + v\alpha = u & \text{at } x = 0 \\ D\alpha_x = 0 & \text{at } x = \ell \end{cases} \\
\beta_t &= \Gamma(\alpha, \beta)\chi \\
\pi_t + \mu v\pi_x &= \hat{D}\pi_{xx} - \psi\chi\beta \\
&\begin{cases} -\hat{D}\pi_x + \mu v\pi = \hat{\pi} & \text{at } x = 0 \\ \hat{D}\pi_x = 0 & \text{at } x = \ell \end{cases}
\end{aligned} \tag{1.1}$$

Here, the coefficients $\varphi, \psi > 0$ represent the respective rates at which (when active) the bacteria consume nutrient and pollutant; we take these as dependent only on α except that ψ becomes 0 at $\pi = 0$. The bacterial growth rate Γ may be taken simply as $\gamma(\alpha)\beta$ or, e.g., as Monod kinetics, limiting β. We assume that $D, \hat{D} \geq 0$, that $0 < \underline{v} \leq v \leq \bar{v}$, and that $0 < \underline{\beta} \leq \beta_0 \leq \bar{\beta}$ at $t = 0$. Etc. Note that $\hat{\pi}(\cdot) \geq 0$ is a known input and that $u(\cdot) \gtrsim 0$ is our control function (although in taking v as time-dependent we anticipate some possibility that this too might be controlled).

In this paper we concentrate on the extreme case of negligible diffusion but first note that the other extreme would be 'perfect mixing' (e.g., for small ℓ), giving a lumped parameter model:

$$\dot{\alpha} = -\varphi\chi\beta - \lambda\alpha + u, \qquad \dot{\beta} = \Gamma\chi, \qquad \dot{\pi} = -\psi\chi\beta \tag{1.2}$$

where the term '$-\lambda\alpha$' corresponds to flowthrough of nutrient, the control u now appears in the equation, and we have taken $\mu = 0$, $\hat{\pi} = 0$. This problem has already been treated in [5], showing: existence of solutions for any non-decreasing $U(t) = \int_0^t u$; computational approximation; existence (with some characterization) of optimal controls, e.g., minimizing $\mathcal{J} = [U(T) + b\pi(T)]$.

2 REFORMULATION

Taking $D, \hat{D} = 0$ in (1.1) to consider pure convection, the system becomes:

$$\begin{aligned} \alpha_t + v\alpha_x &= -\varphi\chi\beta \\ \beta_t &= \Gamma(\alpha,\beta)\chi \\ \pi_t + \mu v\pi_x &= -\psi\chi\beta \end{aligned} \tag{2.1}$$

with $v\alpha = u$ at $x = 0$ and (if $\mu > 0$) $\mu v\pi = \hat{\pi}$ — as well as initial data for α, β, π. To this we adjoin the switching rules: $\chi = W[\alpha]$, independently for each $x \in [0, \ell]$, with $\chi \equiv 0$ at $t = 0$. As noted, this contitutes the significant mathematical novelty of this model — which does not fall within the types treated in [8] but which may be viewed as the free boundary problem of determining the region $\mathcal{R} = \{(t,x) : \chi(t,x) = 1\}$ in which the bacteria are active.

Our first observation is that the final equation in (2.1) decouples, along with (2.2), so the dynamics — in this case, the determination of $\mathcal{R}$ for a fixed control function u — will be given entirely by the first two equations of (2.1) together with the switching rules, although the equation for π is, of course, relevant for the optimization of, e.g.,

$$\mathcal{J} = \int_0^T u(t)\,dt - b\int_{\mathcal{R}} \psi\beta\,dx\,dt. \tag{2.2}$$

In considering the dynamics it is convenient to use the coordinate system $[\tau, x]$ where

$$\tau = \tau(t,x) := \int_0^t \frac{ds}{v(s)} - x. \tag{2.3}$$

This is only relevant for $\tau > 0$ since $\chi = 0$ (so nothing happens) before that. Since τ increases with t at each x, the switching rules can equivalently be expressed in terms of the τ-history of α. Using (2.3), the relevant portion of (2.1) becomes

$$\begin{aligned} v\alpha_x &= -\varphi\chi\beta \quad (v\alpha = u \text{ at } x = 0) \\ v\beta_\tau &= \Gamma(\alpha,\beta)\chi. \end{aligned} \tag{2.4}$$

Assuming $\chi \equiv 1$ on $(0, x)$ for some fixed τ, this gives

$$\alpha(\tau, x) = \left[u - \int_0^x \varphi\beta\right]/v \tag{2.5}$$

whence α is a nonincreasing function of x (with $\alpha \le u/v$). Given a bound on u, we have α uniformly bounded (so $\varphi(\alpha)$ is bounded above and below by compactness) and then have β bounded above and below (away from 0, using that $\beta_0 \ge \underline{\beta} > 0$). We also immediately obtain uniform Lipschitz continuity of α with respect to x and of β with respect to τ.

As $\varphi\beta$ is bounded away from 0, it is possible to define a pair of functions $X^*(\tau)$, $X_*(\tau)$ by solving $\alpha = \alpha^*, \alpha_*$, respectively, in (2.5):

$$\int_0^{X^*} \varphi\beta\, dx = u - \alpha^*, \qquad \int_0^{X_*} \varphi\beta\, dx = u - \alpha_*. \tag{2.6}$$

Imposing a further admissibility requirement on the control $u(\cdot)$: that it satisfy a specified Lipschitz condition (with respect to t and so also τ), it then follows from (2.5) that α is Lipschitz continuous jointly in (τ, x) (although this will also require some analysis of the geometry of $\mathcal{R}$) whence, by the Implicit Function Theorem, X^*, X_* are Lipschitz continuous in τ; we also note that $X_* - X^* \ge$ *const.* > 0. Causally in τ, we now construct a new function $\hat{X}$ as the output of a variant of the hysteretic 'play operator' of [3], taking the double input: $[X_*, X^*]$: imagine an inertial mass point (so it stays stationary where possible) which is 'pushed' up by the graph of X^* and down by the graph of X_* as needed to maintain $X^* \le \hat{X} \le X_*$. We note that this operator: $[X^*, X_*] \mapsto \hat{X}$ is nonexpansive with respect to any weighted sup norm on $C[0,T]$ with nonincreasing weight.

An analysis of our switching rules shows that, since α is nonincreasing in x, the set-valued function $\mathcal{R}(x) := \{\tau : \chi(\tau, x) = 1\}$ must also be nonincreasing. The construction by way of (2.6) is now the heart of our characterization of $\mathcal{R}$, since it is easily seen that this $\hat{X}$ provides the suitable boundary:

$$\mathcal{R} = \{(\tau, x) : x \le \hat{X}(\tau)\}. \tag{2.7}$$

3 CONCLUSIONS

At this point we can state and sketch the proofs of our results: **Theorem 1.** For each non-negative Lipschitzian control function $u(\cdot)$ there is a unique solution of the system (2.1) with the switching rules, depending continuously on u.

Proof We begin, of course, with the reformulation above, both with respect to the recoordinatization and the replacement of the switching rules by taking $\chi(\tau, x) = \{1 \text{ if } x \le \hat{X}(\tau);\ 0 \text{ else}\}$.

Noting that there is no difficulty in obtaining $[\alpha, \beta]$ from (2.4) if χ is given, we can consider the map:

$$\hat{X}\ (\text{giving})\ \chi) \overset{(2.4)}{\mapsto} [\alpha, \beta] \overset{(2.6)}{\mapsto} [X^*, X_*] \mapsto \hat{X}.$$

The nonexpansivity of the play operator, as above, ensures that this map can be made contractive by the usual trick of selecting a suitable exponentially weighted norm for $C[0,T]$. Thus, there is a unique solution which, as we easily verify, depends continuously on $u(\cdot) \in C[0,T]$. **qed**

Theorem 2. Taking the set of admissible controls to be nonnegative functions on $[0,T]$ with a fixed Lipschitz bound, the optimal control problem:

Minimize $\mathcal{J}$ over admissible controls, subject to (2.1).

has a solution.

Proof As our admissibility assumption on $u(\cdot)$ gives compactness, we easily obtain existence of an optimal control for (2.2) as usual: extracting, from any minimizing sequence, one which converges in $C[0,T]$ and applying the previous Theorem to show that the limit control is optimal. **qed**

References

[1] Dawes, E. (1988). *Growth and Survival of Bacteria.* pp. 67–93 in *Bacteria in Nature, vol. 2* (J. Poindexter and E.R. Leadbetter, editors). New York: Plenum Publishers.

[2] National Research Council (1993). *In situ Bioremediation: When does it work?* National Academy Press.

[3] Krasnosel'skiĭ, M.A., and A.V. Pokrovskiĭ (1983). *Systems with Hysteresis.* Moscow: Nauka.
[*English transl.* (1989) Berlin: Springer-Verlag].

[4] S.M. Lenhart, *Optimal control for PDE/ODE systems modeling bioreactors.* this conference.

[5] S.M. Lenhart, T.I. Seidman, and J. Yong, *Optimal control of a hybrid bioreactor.* to appear.

[6] R.D. Norris, et.al., (1994). *Handbook of Bioremediation.* Boca Raton: Lewis Publishers.

[7] T.I. Seidman (1990). *Switching systems, I.* Control and Cybernetics **19**, pp. 63–92.

[8] A. Visintin (1994). *Differential Models of Hysteresis.* Berlin: Springer-Verlag.

OPTIMAL CONTROL PROBLEMS GOVERNED BY AN ELLIPTIC DIFFERENTIAL EQUATION WITH CRITICAL EXPONENT

Gengsheng Wang

Department of Mathematics, Huazhong Normal University

Abstract: This work is concerned with analysis and optimal control of a semilinear elliptic partial differential equation involving critical exponent. Some necessary conditions for optimality are given.

Key words and phrases: Minimal positive solution, state constraint, finite codimensionality, Penalty functional.

1 INTRODUCTION

We discuss the optimal control problem for which the state is governed by a semilinear elliptic partial differential equation with a distributed control.

The system reads

$$\begin{cases} -\Delta y = y^p + u, & \text{in} \quad \Omega \\ y|_{\partial\Omega} = 0. \end{cases} \tag{1.1}$$

Where $\Omega \subset R^N$ (with $N \geq 3$) is a bounded region with $\partial\Omega$ smooth. $p = \frac{N+2}{N-2}$ is the critical sobolev exponent.

The cost functional is given by

$$L(y, u) = G(y) + H(u) \tag{1.2}$$

We assume that

$(\mathbf{H_L})$ G and H: $L^2(\Omega) \longrightarrow \bar{R} = (-\infty, +\infty]$ are proper, convex and lower semicontinuous.

In section 2 , we will see that there is a minimal positive solution $y(x; u) \in H_0^1(\Omega)$ for each $u \in B_r^+(0) \subset L^\infty(\Omega) \subset L^2(\Omega)$, where $B_r^+(0)$ is given by

$$B_r^+(0) = \{u \in L^\infty(\Omega) | \, ||u||_\infty \leq r \quad \text{and}\, u(x) \geq 0, \quad \text{a.e.} \quad x \in \Omega\} \tag{1.3}$$

and $r > 0$ is a constant given in §2.

Note that here we define $y(x;0) \equiv 0$. Thus we may consider $u \in B_r^+(0)$ as the control and $y(x;u)$, the minimal positive solution of (1.1) corresponding to the state.

We assume

($\mathbf{H_F}$) Let Y be a Banach space with strict convex dual Y^* , $F : H_0^1(\Omega) \to Y$ be continuously Frechet differentiable and $Q \subset Y$ be a closed and convex subset.

Set

$$A = \{(y(x;u), u) \in L^2(\Omega) \times L^2(\Omega) \,|\, u \in B_r^+(0),\ y(x;u) \text{ is the minimal positive solution of (1.1) corresponding to } u \text{ and } y(x;0) \equiv 0\}$$

A pair $(y,u) \in A$ is called a feasible pair.

$$A_{ad} = \{(y(x;u), u) \in A | F(y(x;u)) \subset Q\}$$

A pair$(y,u) \in A_{ad}$ is called an admissible pair.

Note that $F(y) \subset Q$ is a kind of state constraint which was given by X.Li and J.Yong(cf.[3]). For its applications, we refer readers to [2] and [3].

We formulate the optimal control problem as follows

$$\textbf{(P)} \qquad \text{Inf } L(y,u) \qquad \text{over all} \quad (y,u) \in A_{ad}$$

We shall study the necessary conditions for the problem (P) in this paper.

2 THE MINIMAL POSITIVE SOLUTION

We first quote a result of [5] as follows.

Theorem A: For any $u \in H^{-1}(\Omega)$ with $||u||_{H^{-1}} \leq C_N S^{\frac{N}{2}}$, problem (2.1) possesses at least one positive solution y with $y \not\equiv 0$ in Ω . Where $C_N = \frac{4}{N-2}(\frac{N-2}{N+2})^{\frac{N+2}{4}}$ and S is the best sobolev constant for the embedding $H_0^1(\Omega) \to L^p(\Omega)$.

From Theorem A and the methods of monotone interation we can prove the existence of minimal positive solution for problem (2.1).

Theorem 2.1 *Under the assumption of Theorem A, Problem (2.1) possesses a unique minimal positive solution $y \in C^\alpha(\Omega)$ for some $\alpha \in (0,1)$ if $u \in L^\infty(\Omega)$.*

In the following we discuss some properties of the minimal positive solution of (1.1).

Lemma 2.1 *Let $y(x;u)$ be the minimal positive solution of (1.1), then the corresponding eigenvalue problem*

$$\begin{cases} -\Delta\varphi = \lambda p[y(x;u)]^{p-1}\varphi, & in \quad \Omega \\ \varphi \in H_0^1(\Omega) \end{cases} \tag{2.1}$$

has the first eigenvalue $\lambda_1(u) > 1$ *for all* $u \in B_R^+(0)$ *and the corresponding eigenfunction* $\varphi_1 > 0$ *in* Ω.

Where

$$B_R^+ = \left\{u \in H_0^{-1}(\Omega) \mid ||u||_{H_0^1(\Omega)} \leq R, u \geq 0 \quad in \quad \Omega\right\}$$

and $R < C_N S^{\frac{N}{2}}$.

Proof: By the standard argument we can prove that the minimum

$$\lambda_1 = \text{Inf}\left\{\int_\Omega |\nabla v|^2\, dx \mid v \in H_0^1(\Omega), \int_\Omega p y^{p-1}(x;u)v^2\, dx = 1\right\} \tag{2.2}$$

can be achieved by some function $\varphi_1 > 0$. Thus eigenvalue problem (2.2) has a solution (λ_1, φ_1). Now we prove that $\lambda_1 > 1$.

Indeed, for any $u \in B_R^+(0)$, we can find a function $w \in H_0^{-1}(\Omega)$ with $||w||_{H_0^{-1}} \leq C_N S^{\frac{N}{2}}$, $w \geq u$, $w \not\equiv u$ a.e. in Ω such that problem (2.1) (corresponding to w) possesses a minimal positive solution $y(x;w)$. Let $y(x;u)$ be the solution of (2.1), we have

$$\lambda_1 \int_\Omega p y^{p-1}(x;u)\varphi_1[y(x;w) - y(x:u)]dx \int_\Omega p y^{p-1}(x;u)\varphi_1(y(x;w) - y(x;u))\, dx \tag{2.3}$$

Which gives $\lambda_1 > 1$ for all $u \in B_R^+(0)$. This completes the proof.

Theorem 2.2 *Assume* $u \in B_R^+(0)$ *and* $y(x;u)$ *be a minimal positive solution of (1.1) corresponding to* u. *Then for any* $g(x) \in H_0^{-1}(\Omega)$, *The problem*

$$\begin{cases} -\Delta\omega = p y^{p-1}(x;u)\omega + g(x) \\ \omega \in H_0^1(\Omega) \end{cases} \tag{2.4}$$

has a unique solution ω *satisfying*

$$||\omega||_{H_0^1(\Omega)} \leq C|g||_{H_0^{-1}(\Omega)} \tag{2.5}$$

for some constant $C > 0$.

Proof: By a standard argument and Lemma 2.1, one can get the existence of the solution of the equation (2.4).

Now we are on the position to prove(2.5).

Let ω be the solution of (2.5). Multiplying (2.5) by ω and integrating by parts we have

$$\int_\Omega |\nabla\omega|^2\, dx = \int_\Omega p y^{p-1}(x;u)\omega^2\, dx + \int_\Omega g\omega\, dx.$$

Now Lemma 2.1 implies

$$(1-\frac{1}{\lambda_1})||\omega||^2_{H^1_0(\Omega)} \le C||g||_{H^{-1}_0(\Omega)}||\omega||_{H^1_0(\Omega)}$$
$$\le \epsilon||\omega||^2_{H^1_0(\Omega)} + C_\epsilon||g||^2_{H^{-1}_0(\Omega)}.$$

From Lemma2.1 we can choose ϵ small enough so that $(1-\frac{1}{\lambda_1}-\epsilon) \ge \lambda_2 > 0$ for some constant $\lambda_2 > 0$.

Thus

$$||\omega||_{H^1_0(\Omega)} \le \frac{C_\epsilon}{\lambda_2}||g||_{H^{-1}_0(\Omega)}$$

This gives (2.5) by taking $C = \frac{C_\epsilon}{\lambda_2}$.

The uniqueness of the solution for (2.4) comes from (2.5).

Corollary 2.1 *Let $u \in B^+_R(0)$ and $y(x;u)$ be the minimal solution of (1.1). Then $y(x,u)$ is continuous in $H^{-1}_0(\Omega)$ with respect to control function u .*

Proof: Define

$$F : H^{-1}_0(\Omega) \times H^1_0(\Omega) \to H^{-1}_0(\Omega) \quad \text{by}$$

$$F(u,y) = \Delta y + y^p + u, \quad \text{for} \quad (u,y) \in H^{-1}_0(\Omega) \times H^1_0(\Omega) \tag{2.6}$$

From Lemma 2.1 and Theorem 2.2, we know that

$$F_y(u,y)\omega = \Delta\omega + py^{p-1} + py^{p-1}(x;u)\omega$$

is an isomorphism of $H^1_0(\Omega)$ onto $H^{-1}_0(\Omega)$.

It follows from Implicit Function Theorem that the solution of $F(u,y) = 0$ near $(u, y(x;u))$ is given by a continuous curve.

Theorem 2.3 *Let $u, v \in B^+_R(0)$ and $y(x,u), y(x,v)$ be the minimal positive solution of (1.1) corresponding to u, v respectively. If $u \to v$ in $H^{-1}_0(\Omega)$ and $u - v$ dosen't change the sign. Then*

$$||y(x;u) - y(x;v)||_{H^1_0(\Omega)} \le C||u-v||_{H^{-1}_0(\Omega)},$$

for $||u-v||_{H^1_0(\Omega)}$ small enough.

Where C is a constant independent of u.

Proof: Without loss of generality, we may assume that $u \ge v$, a,e. in Ω. By Remark 2.1 and (1.1) we have

$$\int_\Omega |\nabla(y(x;u) - y(x;v))|^2\, dx$$
$$\le p\int_\Omega y^{p-1}(x;u)(y(x;u)-y(x;v))^2\, dx + \int_\Omega (u-v)(y(x;u)-y(x;v))\, dx$$

By lemma 2.1, Holder's inequality and Young's inequality, we have

$$(1-\frac{1}{\lambda_1(u)})\int_\Omega |\nabla(y(x;u) - y(x;v))|^2\, dx$$
$$\le \epsilon||y(x;u)-y(x;v)||^2_{H^1_0(\Omega)} + C_\epsilon||u-v||^2_{H^{-1}_0(\Omega)}$$

for any $\epsilon > 0$. Where C_ϵ is a positive constant depending on ϵ.

Note that $\lambda_1(u)$ is the first eigenvalue for the problem (2.2) corresponding to $y(x;u)$. By corollary 2.1, as $u \to v$ in $H_0^{-1}(\Omega)$, $y(x;u) \to y(x;v)$ in $H_0^1(\Omega)$. Then by (2.3), $\lambda_1(u) \to \lambda_1(v)$.

Thus, as $\|u - v\|_{H_0^{-1}(\Omega)}$ small enough, we have

$$(1 - \frac{1}{\lambda_1(v)} - \epsilon_1 - \epsilon)\|y(x;u) - y(x;v)\|_{H_0^1(\Omega)} \leq C_\epsilon \|u - v\|_{H_0^{-1}(\Omega)}$$

for some $\epsilon_1 > 0$ with $1 - \frac{1}{\lambda_1(v)} - \epsilon_1 > 0$.

This completes the proof.

3 FINITE CODIMENSIONALITY

In this section, we will give some results relative to finite codimensionality of a set. For the detail, we refer readers to [3] and [6].

Lemma 3.1 *Let Q_1 and Q_2 be subsets of some Banach space X. Let Q_1 be finite codimensional in X. Then for any $\alpha \in R \setminus \{0\}$, $\beta \in R$,*

$$\alpha Q_1 - \beta Q_2 \equiv \{\alpha x_1 - \beta x_2 | x_1 \in Q_1, x_2 \in Q_2\}$$

is finite codimensional in X.

Lemma 3.2 *Let Q be finite codimensional in X. Let $\{f_n\}_{n\geq 1} \subset X^*$ with $|f_n| \geq \delta > 0$, $f_n \to f \in X^*$ in the weak-star topology, and*

$$\langle f_n, x\rangle \geq -\epsilon_n, \quad \forall x \in Q, \quad n \geq 1$$

when $\epsilon_n \to 0$. Then $f \neq 0$.

4 NECESSARY CONDITION FOR OPTIMALITY

In this section, we discuss the necessary conditions for (y^*, u^*) to be an optimal pair for (P).

Our basic assumptions are given by (H_L) and (H_F) in §1.

Let (y^*, u^*) be optimal for the problem (P).

Consider the variational systems of (1.1) as follows:

$$\begin{cases} -\Delta z = p y^{*(p-1)} z + (v - u^*)^+, & x \in \Omega \\ z|_{\partial\Omega} = 0 \end{cases} \tag{4.1}$$

and

$$\begin{cases} -\Delta z = p y^{*(p-1)} z + (v - u^*)^-, & x \in \Omega \\ z|_{\partial\Omega} = 0 \end{cases} \tag{4.2}$$

By Theorem 2.2, for each $v \in B_r^+(0)$, both (4.1) and (4.2) have a unique solution $z(.;v^+)$ and $z(.;v^-)$ in $H_0^1(\Omega)$.

Thus we may define

$R^+ = \{z(.;v^+) | v \in B_r^+(0), z(.;v^+)$ is the solution of (4.1) corresponding to $v\}$

$R^- = \{z(\cdot\,;v^-) \,|\, v \in B_r^+(0), z(\cdot\,;v^-)$is the solution of (4.2) corresponding to $v\}$

Our another basic assumption which plays a key role in dealing with the state constraint is as follows:

(H_R) Both $F'(y^*)R^+ - Q$ and $F'(y^*)R^- - Q$ have finite codimentionality.

The following results is important for us to introduce our penalty functionals in the proof of our main Theorem 4.1.

Lemma 4.1 *Let H be a Hilbert space, $f : H \to \bar{R}$ be proper convex and lower semicontinuous. Suppose that ∂f (the subdifferential of f) is locally bounded at y^*. Then there exists a neighborhood $O(y^*)$ of y^* such that $f_\lambda(y) \to f(y)$ uniformly in $O(y^*)$, where f_λ is the regularization of f (cf. [4], [7]).*

Proof: It is trivial from [4] and [7].

Our main results on the necessary condition for (y^*, u^*) to be an optimal pair are as follows:

Theorem 4.1 *Let (y^*, u^*) be an optimal pair for P) and $(H_L), (H_F)$ and (H_R) hold. Assume that ∂G and ∂H, the subdifferentials of G and H are locally bounded at y^* and u^* (in $L^2(\Omega)$) respectively. Then there exists a triplet*

$$(\lambda, \psi, q) \in [-1, 0] \times H_0^1(\Omega) \times Y^*, \quad \textit{such that} \quad (\lambda, q) \neq 0,$$

$$\langle q, \eta - F(y^*)\rangle \leq 0, \quad \forall \eta \in Q. \tag{4.3}$$

$$\begin{cases} -\Delta\Psi = py^{*(p-1)}\Psi + \lambda\alpha - [F'(y^*)]^* q & \textit{in} \quad \Omega \\ \Psi|_{\partial\Omega} = 0, \end{cases} \tag{4.4}$$

where $\alpha \in \partial G(y^*)$
and

$$\langle \Psi + \lambda\beta, v - u^*\rangle \leq 0 \quad \textit{for any} \quad v \in B_r^+(0) \tag{4.5}$$

Where $\beta \in \partial H(u^*)$.
In the case $N[F'(y^*)]^* = 0$ (i.e. $[F'(y^*)]^*$ is injective), $(\lambda, \psi) \neq 0$.

Proof: Without lose of generality, assume that

$$L(y^*, u^*) = G(y^*) + H(u^*) = 0.$$

Since ∂G and ∂H are locally bounded at y^* and u^* respectively, by lemma 4.1, we obtain that there exist neighborhoods $O(y^*)$ and $O(u^*)$ of y^* and u^* in $L^2(\Omega)$ respectively, such that

$$\begin{aligned} &G(y) + H(u) \geq G_\lambda(y) + H_\lambda(u) \to G(y) + H(u) \\ &\quad \text{as} \quad \lambda \to 0 \quad \text{uniformly in} \quad y \in O(y^*) \quad \text{and} \quad u \in O(y^*). \end{aligned}$$

Thus for each $\epsilon > 0$, there exists a $\delta(\epsilon) > 0$ ($\delta(\epsilon) \to 0$ as $\epsilon \to 0$), such that

$$G(y) + H(u) + \epsilon \geq G_{\delta(\epsilon)}(y) + H_{\delta(\epsilon)}(u) + \epsilon > G(y) + H(u) \tag{4.6}$$

for all $(y, u) \in O(y^*) \times O(u^*)$.

Let $U = (B_r^+(0), d)$ with $d(u, v) = \|u - v\|$, where the norm is taken in $L^2(\Omega)$. Then U is a metric space. Note that $B_r^+(0) \subset L^\infty(\Omega) \subset L^2(\Omega)$ and one can check easily that $B_r^+(0)$ is closed in $L^2(\Omega)$. Thus U is a complete metric space.

Now we define $L_\epsilon : U \to R$ by

$$L_\epsilon(u) = \{d_Q^2(F(y(u))) + [G_{\delta(\epsilon)}(y(u)) + H_{\delta(\epsilon)}(u) + \epsilon]^2\}^{\frac{1}{2}}.$$

Where $d_Q(w) = \text{Inf}\|w - z\|, z \in Q$ and $y \equiv y(x; u)$ is the unique minimal positive solution of (1.1) corresponding to u.

By Ekeland's variational principle , there exists a $u^\epsilon \in U$ for each $\epsilon > 0$ such that

$$d(u^*, u^\epsilon) \leq \sqrt{\epsilon}, \quad \text{i.e.}: \quad \|u^* - u^\epsilon\| \leq \sqrt{\epsilon} \tag{4.7}$$

and

$$L_\epsilon(u) - L_\epsilon(u^\epsilon) \geq -\sqrt{\epsilon} d(u, u^\epsilon), \quad \forall u \in U. \tag{4.8}$$

Let $v \in U$, we define

$$u_\rho^\epsilon = u^\epsilon + \rho(v - u^\epsilon)^+. \tag{4.9}$$

It's clear that $u_\rho^\epsilon \in U$ and

$$u_\rho^\epsilon - u^\epsilon = \rho(v - u^\epsilon)^+ \to 0 \quad \text{in} \quad \mathrm{L}^\infty(\Omega) \quad \text{as} \quad \rho \to 0^+.$$

Let $y_\rho^\epsilon \equiv y_\rho^\epsilon(\cdot\, ; u_\rho^\epsilon)$ be the minimal positive solution of (1.1) corresponding to u_ρ^ϵ and $z_\rho^\epsilon \equiv \frac{y_\rho^\epsilon - y^\epsilon}{\rho}$.

Consider

$$\begin{cases} -\Delta z^\epsilon = p(y^\epsilon)^{p-1} z^\epsilon + (v - u^\epsilon)^+, & x \in \Omega \\ z^\epsilon|_{\partial\Omega} = 0. \end{cases} \tag{4.10}$$

We have

$$-\Delta(z^\epsilon - z_\rho^\epsilon) - p(y_\epsilon)^{p-1}(z^\epsilon - z_\rho^\epsilon) = (p(y^\epsilon)^{p-1} - a_\rho^\epsilon) z_\rho^\epsilon, \tag{4.11}$$

where $a_\rho^\epsilon = \int_0^1 p[y^\epsilon + t(y_\rho^\epsilon - y^\epsilon)]^{p-1}\, dt$.

Multiplying (4.11) by $(z^\epsilon - z_\rho^\epsilon)$ and interating on Ω, we obtain (note that both y^ϵ and y_ρ^ϵ are positive)

$$\begin{aligned} &\int_\Omega |\nabla(z^\epsilon - z_\rho^\epsilon)|^2\, dx - \int_\Omega p(y^\epsilon)^{p-1} \cdot (z^\epsilon - z_\rho^\epsilon)^2\, dx \\ &= \int_\Omega [p(y^\epsilon)^{p-1} - a_\rho^\epsilon] z_\rho^\epsilon \cdot (z_\rho^\epsilon - z^\epsilon)\, dx \\ &\leq \frac{C_\epsilon}{2} [\int_\Omega |y_\rho^\epsilon - y^\epsilon|^{\frac{N(p-1)}{2}}\, dx]^{\frac{4}{N}} \cdot \|z_\rho^\epsilon\|_{H_0^1}^2 + \frac{\epsilon}{2} \|z^\epsilon - z_\rho^\epsilon\|_{H_0^1}^2 \end{aligned}$$

for any $\epsilon > 0$.

By Lemma 2.1,

$$(1-\frac{1}{\lambda_1}-\epsilon)\|z^\epsilon - z^\epsilon_\rho\|^2_{H^1_0} \le \frac{C_\epsilon}{2}[\int_\Omega |y^\epsilon_\rho - y^\epsilon|^{\frac{N(p-1)}{2}}\, dx]^{\frac{4}{N}} \cdot \|z^\epsilon_\rho\|^2_{H^1_0}.$$

Taking ϵ small enough s.t. $1-\frac{1}{\lambda_1}-\epsilon \le r > 0$, we have

$$\|z^\epsilon - z^\epsilon_\rho\|^2_{H^1_0} \le \frac{1}{r}\cdot\frac{C}{2}[\int_\Omega |y^\epsilon_\rho - y^\epsilon|^{\frac{N(p-1)}{2}}\, dx]^{\frac{4}{N}} \cdot \|z^\epsilon_\rho\|^2_{H^1_0} \tag{4.12}$$

Since $p=\frac{N+2}{N-2}$ and $y^\epsilon_\rho \to y^\epsilon$ in $H^1_0(\Omega)$, we have

$$\left[\int_\Omega |y^\epsilon_\rho - y^\epsilon|^{\frac{N(p-1)}{2}}\right]^{\frac{4}{N}} \le C_1 \cdot \|y^\epsilon_\rho - y^\epsilon\|^{\frac{4}{N-2}}_{H^1_0}$$

On the other hand , by Theorem 2.3,

$$\|z^\epsilon_\rho\|_{H^1_0} = \|\frac{y^\epsilon_\rho - y^\epsilon}{\rho}\|_{H^1_0} \le \frac{C_2\cdot\rho\|(v-u^\epsilon)^+\|_{L^2(\Omega)}}{\rho} \le C_3$$

Thus (4.12) gives us

$$\|z^\epsilon - z^\epsilon_\rho\|_{H^1_0} \le C_4 \cdot \|y^\epsilon_\rho - y^\epsilon\|^{\frac{2}{N-2}}_{H^1_0} \to 0 \quad \text{as} \quad \rho \to 0.$$

Thus we obtain

$$y^\epsilon_\rho = y^\epsilon + \rho z^\epsilon + \rho\, o(1) \quad \text{in} \quad H^1_0(\Omega) \tag{4.13}$$

Next we estimate $G_{\delta(\epsilon)}(y^\epsilon_\rho) - G_{\delta(\epsilon)}(y^\epsilon)$ and $H_{\delta(\epsilon)}(u^\epsilon_\rho) - H_{\delta(\epsilon)}(u^\epsilon)$.

Clearly, we have

$$\begin{aligned} &\tfrac{G_{\delta(\epsilon)}(y^\epsilon_\rho)-G_{\delta(\epsilon)}(y^\epsilon)}{\rho} = \langle \dot G_{\delta(\epsilon)}(y^\epsilon), z^\epsilon_\rho\rangle + \tfrac{1}{\rho}o(\|y^\epsilon_\rho - y^\epsilon\|_{L^2}) \\ &= \langle \dot G_{\delta(\epsilon)}(y^\epsilon), z^\epsilon_\rho\rangle + o(1), \quad \text{as} \quad \rho\to 0 \end{aligned} \tag{4.14}$$

and

$$\tfrac{H_{\delta(\epsilon)}(u^\epsilon_\rho)-H_{\delta(\epsilon)}(u^\epsilon)}{\rho} = \langle \dot H_{\delta(\epsilon)}(u^\epsilon), (v-u^\epsilon)^+\rangle + o(1), \quad \text{as} \quad \rho\to 0 \tag{4.15}$$

We have

$$\frac{L_\epsilon(u^\epsilon_\rho) - L_\epsilon(u^\epsilon)}{\rho} \ge -\sqrt{\epsilon}\|(v-u^\epsilon)^+\|_{L^2} \ge -\sqrt{\epsilon}M. \tag{4.16}$$

It's clear that

$$L_\epsilon(u^\epsilon_\rho) = L_\epsilon(u^\epsilon) + o(1) \quad \text{in} \quad L^2(\Omega) \quad \text{as} \quad \rho \to 0^+. \tag{4.17}$$

One can check that $L_\epsilon(u^\epsilon) \ne 0$ for ϵ small enough.

So we have

$$
\begin{aligned}
&-\sqrt{\epsilon}M \le \tfrac{1}{L_\epsilon(u_p^\epsilon)+L_\epsilon(u^\epsilon)}\{\tfrac{[G_{\delta(\epsilon)}(y_\rho^\epsilon)+H_{\delta(\epsilon)}(u_\rho^\epsilon)+\epsilon]^2}{\rho}\\
&-\tfrac{[G_{\delta(\epsilon)}(y^\epsilon)+H_{\delta(\epsilon)}(u^\epsilon)+\epsilon]^2}{\rho}+\tfrac{d_Q^2(F(y_\rho^\epsilon))-d_Q^2(F(y^\epsilon))}{\rho}\}\\
&\to \tfrac{G_{\delta(\epsilon)}(y^\epsilon)+H_{\delta(\epsilon)}(u^\epsilon)+\epsilon}{L_\epsilon(u^\epsilon)}[\langle \dot{G}_{\delta(\epsilon)}(y^\epsilon), z^\epsilon\rangle + \langle \dot{H}_{\delta(\epsilon)}(u^\epsilon),(v-u^\epsilon)^+]\\
&+\langle \tfrac{d_Q(F(y^\epsilon))\xi^\epsilon}{L_\epsilon(u^\epsilon)}, F'(y^\epsilon)z^\epsilon\rangle,
\end{aligned}
\tag{4.18}
$$

Define

$$
\lambda_0^\epsilon = \frac{G_{\delta(\epsilon)}(y^\epsilon)+H_{\delta(\epsilon)}(u^\epsilon)+\epsilon}{L_\epsilon(u^\epsilon)} \in [0,1], \qquad q^\epsilon = \frac{d_Q(F(y^\epsilon))\xi^\epsilon}{L_\epsilon(u^\epsilon)} \tag{4.19}
$$

Then

$$
-\sqrt{\epsilon}M \le \lambda_0^\epsilon[\langle \dot{G}_{\delta(\epsilon)}(y^\epsilon), z^\epsilon\rangle + \langle \dot{H}_{\delta(\epsilon)}(u^\epsilon),(v-u^\epsilon)^+\rangle] + \langle q^\epsilon, F'(y^\epsilon)z^\epsilon\rangle. \tag{4.20}
$$

Since Y^* is strictly convex, we have

$$
|\lambda_0^\epsilon|^2 + ||q^\epsilon||_{Y^*}^2 = 1 \tag{4.21}
$$

and

$$
\langle q^\epsilon, \eta - F(y^\epsilon)\rangle \le 0 \quad \forall \eta \in Q \tag{4.22}
$$

In order to pass to the limits for $\epsilon \to 0^+$,
We first consider equation as follows:

$$
\begin{cases} -\Delta z = p(y^*)^{p-1}z + (v-u^*)^+, & x \in \Omega \\ z|_{\partial\Omega} = 0 \end{cases} \tag{4.23}
$$

which has a unique solution z in $H_0^1(\Omega)$, by Theorem 2.2.
By the similar arguments in (4.12) and (4.13), We have

$$
z^\epsilon \to z \ \text{in} H_0^1(\Omega). \tag{4.24}
$$

Consider $\{\dot{G}_{\delta(\epsilon)}(y^\epsilon)\}_{\epsilon>0}$ and $\{\dot{H}_{\delta(\epsilon)}(u^\epsilon)\}_{\epsilon>0}$.
By a standard argument in [4], one can get easily that

$$
\begin{aligned}
\dot{G}_{\delta(\epsilon)}(y^\epsilon) \to \alpha \in \partial G(y^*) \quad &\text{weakly in} \quad L^2(\Omega) \quad \text{and} \\
\dot{H}_{\delta(\epsilon)}(u^\epsilon) \to \beta \in \partial H(u^*) \quad &\text{weakly in} \quad L^2(\Omega).
\end{aligned}
\tag{4.25}
$$

Next, by the hypothese (H_F) , we have $F'(y^\epsilon) \to F'(y^*)$ in $L(H_0^1(\Omega); Y)$.
Thus (4.24) implies

$$
\begin{aligned}
\lambda_0^\epsilon[\langle \dot{G}_{\delta(\epsilon)}(y^\epsilon), z^\epsilon\rangle + \langle \dot{H}_{\delta(\epsilon)}(u^\epsilon), (v-u^\epsilon)^+\rangle]+ \\
\langle q^\epsilon, F'(y^*)z - \eta + F(y^*)\rangle \ge -\theta_\epsilon ,
\end{aligned}
\tag{4.26}
$$

$\forall v \in U$ and $\eta \in Q$, and $\theta_\epsilon \to 0$ uniformly on $v \in U$ as $\epsilon \to 0^+$.

Now since $F'(y^*)R^+ - Q$ is finite codimensional in Y , by Lemma 3.2 and (4.21), we can assume (relabelling if necessary) that $(\lambda_0^\epsilon, q^\epsilon) \to (\lambda_0, q) \neq 0$ weakly in $R \times Y^*$.

Thus, by taking the limits for $\epsilon \to 0$, we obtain

$$\begin{array}{l}\lambda_0[\langle \alpha, z\rangle + \langle \beta, (v-u^*)^+\rangle] + \langle q, F'(y^*)z - \eta + F(y^*)\rangle \geq 0 \\ \quad \text{for all} \quad v \in U \quad \text{and} \quad \eta \in Q\end{array} \tag{4.27}$$

Note that z depends on v .

After some simple calculations, we obtain

$$0 \geq \langle \Psi + \lambda\beta, (v-u^*)^+\rangle, \quad \forall v \in U$$

Similarly, by taking consideration of $u_\rho^\epsilon = u^\epsilon + \rho(v-u^\epsilon)^-$, we obtain

$$0 \geq \langle \Psi + \lambda\beta, (v-u^*)^-\rangle, \quad \forall v \in U$$

Thus

$$\langle \Psi + \lambda\beta, v-u^*\rangle \leq 0, \quad v \in U$$

If $(\lambda, \Psi) = 0$, then by (4.4), we have $[F'(y^*)]^*q = 0$. Thus in the case where $N[F'(y^*)]^* = \{0\}$, we must have $(\lambda, \Psi) \neq 0$

This completes the proof.

References

[1] Yinbin Deng and Gengsheng Wang, Optimal Control Problem of some semilinear equations with critical exponent, J. of Nonlinear Analysis TMA, to appear.

[2] Yong J., Pontryagin maximum principle for semilinear second order elliptic partial differential equations and variational inequalities with state constraint, Diff.Int.Eqs. 5 (1992),1307-1334.

[3] Xunjing Li and Jiongmin Yong, Optimal control theory for infinite dimentional systems, Birhauser Boston 1995.

[4] V.Barbu, Analysis and Control of Nonlinear Infinite Dimensional Systems, Academic Press, New York, 1993.

[5] G.Tarantello, On Nonhomogeneous Elliptic Equations Involving Critical Sobolev's Exponent, Ann.Inst.Henri Poincare, 9(3) (1992) 281-304.

[6] Li, X., and Yong, J., Necessary Conditions of Optimal Control for Distributed Parameter Systems, SIAM J. Control Optim., 29(1991), 895-908.

RECONSTRUCTION OF SOURCE TERMS IN EVOLUTION EQUATIONS BY EXACT CONTROLLABILITY

Masahiro Yamamoto*

Department of Mathematical Sciences
The University of Tokyo, Japan
Komaba, Meguro, Tokyo 153, Japan
email: myama@ms.u-tokyo.ac.jp

Abstract: For fixed $\rho = \rho(x,t)$, we consider the solution $u(f)$ to

$$\begin{aligned} u''(x,t) + Au(x,t) &= f(x)\rho(x,t), \quad x \in \Omega,\ t > 0 \\ u(x,0) &= u'(x,0) = 0, \quad x \in \Omega, \\ B_j u(x,t) &= 0, \quad x \in \partial\Omega,\ t > 0,\ 1 \le j \le m, \end{aligned}$$

where $u' = \frac{\partial u}{\partial t}$, $u'' = \frac{\partial^2 u}{\partial t^2}$, $\Omega \subset R^r$, $r \ge 1$ is a bounded domain with smooth boundary, A is a uniformly symmetric elliptic differential operator of order $2m$ with t-independent smooth coefficients, B_j, $1 \le j \le m$, are t-independent boundary differential operators such that the system $\{A, B_j\}_{1\le j\le m}$ is well-posed. Let $\{C_j\}_{1\le j\le m}$ be complementary boundary differential operators of $\{B_j\}_{1\le j\le m}$. We consider a multidimensional linear inverse problem : for given $\Gamma \subset \partial\Omega$, $T > 0$ and $n \in \{1, ..., m\}$, determine $f(x)$, $x \in \Omega$ from $C_j u(f)(x,t)$, $x \in \Gamma$, $0 < t < T$, $1 \le j \le n$.

By exact controllability based on the Hilbert Uniqueness Method, we reduce our inverse problem to an equation of the second kind which gives reconstruction of f. Moreover under extra regularity assumptions on ρ, we can prove that this equation is a Fredholm equation of the second kind. Our methodology is widely applicable to various equations in mathematical physics.

1 INTRODUCTION

We consider an initial - boundary value problem :

$$u''(x,t) + Au(x,t) = f(x)\rho(x,t), \qquad x \in \Omega,\ t > 0 \tag{1.1}$$

*This work is partially supported by Sanwa Systems Development Co., Ltd. (Tokyo, Japan). This paper has been written during the author's stay at Weierstraß Institut für Angewandte Analysis und Stochastik in Berlin and he thanks Dr. Johannes Elschner for the invitation.

$$u(x,0) = u'(x,0) = 0, \qquad x \in \Omega \tag{1.2}$$

$$B_j u(x,t) = 0, \quad x \in \partial\Omega,\ t > 0,\ 1 \le j \le m, \tag{1.3}$$

where $u' = \frac{\partial u}{\partial t}$, $u'' = \frac{\partial^2 u}{\partial t^2}$, $\Omega \subset R^r$, $r \ge 1$ is a bounded domain with C^2- boundary, A is a uniformly symmetric elliptic differential operator of order $2m$ with t-independent smooth coefficients, B_j, $1 \le j \le m$, are boundary differential operators. More precisely, we set $x = (x_1, ..., x_r) \in R^r$, $\alpha = (\alpha_1, ..., \alpha_r) \in (N \cup \{0\})^r$, $|\alpha| = \alpha_1 + ... + \alpha_r$, $D_x^\alpha = \left(\frac{\partial}{\partial x_1}\right)^{\alpha_1} \cdots \left(\frac{\partial}{\partial x_r}\right)^{\alpha_r}$, and

$$(A\phi)(x) = \sum_{|\alpha|,|\beta| \le m} (-1)^{|\alpha|} D_x^\alpha (a_{\alpha\beta}(x) D_x^\beta \phi)(x),$$

which $a_{\alpha\beta} = a_{\beta\alpha} \in C^\infty(\overline{\Omega})$ are real-valued for $|\alpha|$, $|\beta| \le m$, and we assume the uniform ellipticity : there exists a constant $M_0 > 0$ independent of $x \in \overline{\Omega}$ and $\xi \in R^r$ such that

$$M_0^{-1} |\xi|^{2m} \le \left| \sum_{|\alpha|,|\beta| = m} a_{\alpha\beta}(x) \xi^{\alpha+\beta} \right| \le M_0 |\xi|^{2m}, \qquad x \in \overline{\Omega},\ \xi \in R^r,$$

where $\xi = (\xi_1, ..., \xi_r) \in R^r$ and $\xi^\alpha = \xi_1^{\alpha_1} \cdots \xi_r^{\alpha_r}$ with $\alpha = (\alpha_1, ..., \alpha_r)$, $|\xi|^2 = \xi_1^2 + ... + \xi_r^2$. Moreover we put

$$(B_j \psi)(x) = \sum_{|\alpha| \le m_j} b_{j\alpha}(x) D_x^\alpha \psi(x),$$

where $b_{j\alpha} \in C^\infty(\partial\Omega)$, $0 \le m_j < 2m$. Throughout this paper we assume that $\{B_j\}_{1 \le j \le m}$ is normal on $\partial\Omega$ (e.g. Lions and Magenes [17] Vol.I) and that the system $\{A, B_j\}_{1 \le j \le m}$ is well-posed ([17], Vol.II).

Henceforth let $\{C_j\}_{1 \le j \le m}$ be complementary boundary differential operators of $\{B_j\}_{1 \le j \le m}$, whose coefficients are t-independent and smooth in $x \in \partial\Omega$ ([17], Vol.I).

In this paper, assuming that ρ is given while f is unknown to be determined from observations on a part of lateral boundary, we denote the weak solution to (1.1) - (1.3) by $u(f) = u(f)(x,t)$. For the weak solution, we can further refer to [17]. We discuss

Inverse Source Problem:

For given $\Gamma \subset \partial\Omega$, $T > 0$ and $n \in \{1, ..., m\}$, determine $f(x)$, $x \in \Omega$, from $C_j u(f)(x,t)$, $x \in \Gamma$, $0 < t < T$, $1 \le j \le n$.

In (1.1), the non-homogeneous term $f(x)\rho(x,t)$ is considered to cause actions such as vibrations, and the inverse source problem is significant in mathematical physics. Moreover when we discuss determination of spatially varying coefficients in A, we have to do with this type of inverse problem after subtraction or linearization (e.g. Lavrentiev, Romanov and Shishat·skiĭ[14], Romanov [22]). We notice that we want to determine f with a single boundary measurement.

In the case where $\rho = \rho(t)$ is independent of x, by means of Duhamel's principle (e.g. Rauch [21]), we can reduce the inverse problem to an observability problem, namely, determination of initial data. For the inverse problem in the case of x-indepednent $\rho = \rho(t)$, we can refer to Puel and Yamamoto [18], Yamamoto [24], [25], [26]. On the other hand, the inverse problem becomes more difficult for x-dependent ρ. Fot such a case, the method by Bukhgeim and Klibanov [3] is useful and their method is based on a weighted estimate called a Carleman estimate. For the uniqueness, we can refer to Bukhgeim and Klibanov [3], Iskakov [5], [6], [7], Khaïdarov [9], Klibanov [10]. Moreover for similar inverse problems for Lamé systems and Maxwell's equations, we refer to Ikehata, Nakamura and Yamamoto [4], and Yamamoto [27], respectively. As for an inverse problem with many observations for a hyperbolic equation given by (1.1), we can refer to Rakesh and Symes [20]. For general references for these kinds of inverse problems, the readers can consult monographs : Isakov [8], Lavrentiev, Romanov and Shishat·skiĭ[14], Romanov [22].

Most of the papers above-mentioned mainly treat the uniqueness problem. For stability in determining functions in hyperbolic equations from a single boundary measurement, estimation of Hölder type has been proved (Khaïdarov [9]. also see a remark (p.577) in [10]). Recently the author has established the best possible Lipschitz stability by combination of the Carleman estimate and the exact observability (Yamamoto [28]).

Reconstruction of f is practically important, but such discussions are very few (Bukhgeim [2]). The purpose of this paper is to reduce our inverse problem to an equation of the second kind by the exact controllability, which is a Fredholm equation of the second kind under a natural setting. Then our inverse problem is to solve the equation of the second kind. Further study for the equation will be made in a forthcoming paper.

This paper is composed of four sections. Section 2 is devoted to a brief explanation of the Hilbert Uniqueness Method. In Section 3, we state our main result. In Section 4, we prove the main result.

2 BRIEF EXPLANATION OF THE HILBERT UNIQUENESS METHOD

We give a brief explanation of the Hilbert Uniqueness Method, according to Lions [16]. We refer also to Komornik [11], Lasiecka and Triggiani [13], Lions [15]. We set

$$\begin{aligned}\widetilde{F} &= \widetilde{F_1} \times \widetilde{F_2} \\ &= \{(\phi_1, \phi_2) \in C^\infty(\overline{\Omega})^2 \,;\, B_j\phi_1 = 0 \text{ if the order of } B_j \text{ is less than } m\},\end{aligned}$$

and for $(\phi_1, \phi_2) \in \widetilde{F}$, we denote the solution to

$$w''(x,t) + Aw(x,t) = 0, \qquad x \in \Omega,\, 0 < t < T, \tag{2.1}$$

$$w(x,0) = \phi_1(x), \quad w'(x,0) = \phi_2(x), \qquad x \in \Omega \tag{2.2}$$

$$B_j w(x,t) = 0, \qquad x \in \partial\Omega,\ 0 < t < T,\ 1 \le j \le m \tag{2.3}$$

by $w(\phi_1, \phi_2) = w(\phi_1, \phi_2)(x,t)$. We pose

Assumption A (Unicity)

For a given measurable $\Gamma \subset \partial\Omega$, a finite $T > 0$ and $n \in \{1, ..., m\}$, if the solution $w(\phi_1, \phi_2)$ satisfies

$$C_j w(x,t) = 0, \qquad x \in \Gamma,\ 0 < t < T,\ 1 \le j \le n$$

for $(\phi_1, \phi_2) \in \widetilde{F}$, then $w(\phi_1, \phi_2)(x,t) = 0$, $x \in \Omega$, $0 < t < T$ follows.

This is unicity in a Cauchy problem for $w'' + Aw = 0$, for which we refer to Bardos, Lebeau and Rauch [1] and Tataru [23] for example. On Assumption A, we can define a norm $||(\phi_1, \phi_2)||_F$ by

$$||(\phi_1, \phi_2)||_F \equiv \left(||\phi_1||^2_{F_1} + ||\phi_2||^2_{F_2}\right)^{\frac{1}{2}} = \left(\sum_{j=1}^{n} ||C_j w(\phi_1, \phi_2)||^2_{L^2(\Gamma \times (0,T))}\right)^{\frac{1}{2}},$$

for any $(\phi_1, \phi_2) \in \widetilde{F}$, where $||\eta||_{L^2(\Gamma\times(0,T))} = \left(\int_\Gamma \int_0^T |\eta(x,t)|^2 dt dS_x\right)^{\frac{1}{2}}$. Let a Hilbert space $F \equiv F_1 \times F_2$ be the *completion* of $\widetilde{F}$ by the norm $|| \cdot ||_F$. Let $F' = F_1' \times F_2'$ be its dual. Throughout this paper, $\cdot'$ denotes the dual space and we identify the dual spaces $L^2(\Gamma \times (0,T))'$ of $L^2(\Gamma \times (0,T))$ and $L^2(\Omega)'$ of $L^2(\Omega)$ respectively with itself. The space F' is related to the exactly controllable set and the essence of the Hilbert Uniqueness Method is construction of the Hilbert space F'.

Next let us consider

$$\psi''(x,t) + A\psi(x,t) = 0, \quad x \in \Omega,\ 0 < t < T \tag{2.4}$$

$$\psi(x,T) = \psi'(x,T) = 0, \quad x \in \Omega \tag{2.5}$$

$$B_j\psi(x,t) = \begin{cases} v_j(x,t), & x \in \Gamma,\ 0 < t < T\ :\ 1 \le j \le n \\ 0, & x \in \partial\Omega \setminus \Gamma,\ 0 < t < T\ :\ 1 \le j \le n \\ 0, & x \in \partial\Omega,\ 0 < t < T\ :\ n+1 \le j \le m. \end{cases} \tag{2.6}$$

For the system (2.4) - (2.6) with a uniformly symmetric elliptic operator A of order $2m$, a general treatment (Theorem 4.1 (p.107 : Vol.II) in [17]) tells that for any $v = (v_1, ..., v_n) \in L^2(\Gamma \times (0,T))^n$, there exists a unique weak solution $\psi(v) \in H^{0,-1}(\Omega \times (0,T)) \equiv \left(H^1_0(0,T; L^2(\Omega)) \cap L^2(0,T; L^2(\Omega))\right)'$, where $H^1_0(0,T; L^2(\Omega)) = \{u \in H^1(0,T; L^2(\Omega))\,;\, u(\cdot,0) = u(\cdot,T) = 0\}$. Furthermore we refer to Theorems 6.1 and 6.2 (pp.118-119 : Vol.II) in [17], and especially for a wave equation, we also quote Lasiecka, Lions and Triggiani [12], Lions [16].

In applying a result (Theorem 0 below) on exact controllability, we however pose a stronger assumption for the regularity of $\psi(v)$.

Assumption B (Regularity in the control system)
For $v \in L^2(\Gamma \times (0,T))^n$, the weak solution $\psi(v)$ satisfies

$$\psi(v) \in C^0([0,T]; F_2'), \quad \psi(v)' \in C^0([0,T]; F_1')$$

$$||\psi(v)||_{C^0([0,T];F_2')} \leq M_1 ||v||_{L^2(\Gamma\times(0,T))^n}$$

where $M_1 = M_1(\Omega, \Gamma, T) > 0$ is independent of v.

Example 1 : wave equation ([11], [13], [16])
For an arbitrarily given $x_0 \in R^r$, we set

$$\begin{aligned} \Gamma_+(x_0) &= \{x \in \partial\Omega; (x - x_0, \nu(x)) > 0\} \\ R_0 &= R_0(x_0) = \sup_{x\in\partial\Omega} |x - x_0|, \end{aligned} \tag{2.7}$$

where $\nu(x)$ is the outward unit normal to $\partial\Omega$ and $(\cdot,\cdot)$ is the inner product in R^r. We consider: $A = -\Delta$ (the Laplacian), $m = 1$,

$$B_1 u = u|_{\partial\Omega}, \qquad C_1 u = \frac{\partial u}{\partial n}_{|\Gamma}.$$

If

$$T > 2R_0$$

and a measurable set $\Gamma \subset \partial\Omega$ satisfies

$$\Gamma \supset \Gamma_+(x_0), \tag{2.8}$$

then

$$F_1 = H_0^1(\Omega), \qquad F_2 = L^2(\Omega), \tag{2.9}$$

and Assumptions A and B hold true.

Example II: plate equation (e.g. [11], [16]).
Let $A = \Delta^2$, $m = 2$ and

$$B_1 u = u|_{\partial\Omega}, \quad B_2 u = \frac{\partial u}{\partial n}, \quad C_1 u = \Delta u_{|\Gamma}, \quad C_2 u = \frac{\partial \Delta u}{\partial n}_{|\Gamma}.$$

We set $n = 1$. If we choose Γ satisfying (2.8), then for any $T > 0$, $F_2 = L^2(\Omega)$ holds, and Assumptions A and B hold true.

By the Hilbert Uniqueness Method, we show boundary exact controllability:
Theorem 0 (Théorème 3.2 (p.119) in [16]) On Assumptions A and B, for any $(\phi_1, \phi_2) \in F_2' \times F_1'$, there exists $v = (v_1, ..., v_n) \in L^2(\Gamma \times (0,T))^n$ such that the weak solution $\psi = \psi(v)$ to (2.4) - (2.6) satisfies

$$\psi(v)(\cdot, 0) = \phi_1, \quad \psi(v)'(\cdot, 0) = \phi_2. \tag{2.10}$$

Moreover we can construct a map from (ϕ_1, ϕ_2) to v such that

$$||v||_{L^2(\Gamma\times(0,T))^n} \le M_1(||\phi_1||_{F_2'} + ||\phi_2||_{F_1'}), \qquad (\phi_1, \phi_2) \in F_2' \times F_1',$$

where $M_1 = M_1(\Omega, \Gamma, T) > 0$ is independent of (ϕ_1, ϕ_2).

This theorem defines a bounded linear operator $g : F_2' \longrightarrow L^2(\Gamma \times (0,T))^n$ which maps $\phi_1 \in F_2'$ to $v \in L^2(\Gamma \times (0,T))^n$ realizing $\psi(v)(\cdot, 0) = \phi_1$ and $\psi(v)'(\cdot, 0) = 0$, and

$$||g(\phi_1)||_{L^2(\Gamma\times(0,T))^n} \le M_1||\phi_1||_{F_2'}. \tag{2.11}$$

In (2.6), v_j, $1 \le j \le n$, are regarded as boundary controls which steer the system described by (2.4) - (2.5) to the equilibrium at time T starting from the initial state given by (ϕ_1, ϕ_2).

3 MAIN RESULT: REDUCTION OF THE GENERAL INVERSE SOURCE PROBLEM TO AN EQUATION OF THE SECOND KIND

We discuss the initial - boundary value problem (1.1) - (1.3) with $\rho = \rho(x,t)$ satisfying

$$\left\| \int_0^T \rho'(\cdot, t)\psi(\cdot, t)dt \right\|_{F_2'} \le M_2||\psi||_{C^0([0,T];F_2')}, \quad \psi \in C^0([0,T]; F_2') \tag{3.1}$$

$$||f\rho(\cdot, 0)||_{F_2} \le M_2||f||_{F_2}, \quad f \in F_2 \tag{3.2}$$

$$\rho \in H^1(0,T; L^\infty(\Omega)) \tag{3.3}$$

$$||f\rho'||_{L^2(0,T;F_2)} \le M_2||f||_{F_2}, \quad f \in F_2 \tag{3.4}$$

Here $M_2 > 0$ is independent of ψ and f. We always pose Assumptions A and B.

Remark If we can characterize F_2, for example, as $F_2 = L^2(\Omega)$ (cf. Examples in Section 2), then the conditions (3.1) - (3.4) are equivalent to

$$\rho \in H^1(0,T; L^\infty(\Omega)), \quad \rho(\cdot, 0) \in L^\infty(\Omega). \tag{3.5}$$

We recall that a linear operator $g : F_2' \longrightarrow L^2(\Gamma \times (0,T))^n$ is defined in Theorem 0 in Section 2 and satisfies (2.11). We define a linear operator S in F_2' by

$$(S\phi_1)(x) = \int_0^T \rho'(x,t)\psi(g(\phi_1))(x,t)dt, \qquad \phi_1 \in F_2'. \tag{3.6}$$

Then we are ready to state the main result:

Theorem Under Assumptions A and B, (3.1) - (3.4);

(1) $S : F_2' \longrightarrow F_2'$ is a bounded linear operator.

(2) Let $v \in H^1(0,T;L^2(\Gamma))^n$. Then $f \in F_2$ satisfies

$$g^*\left(v' - (C_1u(f)', ..., C_nu(f)')\right) = 0 \tag{3.7}$$

if and only if $f \in F_2$ satisfies

$$\rho(\cdot,0)f + S^*f = g^*v'. \tag{3.8}$$

Here $S^* : F_2 \longrightarrow F_2$ is the adjoint of $S : F_2' \longrightarrow F_2'$, and g^* is the one of a bounded linear operator $g : F_2' \longrightarrow L^2(\Gamma \times (0,T))^n$. The operator equation (3.7) is our desired one of the second kind.

Corollary 3.1 *If f is a solution of our inverse problem, that is, $f \in F_2$ satisfies*

$$(C_1u(f), ..., C_nu(f)) = v \tag{3.9}$$

for $v \in H^1(0,T;L^2(\Gamma))^n$, then f solves (3.7).

Remark In general, $\mathcal{R}(g)$ is not dense in $L^2(\Gamma \times (0,T))^n$, so that g^* is not injective. Thus in Theorem, we can not replace (3.6) by (3.6').

Henceforth we assume

$$\rho(x,0) \neq 0, \qquad x \in \overline{\Omega}. \tag{3.10}$$

Then (3.7) is an equation of the second kind:

$$f + \frac{1}{\rho(\cdot,0)}S^*f = \frac{1}{\rho(\cdot,0)}g^*v'. \tag{3.11}$$

Moreover Corollary 1 asserts that it is sufficient to consider (3.9) for reconstructing f. For similar linear inverse problems with singular data such as Dirac delta functions in multidimensional cases and similar ones with smooth data in one-dimensional cases, we can reduce the problems to a Volterra equation of the second kind (e.g. Chapter 2 and Section 3 of Chapter 4 in [22]). However in multidimensional cases with not necessarily singular data, a general way for such reduction has not been published (cf. Bukhgeim [2]).

Here we do not give direct expression of S^*. In special cases, direct expression of S^* is not difficult. For example, in Example 1 in Section 2, let $r = 1$ (i.e., the spatial dimension is 1), $\Omega = (0,1)$, $\Gamma = \{0\}$ (one end point) and $T = 2$. Then we can construct the control operator $g : L^2(0,1) \longrightarrow L^2(0,2)$ by consideration of the dependency domain of the one-dimensional wave equation and D'Alembert's formula.

Next we have to study the unique solvability of the equation (3.9). First by the contraction mapping principle, we can readily see

Corollary 3.2 *Let*

$$\left\| \frac{\rho'(\cdot,\cdot)}{\rho(\cdot,0)} \right\|_{L^1(0,T;L^\infty(\Omega))} \tag{3.12}$$

be sufficiently small and let $v = (C_1u(f), ..., C_nu(f))$. *Then* f *is given as a unique solution of (3.9) by iteration.*

We consider a hyperbolic equation of the second order and we take $C_1u = \frac{\partial u}{\partial n}_{|\Gamma}$ as the boundary observation where the subboundary Γ satisfies (2.8):

$$u''(x,t) = \Delta u(x,t) - p(x)u(x,t) + f(x)\rho(x,t), \quad x \in \Omega,\, t > 0 \tag{3.13}$$

$$u(x,0) = u'(x,0) = 0, \qquad x \in \Omega \tag{3.14}$$

$$u(x,t) = 0, \qquad x \in \partial\Omega, t > 0. \tag{3.15}$$

Moreover in addition to (3.1') we assume

$$p \in L^\infty(\Omega) \tag{3.16}$$

$$\rho, \frac{\rho}{\rho(\cdot,0)} \in H^2(0,T;L^\infty(\Omega)) \tag{3.17}$$

$$T > 2R_0 \tag{3.18}$$

where R_0 is given by (2.7). Then by the argument in the proof of Lemma 5.5 in Puel and Yamamoto [19], we can prove

Corollary 3.3 *Under the assumptions (3.14) - (3.16), the operator* $S^* : L^2(\Omega) \longrightarrow L^2(\Omega)$ *is compact. Therefore the equation (3.9) is a Fredholm equation of the second kind in* $L^2(\Omega)$.

In Corollary 3, for the unique solvability, it suffices to verify that $f + \frac{1}{\rho(\cdot,0)}S^*f = 0$ implies $f = 0$. This is equivalent to the uniqueness in some inverse problem and the method in Bukhgeim and Klibanov [3] may be helpful. In a forthcoming paper, we will treat details of the unique solvability.

References

[1] Bardos, C., Lebeau, G. and Rauch, J., Sharp sufficient conditions for the observation, control, and stabilization of waves from the boundary. SIAM J. Control and Optimization V.30 (1992), 1024 – 1065.

[2] Bukhgeim, A.L., private communication.

[3] Bukhgeim, A.L. and Klibanov, M.V., Global uniqueness of a class of multidimensional inverse problems. (English translation) Soviet Math. Dokl. V.24 (1981), 244–247.

[4] Ikehata, M., Nakamura, G. and Yamamoto, M., Uniqueness in inverse problems for the isotropic Lamé system. to appear in Journal of Mathematical Sciences of The University of Tokyo (1998).

[5] Isakov, V., A nonhyperbolic Cauchy problem for $\Box_b \Box_c$ and its applications to elasticity theory. Comm. Pure and Applied Math. V.34 (1986), 747–767.

[6] Isakov, V., Inverse Source Problems. American Mathematical Society, Providence, Rhode Island, 1990

[7] Isakov, V., Uniqueness and stability in multi-dimensional inverse problems. Inverse Problems V.9 (1993), 579–621.

[8] Isakov, V., Inverse Problems for Partial Differential Equations. Springer-Verlag, Berlin 1998

[9] Khaĭdarov, A., On stability estimates in multidimensional inverse problems for differential equations. aperinfo (English translation) Soviet Math. Dokl. V.38 (1989), 614–617.

[10] Klibanov, M.V., Inverse problems and Carleman estimates. Inverse Problems V.8 (1992), 575–596.

[11] Komornik, V., Exact Controllability and Stabilization : the Multiplier Method. John Wiley & Sons, Chichester 1994.

[12] Lasiecka, I., Lions, J.-L. and Triggiani, R., Non homogeneous boundary value problems for second order hyperbolic operators. J. Math. Pures Appl. V.65 (1986), 149–192.

[13] Lasiecka, I. and Triggiani, R., Exact controllability of the wave equation with Neumann boundary control. Appl. Math. Optim. V.19 (1989), 243–290.

[14] Lavrent'ev, M.M., Romanov, V.G. and Shishat·skiĭ, S.P., Ill-posed Problems of Mathematical Physics and Analysis. American Mathematical Society, Providence, Rhode Island (English translation) 1986.

[15] Lions, J.-L., Exact controllability, stabilization and perturbations for distributed systems. SIAM Review V.30 (1988), 1–68.

[16] Lions, J.-L., Contrôlabilité Exacte Perturbations et Stabilisation de Systèmes Distribués. Vol.1. Masson, Paris 1988.

[17] Lions, J.-L. and Magenes, E., Non-homogeneous Boundary Value Problems and Applications. Volumes I and II. (English translation) Springer-Verlag, Berlin 1972.

[18] Puel, J.-P. and Yamamoto, M., Applications de la contrôlabilité exacte à quelques problèmes inverses hyperboliques. C. R. Acad. Sci. Paris Sér.I Math. V.320 (1995), 1171–1176.

[19] Puel, J.-P. and Yamamoto, M., Generic well-posedness in a multidimensional hyperbolic inverse problem. Journal of Inverse and Ill-posed Problems V.5 (1997), 55-83.

[20] Rakesh and Symes, W.W., Uniqueness for an inverse problem for the wave equation. Commun. in Partial Differential Equations V.13 (1988), 87–96.

[21] Rauch, J., Partial Differential Equations. Springer-Verlag, Berlin 1991.

[22] Romanov, V.G., Inverse Problems of Mathematical Physics. (English translation) VNU Science Press, Utrecht 1987

[23] Tataru, D., Unique continuation for solutions to PDE's; between Hörmander and Holmgren's theorems. Commun. in Partial Differential Equations V. 20 (1995), 855–884.

[24] Yamamoto, M., Stability, reconstruction formula and regularization for an inverse source hyperbolic problem by a control method. Inverse Problems V.11 (1995), 481–496.

[25] Yamamoto. M., Application of the Hilbert uniqueness method to inverse source problems: stability and regularization. Z. Angew. Math. Mech. V.75, Suppl. 2 (1995), 515–516.

[26] Yamamoto, M., On ill-posedness and a Tikhonov regularization for a multidimensional inverse hyperbolic problem. J. Math. Kyoto University V.36 (1996), 825–856.

[27] Yamamoto, M., On an inverse problem of determining source terms in Maxwell's equations with a single measurement. in "Inverse Problems, Tomography and Image Processing", edited by A.G. Ramm, Plenum Press, New York (1998), 241–256

[28] Yamamoto, M., Uniqueness and stability in multidimensional hyperbolic inverse problems. to appear in J. Math. Pures Appl. (1998).

NECESSARY OPTIMALITY CONDITIONS FOR CONTROL OF STRONGLY MONOTONE VARIATIONAL INEQUALITIES

Jane J. Ye

Department of Mathematics and Statistics, University of Victoria
Victoria, British Clumbia, Canada V8W 3P4*

janeye@uvic.ca

1 INTRODUCTION

The optimal control problem for a system governed by an elliptic variational inequality, first proposed by J.L. Lions (1969,1972) and studied in Barbu (1984) is as follows: Let V and H be two Hilbert spaces (*state spaces*) such that

$$V \subseteq H = H^* \subseteq V^*$$

where V^* is the dual of V, H^* is the dual of H which is identified with H, and injections are dense and continuous. Let U be another Hilbert space (*control space*). Suppose that $A \in L(V, V^*)$ is *coercive*, i.e., there exists a constant $\mu > 0$ such that

$$\langle Av, v \rangle \geq \mu ||v||_V^2 \quad \forall v \in V$$

where $\langle \cdot, \cdot \rangle$ is the duality pairing on $V^* \times V$ and identified with the inner product on H. The optimal control problem for an elliptic variational inequality is the following minimization problem:

$$\begin{array}{llll} (P) & \min & g(y) + h(u) & \\ & \text{s.t.} & y \in K, u \in U_{ad} & \\ & & \langle Ay, y' - y \rangle \geq \langle Bu + f, y' - y \rangle & \forall y' \in K, \end{array}$$

where $B \in L(U, V^*)$ is compact, $K \subset V$ and $U_{ad} \subseteq U$ are two closed convex subsets in V and U respectively, $f \in V^*$, $g : K \to R_+$ and $h : U_{ad} \to R_+$ are two given functions.

*The research of this paper is partially supported by NSERC.

In this paper we study the following optimal control of strongly monotone variational inequality which is more general than the one proposed by Lions:

$$
\begin{array}{lll}
(OCVI) & \min & J(y,u) \\
 & \text{s.t.} & y \in K, u \in U_{ad} \\
 & & \langle F(y,u), y'-y \rangle \geq 0 \quad \forall y' \in K,
\end{array}
$$

where the following assumptions are satisfied

(A1) K and U_{ad} are closed convex subsets of Asplund spaces (which include all reflexive Banach spaces) V and U respectively. There is a finite codimensional closed subspace M such that $U_{ad} \subseteq M$ and the relative interior of U_{ad} with respect to the subspace M is nonempty.

(A2) $J : V \times U_{ad} \to R$ is Lipschitz near $(\bar{y}, \bar{u})$.

(A3) $F : V \times U_{ad} \to V^*$ is strictly differentiable at $(\bar{y}, \bar{u})$ (see definition given in Remark 2) and locally strongly monotone in y uniformly in u, i.e., there exist $\mu > 0$ and $U(\bar{y}, \bar{u})$, a neighborhood of $(\bar{y}, \bar{u})$ such that

$$\langle F(y',u)-F(y,u), y'-y \rangle \geq \mu \|y'-y\|^2 \;\; \forall (y,u),(y',u) \in U(\bar{y},\bar{u}) \cap (K \times U_{ad}).$$

Our main result is the following theorem:

Theorem 1 *Let* $(\bar{y}, \bar{u})$ *be a local solution of problem* (OCVI). *Then there exists* $\eta \in V$ *such that*

$$0 \in \partial J(\bar{y},\bar{u}) + F'(\bar{y},\bar{u})^* \eta + D^* N_K(\bar{y}, -F(\bar{y},\bar{u}))(\eta) \times \{0\} + \{0\} \times N(\bar{u}, U_{ad}) \tag{1.1}$$

where ∂ *denotes the limiting subgradient (see Definition 2),* F' *denotes the strict derivative (see Remark 2),* $N(\bar{u}, U_{ad})$ *denotes the normal cone of the convex set* U_{ad} *at* $\bar{u}$ *and* N_K *denotes the normal cone operator defined by*

$$N_K(y) := \begin{cases} \text{the normal cone of } K \text{ at } y & \text{if } y \in K \\ \emptyset & \text{if } y \notin K \end{cases}$$

and D^* *denotes the coderivative of a set-valued map (see Definition 5).*

This is in fact in the form of the optimality condition given by Shi (1988, 1990) with the paratingent coderivative of the the set-valued map N_K replaced by the Mordukhovich coderivative.

In the case where $J(y,u) = g(y) + h(u)$ and $F(y,u) = Ay - Bu - f$ as in problem (P), Inclusion (1.1) becomes

$$0 \in \partial g(\bar{y}) + A^* \eta + D^* N_K(\bar{y}, -F(\bar{y},\bar{u}))(\eta) \tag{1.2}$$

$$0 \in \partial h(\bar{u}) - B^* \eta + N(\bar{u}, U_{ad}). \tag{1.3}$$

Notice that $N_K(y) = \partial\psi_K(y)$, the coderivative of the set-valued map N_K can be considered as a second order generalized derivative of ψ_K. Hence inclusions (1.2) and (1.3) are in the form of the necessary optimality condition given in Theorem 3.1 of Barbu (1984) with the Clarke subgradient replaced by the limiting subgradient which is in general a smaller set than the Clarke subgradient and with the *notational* second order generalized derivative replaced by the *true* second order generalized derivative D^*N_K.

We organize the paper as follows. §2 contains background material on nonsmooth analysis and preliminary results. In §3 we derive necessary optimality conditions for (OCVI).

2 PRELIMINARIES

This section contains some background material on nonsmooth analysis which will be used in the next section. We only give concise definitions that will be needed in the paper. For more detail information on the subject, our references are Clarke (1983), Mordukhovich and Shao (1996a,b).

First we give some concepts for various normal cones.

Definition 1 *Let Ω be a nonempty subset of a Banach space X and let $\epsilon \geq 0$.*

(i) *Given $\bar{x} \in cl\Omega$, the closure of set Ω, the set*

$$\hat{N}_\epsilon(\bar{x},\Omega) := \{x^* \in X^* : \limsup_{x\to\bar{x},x\in\Omega} \frac{\langle x^*, x-\bar{x}\rangle}{\|x-\bar{x}\|} \leq \epsilon\} \tag{1.4}$$

is called the set of Fréchel $\epsilon-$normal to set Ω at point $\bar{x}$. When $\epsilon = 0$, the set (1.4) *is a cone which is called the Fréchel normal cone to Ω at point $\bar{x}$ and is denoted by $\hat{N}(\bar{x},\Omega)$.*

(ii) *The following nonempty cone*

$$\begin{aligned} N(\bar{x},\Omega) \quad := \quad & \{x^* \in X^* | \exists x_k \to \bar{x}, \epsilon_k \downarrow 0, x_k^* \xrightarrow{w^*} x^*, x_k^* \in \hat{N}_{\epsilon_k}(x_k,\Omega) \\ & \text{as } k\to\infty\} \end{aligned} \tag{1.5}$$

is called the limiting normal cone to Ω at point $\bar{x}$,

As proved in Mordukhovich and Shao (1996a), in Asplund spaces X the normal cone (1.5) admits the simplified representation

$$N(\bar{x},\Omega) = \{x^* \in X^* | \exists x_k \to \bar{x}, x_k^* \xrightarrow{w^*} x^*, x_k^* \in \hat{N}(x_k,\Omega) \text{ as } k\to\infty\}$$

Using the definitions for normal cones, we now give definitions for subgradients of a single-valued map.

Definition 2 *Let X be a Banach space and $f : X \to R \cup \{+\infty\}$ be lower semicontinuous and finite at $\bar{x} \in X$. The limiting subgradient of f at $\bar{x}$ is defined by*

$$\partial f(\bar{x}) := \{x^* \in X^* : (x^*,-1) \in N((\bar{x},f(\bar{x})), epi(f))\}$$

and the singular subdifferential of f at $\bar{x}$ is defined by

$$\partial^{\infty} f(\bar{x}) := \{x^* \in X^* : (x^*, 0) \in N((\bar{x}, f(\bar{x})), epi(f))\},$$

where $epi(f) := \{(x, r) \in X \times R : f(x) \leq r\}$ is the epigraph of f.

Remark 1 *Let Ω be a closed set of a Banach space and ψ_Ω denote the indicator function of Ω. Then it follows easily from the definition that*

$$\partial \psi_\Omega(\bar{x}) = \partial^{\infty} \psi_\Omega(\bar{x}) = N_\Omega(\bar{x}).$$

The following fact is also well-known and follows easily from the definition:

Proposition 1 *Let X be a Banach space and $f : X \to R \cup \{+\infty\}$ be lower semicontinuous. If f has a local minimum at $\bar{x} \in X$, then*

$$0 \in \partial f(\bar{x}).$$

To ensure that the sum rule holds in an infinite dimensional Asplund space, we need the following definitions.

Definition 3 *Let X be a Banach space and Ω a closed subset of X. Ω is said to be sequentially normally compact at $\bar{x} \in \Omega$ if any sequence (x_k, x_k^*) satisfying*

$$x_k^* \in \hat{N}(x_k, \Omega), x_k \to \bar{x}, x_k^* \xrightarrow{w^*} 0 \text{ as } k \to \infty$$

contains a subsequence with $\|x_{k_\nu}^*\| \to 0$ as $\nu \to 0$.

Definition 4 *Let Let X be a Banach space and $f : X \to R \cup \{+\infty\}$ be lower semicontinuous and finite at $\bar{x} \in X$. f is said to be sequentially normally epi-compact around $\bar{x}$ if its epigraph is sequentially normally compact at $\bar{x}$.*

Proposition 2 *Let Let X be a Banach space and $f : X \to R \cup \{+\infty\}$ be directionally Lipschitz in the sense of Clarke (1983) at $\bar{x} \in X$. Then f is sequentially normally epi-compact around $\bar{x}$.*

Proof. By Proposition 3.1 of Borwein (1987), if f is directionally Lipschitz at $\bar{x}$, then it is compactly Lipschitz at $\bar{x}$, i.e., its epigragh is compactly epi-Lipschitz at $(\bar{x}, f(\bar{x}))$ in the sense of Borwein and Strojwas (1985). By Proposition 3.7 of Loewen (1992), a compactly epi-Lipschitz set is sequentially normally compact. Hence the proof of the proposition is complete.

■

The following is the sum rule for limiting subgradients.

Proposition 3 *[Corollary 3.4 of Mordukhovich and Shao (1996b)] Let X be an Asplund space, functions $f_i : X \to R \cup \{\infty\}$ be lower semicontinuous and finite at $\bar{x}$, $i = 1, 2$ and one of them be sequentially normally epi-compact around $\bar{x}$. Then one has the inclusion*

$$\partial(f_1 + f_2)(\bar{x}) \subset \partial f_1(\bar{x}) + \partial f_2(\bar{x})$$

provided that

$$\partial^\infty f_1(\bar{x}) \cap (-\partial^\infty f_2(\bar{x})) = \{0\}.$$

For set-valued maps, the definition for limiting normal cone leads to the definition of coderivative of a set-valued map introduced by Mordukhovich.

Definition 5 *Let $\Phi : X \rightrightarrows Y$ be an arbitrary set-valued map (assigning to each $x \in X$ a set $\Phi(x) \subset Y$ which may be empty) and $(\bar{x}, \bar{y}) \in$ cl gphΦ where gphΦ is the graph of the set-valued Φ defined by*

$$gph\Phi := \{(x, y) \in X \times Y : y \in \Phi(x)\}$$

and clΩ denotes the closure of the set Ω. The set-valued map $D^*\Phi(\bar{x}, \bar{y})$ from Y^* into X^* defined by

$$D^*\Phi(\bar{x}, \bar{y})(y^*) = \{x^* \in X^* : (x^*, -y^*) \in N((\bar{x}, \bar{y}), gph\Phi)\},$$

is called the coderivative of Φ at point $(\bar{x}, \bar{y})$. By convention for $(\bar{x}, \bar{y}) \notin$ clgphΦ we define $D^*\Phi(\bar{x}, \bar{y})(y^*) = \emptyset$. The symbol $D^*\Phi(\bar{x})$ is used when Φ is single-valued at $\bar{x}$ and $\bar{y} = \Phi(\bar{x})$.

Remark 2 *Recalled that a single-valued mapping $\Phi : X \to Y$ is called strictly differentiable at $\bar{x}$ with the derivative $\Phi'(\bar{x})$ if*

$$\lim_{x, x' \to \bar{x}} \frac{\Phi(x) - \Phi(x') - \Phi'(\bar{x})(x - x')}{\|x - x'\|} = 0$$

In the special case when a set-valued map is single-valued and $\Phi : X \to Y$ is strictly differentiable at $\bar{x}$, the coderivative coincides with the adjoint linear operator to the classical strict derivative, i.e.,

$$D^*\Phi(\bar{x})(y^*) = \Phi'(\bar{x})^* y^* \quad \forall y^* \in Y^*,$$

where $\Phi'(\bar{x})^*$ denotes the adjoint of $\Phi'(\bar{x})$.

The following proposition is a sum rule for coderivatives when one mapping is single-valued and strictly differentiable.

Proposition 4 [Theorem 3.5 of Mordukhovich and Shao (1996b)] *Let X, Y be Banach spaces, $f : X \to Y$ be strictly differentiable at $\bar{x}$ and $\Phi : X \rightrightarrows Y$ be an arbitrary closed set-valued map. Then for any $\bar{y} \in f(\bar{x}) + \Phi(\bar{x})$ and $y^* \in Y^*$ one has*

$$D^*(f + \Phi)(\bar{x}, \bar{y})(y^*) = f'(\bar{x})^* y^* + D^*\Phi(\bar{x}, \bar{y} - f(\bar{x}))(y^*).$$

3 PROOF OF THE NECESSARY OPTIMALITY CONDITION

The purpose of this section is to derive the necessary optimality conditions involving coderivatives for (OCVI) as stated in Theorem 1.

Proof of Theoerm 1. Since K is a convex set, by the definition of a normal cone in the sense of convex analysis, it is easy to see that problem (OCVI) can be rewritten as the optimization problem with generalized equation constraints (GP):

$$\begin{array}{lll} \text{(GP)} & \min & J(y,u) \\ & \text{s.t.} & (y,u) \in V \times U_{ad}. \\ & & 0 \in F(y,u) + N_K(y), \end{array}$$

where

$$N_K(y) := \begin{cases} \text{the normal cone of } K \text{ at } y & \text{if } y \in K \\ \emptyset & \text{if } y \notin K \end{cases}$$

is the normal cone operator.

Let $\Phi(y,u) : V \times U \rightrightarrows V^*$ be the set-valued map defined by

$$\Phi(y,u) := F(y,u) + N_K(y).$$

By local optimality of the pair $(\bar{y},\bar{u})$ we can find $U(\bar{y},\bar{u})$, a neighborhood of $(\bar{y},\bar{u})$, such that

$$\begin{aligned} J(\bar{y},\bar{u}) &\leq J(y^*,u) \quad \forall (y^*,u) \in U(\bar{y},\bar{u}) \cap (V \times U_{ad}) \text{ s.t. } 0 \in \Phi(y^*,u), \\ &\leq J(y,u) + \frac{L_J}{\mu} \frac{\langle F(y^*,u) - F(y,u), y^* - y\rangle}{\|y^* - y\|} \\ &\qquad \forall (y^*,u),(y,u) \in U(\bar{y},\bar{u}) \cap (K \times U_{ad}) \text{ s.t. } 0 \in \Phi(y^*,u). \end{aligned}$$

Let $y, y^* \in K, u \in U_{ad}$ be such that $0 \in \Phi(y^*,u)$ and $v \in \Phi(y,u)$. Then by definition of the normal cone, we have

$$\begin{aligned} \langle v - F(y,u), y' - y\rangle &\leq 0 \quad \forall y' \in K \\ \langle -F(y^*,u), y' - y^*\rangle &\leq 0 \quad \forall y' \in K. \end{aligned}$$

In particular one has

$$\langle v - F(y,u), y^* - y\rangle \leq 0, \quad \langle -F(y^*,u), y - y^*\rangle \leq 0$$

which implies that

$$\langle v + F(y^*,u) - F(y,u), y^* - y\rangle \leq 0.$$

Hence we have

$$J(\bar{y},\bar{u}) \leq J(y,u) + \frac{L_J}{\mu}\|v\| \quad \forall (y,u,v) \in Gr\Phi, (y,u) \in U(\bar{y},\bar{u}) \cap (V \times U_{ad}).$$

That is, $(\bar{y},\bar{u},0)$ is a local solution to the penalized problem of (GP):

$$\begin{array}{ll} \min & J(y,u) + \dfrac{L_J}{\mu}\|v\| \\ \text{s.t.} & (y,u) \in V \times U_{ad}, \quad (y,u,v) \in Gr\Phi. \end{array}$$

Let $\psi_\Omega(x)$ denote the indicate function of Ω. Then it is easy to see that $(\bar{y},\bar{u},0)$ is a local minimizer of the lower semicontinuous function

$$f(y,u,v) := J(y,u) + \frac{L_J}{\mu}\|v\| + \psi_{Gr\Phi}(y,u,v) + \psi_{U_{ad}}(u).$$

It follows from Propositions 1 that

$$0 \in \partial f(\bar{y},\bar{u},0). \tag{1.6}$$

Since J is Lipschitz near $(\bar{y},\bar{u})$, $g(y,u,v) := J(y,u) + \frac{L_J}{\mu}\|v\|$ is Lipschitz at $(\bar{y},\bar{u},0)$. Hence it is directionally Lipschitz by Theorem 2.9.4 of Clarke (1983) and $\partial^\infty g(\bar{y},\bar{u},0) = \{0\}$ by Proposition 2.5 of Mordukhovich and Shao (1996a). Consequently by Proposition 3 we have

$$\begin{aligned}\partial f(\bar{y},\bar{u},0) &\subseteq \partial g(\bar{y},\bar{u},0) + \partial(\psi_{Gr\Phi} + \psi_{U_{ad}})(\bar{y},\bar{u},0)\\ &\subseteq \partial J(\bar{y},\bar{u}) \times \frac{L_J}{\mu}B + \partial(\psi_{Gr\Phi} + \psi_{U_{ad}})(\bar{y},\bar{u},0),\end{aligned} \tag{1.7}$$

where B is the closed unit ball of V. Next we shall prove that

$$\partial(\psi_{Gr\Phi} + \psi_{U_{ad}})(\bar{y},\bar{u},0) \subseteq \partial\psi_{Gr\Phi}(\bar{y},\bar{u},0) + \{0\} \times \partial\psi_{U_{ad}}(\bar{u}) \times \{0\}$$

by using the sum rule Propostion 3. By (vii) of Theorem 1 and Remark 3 of Borwein, Lucet and Mordukhovich (1998), the assumption (A1) implies that U_{ad} is compactly Epi-Lipschitz. Hence the epigragh of the function $\psi_{U_{ad}}$ is also compactly Epi-Lipschitz. By Proposition 3.7 of Loewen (1992), a compactly epi-Lipschitz set is sequentially normally compact. Therefore the function $\psi_{U_{ad}}$ is sequentially normlly epi-compact around every point in U_{ad}. Now we check the condition

$$\partial^\infty\psi_{Gr\Phi}(\bar{y},\bar{u},0) \cap (-\{0\} \times \partial^\infty\psi_{U_{ad}}(\bar{u}) \times \{0\}) = \{0\}.$$

Let $(0,\xi_2,0) \in \partial^\infty\psi_{Gr\Phi}(\bar{y},\bar{u},0) \cap (-\{0\} \times \partial^\infty\psi_{U_{ad}}(\bar{u}) \times \{0\})$. Then

$$(0,\xi_2,0) \in \partial^\infty\psi_{Gr\Phi}(\bar{y},\bar{u},0) = N_{Gr\Phi}(\bar{y},\bar{u},0)$$

So by definition of coderivatives,

$$(0,\xi_2) \in D^*\Phi(\bar{y},\bar{u},0)(0).$$

By the sum rule for coderivatives Proposition 4, we have

$$D^*\Phi(\bar{y},\bar{u},0)(0) \subset F'(\bar{y},\bar{u})^*0 + D^*N_K(\bar{y},-F(\bar{y},\bar{u}))(0) \times \{0\}$$

which implies that $\xi_2 = 0$. Hence by Proposition 3 we have

$$\begin{aligned}\partial(\psi_{Gr\Phi} + \psi_{U_{ad}})(\bar{y},\bar{u},0) &\subseteq \partial\psi_{Gr\Phi}(\bar{y},\bar{u},0) + \{0\} \times \partial\psi_{U_{ad}}(\bar{u}) \times \{0\}\\ &= N_{Gr\Phi}(\bar{y},\bar{u},0) + \{0\} \times N(\bar{u},U_{ad}) \times \{0\}.\end{aligned} \tag{1.8}$$

By (1.6), (1.7) and (1.8), we have

$$0 \in \partial J(\bar{y}, \bar{u}) \times \frac{L_J}{\mu} B + N_{Gr\Phi}(\bar{y}, \bar{u}, 0) + \{0\} \times N(\bar{u}, U_{ad}) \times \{0\}.$$

That is, there exist $\eta \in \frac{L_J}{\mu} B$, $(\xi_1, \xi_2) \in \partial J(\bar{y}, \bar{u})$ and $\zeta \in N(\bar{u}, U_{ad})$ such that

$$(-\xi_1, -\xi_2 - \zeta, -\eta) \in N_{Gr\Phi}(\bar{y}, \bar{u}, 0).$$

Hence by the definition of coderivatives and the sum rule for coderivatives Proposition 4, we have

$$\begin{array}{rcl} (-\xi_1, -\xi_2 - \zeta) & \in & D^*\Phi(\bar{y}, \bar{u}, 0)(\eta) \\ & \subset & F'(\bar{y}, \bar{u})^* \eta + D^* N_K(\bar{y}, -F(\bar{y}, \bar{u}))(\eta) \times \{0\} \end{array}$$

The proof of the theorem is complete. ■

References

Barbu, V. (1984). *Optimal Control of Variational Inequalities* Pitman, Boston.

Borwein, J.M. (1987). Epi-Lipschitz-like sets in Banach space:theorems and examples, *Nonlinear Analysis: TMA.* **11** 1207-1217.

Borwein, J.M., Lucet, Y. and Mordukhovich, B. (1998). Compactly Epi-Lipschitz convex sets, Preprint.

Borwein, J.M. and Strojwas (1985), H.M., Tangential approximations, *Nonlinear Analysis TMA* **9** 1347-1366.

Clarke, F.H. (1983). *Optimization and Nonsmooth Analysis*, Wiley-Interscience, New York.

Lion, J.L. (1969). *Quelwues méthodes de résolution des probleémes aux limites non-linéaires*, Dunod-Gauthier-Villars, Paris.

Lion, J.L. (1972). Some aspects of the optimal control of distributed parameter systems CBMS-NSF Regional Conference Series in Applied Mathematics, Society for Industrial and Applied Mathematics, Philadelphia, PA.

Loewen, P.D. (1992). Limits of Fréchet Normals in nonsmooth analysis, in: A.D. Ioffe *et al.* (eds.), Optimization and Nonlinear Analysis, Pitman Research Notes in Math. Series **244**, Longman, Harlow, Essex, 178-188.

Mordukhovich, B.S. and Shao, Y. (1996a). Nonsmooth sequential analysis in Asplund spaces *Trans. Amer. Math. Soc.* **348** 1235-1280.

Mordukhovich, B.S. and Shao, Y. (1996b). Nonconvex differential calculus for infinite-dimensional multifunctions, *Set-Valued Analysis* **4** 205-236.

Shi, S. (1988). Optimal control of strongly monotone variational inequalities, *SIAM J. on Control Optim.* **26** 274-290.

Shi, S. (1990). Erratum: Optimal control of strongly monotone variational inequalities, *SIAM J. on Control Optim.* **28** 243-249.

OPTIMAL CONTROL OF A CLASS OF STRONGLY NONLINEAR EVOLUTION SYSTEMS *

Z. H. Zhang & X. Xiang

Department of Mathematics
Guizhou University
Guiyang, Guizhou, P.R. China.

Abstract: In this paper, we investigate the existence of optimal controls of the first-order nonlinear evolution systems whose principle operator is pseudomonotone operator with nonmonotone nonlinearities perturbation.

Keywords. Nonmonotone nonlinearities, Pseudomonotone operator, Optimality.

1. Introduction

Recently, optimal control problems of nonlinear distributed parameter systems have been extensively studied. Many authors pay great attention to several class of semilinear and nonlinear evolution equations, such as, Papageorgiou ([1]-[2]), Ahmed ([4]) and Xiang ([5]). Meanwhile, to our knowlege, the class of nonlinear control systems with pseudomonotone operator has rarely been studied.

As well-known, in studying existence of solutions of nonlinear evolution equations, we usually consider the evolution triples ([3]), $X \hookrightarrow H \hookrightarrow X^*$, or more generally, consider the chairs of spaces ([1]),

$$X \hookrightarrow Y \hookrightarrow H \hookrightarrow Y^* \hookrightarrow X^*.$$

The nonlinear perturbation is assumed by some authors as following:

$f : I \times Y \to Y^*$([1]), here I denotes the time horizon, or $f : I \times H \to H$([2]), or $f : I \times H \to X^*$([5]). In this paper, by developing the concept of pseudomonotone operator and using some technique of convergence, we deal with the existence of solutions of nonlinear evolution equation whose principle operator is pseudomonotone operator with nonmonotone nonlinearities perturbation being from $I \times Y$ into X^*, which includes the results of [1] and [2]. Further,

*Projects Supported by Chinese National Natural Science Foundation 29362003 .

we solve the existence of optimal controls of nonlinear control system.

2. Preliminaries

Let T be a fixed number, and $I = [0, T]$ the time horizon, H a separable Hilbert space, and X, Y subspaces of H carrying the structure of a separable reflexive Banach space. We will assume that $X \hookrightarrow Y \hookrightarrow H$, with all embeddings being continuous, dense, and X into Y is also compact. Identifying H with its dual, we have

$$X \hookrightarrow Y \hookrightarrow H \hookrightarrow Y^* \hookrightarrow X^*,$$

with all embeddings being continuous, dense, and the first and the last are also compact. In the literature, triples like (X, H, X^*) and (Y, H, Y^*) are usually called evolution triples ([3]). For simplicity, we denote by $< .,. >$ and $(.,.)$ the duality brackets of the pair (X, X^*) and the inner product of H, respectively. They are compatible in the sense that $< .,. > |_{X\times H} = (.,.)$ (proposition 23.13, [3]). Also, by $||.||$ (*resp.*, $||.||_Y, |.|, ||.||_{Y^*}, ||.||_*$), we denote the norm of X (resp.,Y, H, Y^*, X^*), and by $\longrightarrow^s$ ($\longrightarrow^w$), we denote the strongly (weakly) convergence. We set

$$W_{pq} = \{x \in L^p(I, X),\ \dot{x} \in L^q(I, X^*)\}(1 < p, q < \infty,\ \tfrac{1}{p} + \tfrac{1}{q} = 1),$$

where the derivative in the definition should be understood in the sense of vector-valued distributions, equiped with the norm

$$||x||_{W_{pq}} = (||x||^2_{L^p(I,X)} + ||\dot{x}||^2_{L^q(I,X^*)})^{\frac{1}{2}}.$$

The space W_{pq} becomes a separable reflexsive Banach space. Moreover, $W_{pq} \hookrightarrow L^p(I, X)$ and $W_{pq} \hookrightarrow C(I, H)$ continuously, and $W_{pq} \hookrightarrow L^p(I, Y)$ compactly (pp. 650, [1]). For convenience, we write the spaces

$V = L^p(I, X)$, $V^* = L^q(I, X^*)$, $\mathcal{H} = L^2(I, H)$, $Z = L^p(I, Y)$, $Z^* = L^q(I, Y^*)$.

By $||.||_V$ (resp., $||||_{V^*}$, $||.||_{\mathcal{H}}$, $||.||_Z$, $||.||_{Z^*}$), we denote the norm of the space V (resp., V^*, $\mathcal{H}$, Z, Z^*). Let $<< .,. >>$ and $((.,.))$ denote the duality brackets for the pair (V, V^*) and the inner product of $\mathcal{H}$, respectively.

Next we model the control space by a reflexive separable B-space E with the norm $||.||_E$. By $P_{f(c)}(E)$, we denote a class of nonempty, closed (convex) subsets of E. Recall that a multi-function $U : I \to P_f(E)$ is called measurable if

$$GrU = \{(t, v) \in I \times E : v \in U(t)\} \in B(I) \times B(E),$$

with $B(I)(B(E))$ being the Borel σ–field of $I(E)$. By $\mathcal{U}_{ad}$, we denote the admissible control set of all selectors of U(.) that belong to Lebesgue-Bochner space $L^q(I, E)$, $1 < q < \infty$; that is,

$$\mathcal{U}_{ad} = \{u \in L^q(I, E) : u(t) \in U(t),\ \text{a.e. on } I\}.$$

Note that

$$\mathcal{U}_{ad} \neq \phi, \text{ if } t \to |U(t)| = \sup\{||u||_E,\ u \in U(t)\} \in L^q_+,$$

in which case the multi-function is said to be L^q–bounded.

Finally recall that an operator $A : X \to X^*$ is said to be pseudomonotone if, for each $x \in X$ and each sequence (x_n) in X,

$$x_n \longrightarrow^w x \text{ in X as } n \to \infty \text{ and } \overline{\lim}_{n\to\infty} < Ax_n, x_n - x > \leq 0$$

imply

$$< Ax, x - w > \leq \underline{\lim}_{n\to\infty} < Ax_n, x_n - w >, \text{ for all } w \in X.$$

In order to verify the existence of solutions of System (3.1)(See next section), we develop the concept of pseudomonotone.

Definition An operator $A : M \subseteq X \to X^*$ on the complete subspace M of the real Banach space X is called quasi-pseudomonotone if, for each $x \in M$ and each sequence (x_n) in M,

$$x_n \longrightarrow^w x \text{ in } M \text{ as } n \to \infty \text{ and } \overline{\lim}_{n\to\infty} < Ax_n, x_n - x > \leq 0,$$

there exists a subsequence (x_{n_k}) of (x_n), s.t.

$$< Ax, x - w > \leq \underline{\lim}_{k\to\infty} < Ax_{n_k}, x_{n_k} - w >, \text{ for all } w \in X.$$

3. Existence of Solutions of Nonlinear Evolution Equation

We consider the following initial value problem:

$$\begin{cases} \dot{x}(t) + A(t, x(t)) = f(t, x(t)), & a.e.on I \qquad (3.1) \\ x(0) = x_0 \end{cases}$$

Recall that a function $x \in W_{pq}$ is a solution of (3.1) if, for all $v \in X$, x satisfies

$$\begin{cases} \frac{d}{dt}(x(t), v) + < A(t, x(t)), v > = < f(t, x(t)), v >, & a.e.on I \\ x(0) = x_0 \end{cases}$$

where $\dot{x}$ means the generalized distributed derivatives.

In order to investigate the existence of solutions of (3.1), we need the following hypotheses.

Hypothesis (A). $A : I \times X \to X^*$ is an operator s.t.

(a) $t \to A(t, x)$ is measurable;

(b) $x \to A(t, x)$ is pseudomonotone;

(c) $\|A(t, x)\|_* \leq a_1(t) + b_1\|x\|^{p-1}$,

for all $x \in X$, with $a_1 \in L^q_+(0, T)$, $b_1 > 0$, $2 \leq p < \infty$, $\frac{1}{p} + \frac{1}{q} = 1$;

(d) $< A(t, x) - A(t, y), x - y > \geq c\|x - y\|^p - d|x - y|^2$,

for all x, $y \in X$, with $c > 0$ and $d \geq 0$.

Hypothesis (F). $f : I \times Y \to X^*$ is also an operator s.t.

(a) $t \to f(t, x)$ is measurable;

(b) $x \to f(t, x)$ is continuous from Y into X*;

(c) $\|f(t, x)\|_* \leq a_2(t) + b_2\|x\|_Y^{p-1}$,

for all $x \in Y$, with $a_2 \in L^q_+(0, T)$ and $b_2 > 0$;

(d) $< f(t, x), x > \leq \phi(t)$, for all $x \in X$, with $\phi \in L^1(0, T)$.

Hypothesis ($\mathbf{H_0}$). $x_0 \in H$.

Next we define the Nemyckii operators $\widehat{A}$ and $\widehat{f}$, given by

$$(\widehat{A}x)(t) = A(t, x(t)), \text{ for } x \in V; \ (\widehat{f}x)(t) = f(t, x(t)), \text{ for } x \in Z.$$

Let $\widehat{A}|_{W_{pq}}$ denote the largest restriction of $\widehat{A}$ on W_{pq}. We have some important properties on the operators $\widehat{A}$, $\widehat{f}$ and $\widehat{A}|_{W_{pq}}$.

Lemma 1. Under Hypothesis (A), $\widehat{A} : V \to V^*$ is bounded and satisfies

(a) $\quad ||\widehat{A}x||_{V^*} \leq const(||a_1||_q + b_1||x||_V^{p-1})$, for all $x \in V$;

(b) $\quad << \widehat{A}x - \widehat{A}y,\, x - y >> \geq c||x - y||_V^p - d||x - y||_{\mathcal{H}}^2$,

for all $x,\, y \in V$;

(c) (i) $\widehat{A}|_{W_{pq}}$ is quasi-pseudomonotone from $W_{pq} \subseteq V$ into V^*.

(ii) $\widehat{A}|_{W_{pq}}$ satisfies the condition(M), i.e.

$x_n \longrightarrow^w x$ in W_{pq}, $\widehat{A}x_n \longrightarrow^w b$ in V^*, and $\overline{\lim}_{n\to\infty} << \widehat{A}x_n,\, x_n >> \leq << b,\, x >>$,

imply

$$\widehat{A}x = b.$$

Proof. One can easily verify that (a) and (b) hold true. Now we show (c)(i).

Let $x_n \longrightarrow^w x$ in W_{pq} and $\overline{\lim}_{n\to\infty} << \widehat{A}x_n, x_n - x >> \leq 0$.

Thanks to (b), we have

$$c\, \overline{\lim}_{n\to\infty}||x_n - x||_V^p \leq d\, \overline{\lim}_{n\to\infty}||x_n - x||_{\mathcal{H}}^2$$
$$+ \overline{\lim}_{n\to\infty} << \widehat{A}x_n,\, x_n - x >>$$
$$- \lim_{n\to\infty} << \widehat{A}x, x_n - x >> \leq 0.$$

Since $W_{pq} \hookrightarrow \mathcal{H} = L^2(I, H)$ compactly, we get that $x_n \longrightarrow^s x$ in $\mathcal{H}$. Hence, it follows that $x_n \longrightarrow^s x$ in $V = L^p(I, X)$. Thus, there exists a subsequence of (x_n), denoted by (x_{n_k}), such that $x_{n_k}(t) \longrightarrow^s x(t)$ in X, a.e.on I, and there exists a real function $v \in L^p(0,T)$ satisfying

$$||x_{n_k}(t)|| \leq v(t), \text{ for } a.e.\ t \in I$$

(See pp. 579, [3]).

By virtue of Hypothesis (A)(c), we can show that

$$\lim_{k\to\infty} < A(t, x_{n_k}(t)),\, x_{n_k}(t) - x(t) >= 0, \text{ a.e. on } I.$$

Since A is pseudomonotone, we have

$$< A(t, x(t)),\, x(t) - w(t) > \leq \underline{\lim}_{k\to\infty} < A(t, x_{n_k}(t)),\, x_{n_k}(t) - w(t) > dt,$$

a.e. on I, for all $w \in V$. And consequently,

$$<< \widehat{A}x, x - w >> \leq \int_I \underline{\lim}_{k\to\infty} < A(t, x_{n_k}(t), x_{n_k}(t) - w(t) > dt.$$

Invoking Hypothesis (A) and Theorem 7 of [7], we obtain that

$$\int_I \underline{\lim}_{k\to\infty} < A(t, x_{n_k}(t)), x_{n_k}(t) - w(t) > dt$$
$$\leq \underline{\lim}_{k\to\infty} \int_I < A(t, x_{n_k}(t), x_{n_k}(t) - w(t) > dt.$$

Hence,

$$<< \widehat{A}x,\, x - w >> \leq \underline{\lim}_{k\to\infty} << \widehat{A}x_{n_k},\, x_{n_k} - w >> .$$

Next we show that $\widehat{A}|_{W_{pq}}$ satisfies the condition(M). Let $x_n \longrightarrow^w x$ in W_{pq}, $\widehat{A}x_n \longrightarrow^w b$ in V^* and $\overline{\lim}_{n\to\infty} << \widehat{A}x_n,\, x_n >> \leq << b,\, x >>$.

Due to (c)(i) and

$$\overline{\lim}_{n\to\infty} << \widehat{A}x_n,\, x_n - x >> \leq \overline{\lim}_{n\to\infty} << \widehat{A}x_n, x_n >>$$
$$- \lim_{n\to\infty} << \widehat{A}x_n,\, x >> \leq 0,$$

there exists a subsequence of (x_n), denoted by (x_{n_k}), such that

$$<< \widehat{A}x, x - w >> \leq \underline{\lim}_{k\to\infty} << \widehat{A}x_{n_k}, x_{n_k} - w >>$$

$$\leq \overline{\lim}_{k\to\infty} << \widehat{A}x_{n_k}, x_{n_k} >> - \lim_{k\to\infty} << \widehat{A}x_{n_k}, w >>$$
$$\leq << b, x - w >>,$$

for all $w \in V$. Thus,

$$\widehat{A}x = b.$$

Similar to the proofs of Theorem 23.A and Lemma 2.5.1 of [4], one can verify that the following Lemmas.

Lemma 2. Under Hypothsis(F), $\widehat{f} : Z \to V^*$ is bounded and satisfies

(a) $\quad ||\widehat{f}x||_{V^*} \leq const(||a_2||_q + b_2||x||_Z^{p-1})$;

(b) $\quad << \widehat{f}x, x >> \leq ||\phi||_1$, for all $x \in V$;

(c) $\quad \widehat{f} : Z \to V^*$ is continuous.

Lemma 3. If Hypotheses (A), (F), (H_0) are satisfied, and $x \in W_{pq}$ is a solution of (3.1), then there exist positive constants M_1, M_2, M_3, such that

$$\max_{t\in I} |x(t)| \leq M_1;$$
$$||x||_{W_{pq}} \leq M_2;$$
$$||\widehat{A}x||_{V^*} \leq M_3.$$

Note that (3.1) is equvalent to the following operator equation

$$\begin{cases} \dot{x} + \widehat{A}x = \widehat{f}x \\ x(0) = x_0 \end{cases} \qquad (3.2)$$

(See pp. 767-770, [3]).

Theorem 1. Suppose that Hypotheses of (A), (F), and (H_0) hold, then the problem(3.1) admits a solution $x \in W_{pq} \subseteq C(I, H)$.

Proof. Let $\{w_1, w_2, ..., w_n, ...\}$ be a basis in X. We set $X_n = span\{w_1, w_2, ..., w_n\}$, and introduce on the n-dimensional space X_n the scalar product of the H-space H. Note that $X_n \subseteq X \subseteq H$.

Suppose that $x_{n_0} \longrightarrow^s x_0$ in H. The Galerkin equation of (3.1) can be constructed by

$$\begin{cases} < \dot{x}_n(t), w_j > + < A(t, x_n(t)), w_j > = < f(t, x_n(t)), w_j >, \quad a.e. on I \\ x_n(0) = x_{n_0}, \qquad j = 1, 2, ..., n \\ w_j \in X_n, x_n \in X_n \end{cases} \qquad (3.3)$$

By using Lemma 3, Hypotheses (A)(a)-(d), (F)(a)-(c), Proposition 27.7 of [3] and Carátheodory theorem ([7]), the Galerkin equation(3.3) must have a solution x_n (n=1, 2, ...).

By Lemma 3, we know that there exists $M^* > 0$ s.t.

$$\max_{t\in I} |x_n(t)| + ||x_n||_{W_{pq}} + ||\widehat{A}x_n||_{V^*} \leq M^*.$$

Hence, we can find a subsequence, again denoted by $\{x_n\}$, such that $x_n \longrightarrow^w x$ in W_{pq}, $\widehat{A}x_n \longrightarrow^w b$ in V^*, and $x_n(T) \longrightarrow^w z$ in H. Using the same procedure of the proof of Lemma 30.4 of [3], we can verify that the limit elements x, w, and z satisfy

$$\begin{cases} \dot{x} + b = \widehat{f}x \\ x(0) = x_0, \quad x(T) = z, \quad x \in W_{pq} \end{cases}$$

It follows from the Galerkin equation that

$$<< \widehat{A}x_n, x_n >> = << \widehat{f}x_n, x_n >> - \tfrac{1}{2}(|x_n(T)|^2 - |x_{n0}|^2).$$

Since $x_n(T) \longrightarrow^w x(T)$ in H, we have

$$|x(T)| \leq \underline{\lim}_{n\to\infty}|x_n(T)|.$$

Hence

$$\overline{\lim}_{n\to\infty} << \widehat{A}x_n, x_n >> \leq << b, x >> .$$

Using Lemma 1, we assert that

$$\widehat{A}x = b.$$

This shows that x is a solution of (3.2), and hence a solution of (3.1).

4. Existence of Admissible Trajectories and Optimal Controls

We now consider the following controlled evolution system:

$$\begin{cases} \dot{x}(t) + A(t, x(t)) = f(t, x(t), u(t)), & a.e.onI \\ x(0) = x_0, & u \in \mathcal{U}_{ad} \end{cases}$$

We impose some hypotheses.

.**Hypothsis (F*).** $f : I \times Y \times E \to X^*$ satisfies:

(a) $(t, x, u) \to f(t, x, u)$ is measurable;

(b) $(x, u) \longrightarrow f(t, x, u)$ is continuous from $Y \times E_w$ into X^*,

here E_w denotes the space E with its weak topology;

(c) $\| f(t, x, u)\|_* \leq a_3(t) + b_3(\|x\|_Y^{p-1} + \|u\|_E)$,

for all $x \in Y$ and $u \in E$, with $a_3 \in L_+^q(0, T)$ and $b_3 > 0$;

(d) $< f(t, x, u), x > \leq \widetilde{\phi}(t) + k\|u\|_E^{\frac{p}{p-1}}$,

for all $x \in X$ and $u \in E$ with $\widetilde{\phi} \in L^1(0, T)$ and $k \in R$.

Hypothsis (U). $U : I \to P_f(E)$ is L^q-bounded multifunction.

By virtue of Theorem 1, we can conclude the existence of admissible trajectories of (4.1).

Theorem 2. Assume that Hypotheses $(A), (F^*), (H_0)$, and (U) hold, then the problem (4.1) admits a solution corresponding to each date pair $(x_0, u) \in H \times \mathcal{U}_{ad}$.

By Hypothesis (U), $\mathcal{U}_{ad}$ is a bounded subset in $L^q(I, E)$. Set $P(x_0) = \{x_u \in W_{pq}$, x_u is a solution of (4.1) corresponding to $u \in \mathcal{U}_{ad}$ and $x_0 \in H\}$, and $K(T) = \{y = x(T) : x \in P(x_0)\}$. Utilizing the similar method of Lemma 3, we obtain

Lemma 5. Under Hypotheses of Theorem 2, there exists positive constants M_1, M_2, and M_3, such that for all $x \in P(x_0)$,

$$\max_{t\in I} |x(t)| \leq M_1;$$
$$\|x\|_{W_{pq}} \leq M_2.$$

Consider the optimal control problem (P):

$$J(x, u) = \phi_0(x(T)) + \int_0^T L(t, x(t), u(t))dt \longrightarrow \inf$$

s.t.

$$\begin{cases} \dot{x}(t) + A(t, x(t)) = f(t, x(t), u(t)), & a.e.onI \\ x(0) = x_0, \quad u \in \mathcal{U}_{ad} \end{cases}$$

Suppose

Hypothesis (U*). $U : I \to P_{fc}(E)$ is L^q-bounded multi-function.

Hypothesis (F).** $f(.,x_n(.),u_n(.)) \longrightarrow^s f(.,x(.),u(.))$ in V^* as $x_n \longrightarrow^s x$ in $Z = L^p(I,Y)$ and $u_n \longrightarrow^w u$ in $L^q(I,E)$.

Hypothesis (L). $L : I \times X \times E \to R \cup \{+\infty\}$ s.t.

(a) $(t,x,u) \to L(t,x,u)$ is measurable;

(b) $(x,u) \to L(t,x,u)$ is l.s.c.;

(c) $u \to L(t,x,u)$ is convex;

(d) $\phi(t) - c\,(\|x\| + \|u\|_E) \leq L(t,x,u)$, a.e. on I,

with $\phi \in L^1(0,T)$ and $c \geq 0$.

Hypothesis (P). $P \equiv K(T) \cap Q \neq \phi$, where Q denotes a weakly closed subset of H-space H.

Hypothesis (Φ_0). $\phi_0 : P \subseteq H \to R$ is w.l.s.c.

Hypothesis (J). There exists an admissible state-control pair (x,u) such that

$J(x,u) < +\infty$.

Theorem 3. If Hypotheses $(A), (F^*), (F^{**}), (H_0), (U^*), (L), (P), (\Phi_0)$, and (J) hold, then the problem (P) has an optimal pair (x,u).

Proof Suppose that (x_n,u_n) is the minimizing sequence, then it satisfies the following initial value problem:

$$\begin{cases} \dot{x}_n(t) + A(t,x_n(t)) = f(t,x_n(t),u_n(t)) \\ x_n(0) = x_0 \end{cases} \tag{4.2}$$

Recalling Lemma 5 and boundedness of $\mathcal{U}_{ad}$ in $L^q(I,E)$, by passing to a subsequence, if necessary, we may assume that $x_n \longrightarrow^w x$ in W_{pq}, $\widehat{A}x_n \longrightarrow^w b$ in V^*, $x_n(T) \longrightarrow^w z \in Q$ in H, and $u_n \longrightarrow^w u$ in $L^q(I,E)$. By virtue of Mazur Lemma and Hypothesis (U*), we have

$$u(t) \in \overline{co}w - \lim_n \sup\{u_n(t)\}_{n\geq 1} \subset U(t), \text{a.e. on } I.$$

Hence, $u \in \mathcal{U}_{ad}$. As in the proof of Theorem 1, we can prove that x satisfies

$$\begin{cases} \dot{x} + \widehat{A}x = f(.,x(.),u(.)) \\ x(0) = x_0 \\ x(T) = z \in Q \end{cases}$$

Therefore, $x(T) \in P \subseteq H$. Next, it follows from (4.2) that $<< \dot{x}_n - \dot{x}, x_n - x >> + << \widehat{A}x_n - \widehat{A}x, x_n - x >> = << f(.,x_n(.),u_n(.)) - f(.,x(.),u(.)), x_n(.) - x(.) >>$.

By Lemma 1, we obtain that

$$c\|x_n - x\|_V^p \leq d\|x_n - x\|_{\mathcal{H}}^2 + << f(.,x_n(.),u_n(.)) - f(.,x(.),u(.)), x_n(.) - x(.) >> .$$

Therefore, by Hypothesis (F**), $W_{pq} \hookrightarrow L^p(I,Y)$compactly, we claim that $x_n \longrightarrow^s x$ in V. Since $x_n(T) \longrightarrow^w x(T)$in H, Hypothesis (Φ_0) infers that

$$\phi_0(x(T)) \leq \underline{\lim}_{n\to\infty} \phi_0(x_n(T)).$$

By applying Hypothesis (L) and Balder theorem 2.1 ([6]), we conclude that

$$\int_0^T L(t,x(t),u(t))dt \leq \underline{\lim}_{n\to\infty} \int_0^T L(t,x_n(t),u_n(t))dt$$

Consequently,

$J(x,u) \leq \underline{\lim}_{n\to\infty} J(x_n,u_n) = \lim_{n\to\infty} J(x_n,u_n)$.

This shows that $(x,u)$is the desired optimal pair.

5. Examples

Our results can be used to the following the initial boundary value problem of parabolic system of order 2m

$$\begin{cases} \frac{\partial}{\partial t}x(t,z) + L(t)x(t,z) = f(t,z,x(t,z)), & on I\times G \\ D^\beta x = 0 \quad on I\times \partial G, & |\beta| \leq m-1 \\ x(0,z) = x_0(z), & on G \end{cases}$$

where $L(t)x(t,z) = \sum_{|\beta|\leq m}(-1)^{|\beta|}D^\beta A_\beta(t,z,\theta(x(t,z)))$,

$f(t,z,x(t,z)) = \sum_{|\beta|\leq k}(-1)^{|\beta|}D^\beta f_\alpha(t,z,\eta(x(t,z)))$, $\beta = (\beta_1,\beta_2,...,\beta_n)$ is n-tuple of nonnegative integers, $|\beta| = \sum_{i=1}^m \beta_i$, $D^\beta = D_1^{\beta_1}D_2^{\beta_2}...D_n^{\beta_n}$, $D_i = \frac{\partial}{\partial x_i}$, $\theta(x) = \{D^\beta x : |\beta| \leq m\}$, $\eta(x) = \{D^\beta x, |\beta| \leq k\}$,and $k < m$. Take

$X = W_0^{m,p}(G)$, $Y = W_0^{k,p}(G)$, $H = L^2(G)$.

Their dual spaces are

$X^* = W^{-m,q}(G)$, $Y = W^{-k,q}(G)$, $H^* = H$,

respectively. From Soblev embedding theorem, we know that

$$X \hookrightarrow Y \hookrightarrow H \hookrightarrow Y^* \hookrightarrow X^*,$$

with all embedings being continuous, dense, and compact([1]). Define,

$a_1(t,x,y) = \int_G \sum_{|\beta|<m} \xi_\beta(t,z,\theta(x))D^\beta y dz$,

$a_2(t,x,y) = \int_G \sum_{|\beta|=m} \xi_\beta(t,z,\theta(x))D^\beta y dz$,for all x,$y \in X$,

$b(t,x,y) = \int_G \sum_{|\beta|\leq k} f_\beta(t,z,\eta(x))D^\beta y dz$.

Under some reasonable assumptions similar to Section 30.4 of [3], one can verify that there exist operaters $A_i(.,.) : I\times X \to X^*(i=1,2)$ and $f(.,.) : I\times Y \to X^*$ satisfying

$< A_i(t,x), y >= a_i(t,x,y)$,$i = 1,2$;

$< f(t,x), y >= b(t,x,y)$.

Define

$A(t,x) = A_1(t,x) + A_2(t,x)$.

By Proposition 26.16 of [3] and Corrollary 26.14 of [3], we can check that the operaters A and f satisfy the assumptions of Theorem 1. Hence, the existence of weak solutions can be proved. Futher, under some assumptions of Example 1 of [1], we can solve the following optimal control problem:

$$J(x,u) = \int_G |x(T,z)|dz + \int_0^T \int_G \frac{1}{2}|x(t,z) - y_0(z)|^2 dtdz$$
$$+ \frac{\theta}{2}\int_0^T \int_G |u(t,z)|^2 dtdz \longrightarrow \inf,$$

s.t.

$$\begin{cases} \frac{\partial}{\partial t}x(t,z) + L(t)x(t,z) = f(t,z,x(t,z)) + b(t)u(t,z), & on I\times Q \\ D^\beta x = 0, \quad on I\times \partial G, & |\beta| \leq m-1 \\ x(0,z) = x_0(z) & on G \end{cases}$$

References

[1] N.S.Papgeorgiou, Optimal control of nonlinear evolution equations with nonmonotone nonlinearities, Journal of optimization theory and application, Vol. 77, No. 3, pp.643-660, 1993.

[2] N.S.Papageorgiou. Properties of the relaxd trajectories of evolution equations and optimal control. SIAM Journal on control and optimization, Vol. 27, No.2, pp.267-288, 1989.

[3] E.Zeidler, Nonlinear Functional Analysis and its Applications II, Springer-Verlag, New York, Berlin Heidelberg London Paris,Tokyo, 1992.

[4] N.U.Ahmed and K.L.Teo, Optimal control of distributed parameter systems. North-Holland, New

York, Oxford, pp.368-391, 1981.

[5] N.U. Ahmed & X.L. Xiang, Nonlinear Uncertain Systems and Necessary Conditions of Optimality, SIAM Journal on control and optimization, Vol. 35, No. 5, Sept. 1997.

[6] E.J.Balder, Necessary and sufficient conditions for L^1−strong-weak lower semicontinuity of integral functions, nonlinear analysis, Vol.11, No. 12, pp.1399-1403, 1987.

[7] R.F.Curtain, Functional analysis in modern applied mathematics, United States edition published by academic press inc. 111 Fifth Avenue, New York, New York 100003, 1977.

[8] Xia Daoxing, Real varation function and functional analysis, published by Education House, Book 2, 1987.

Part II. Stochastic Systems

ROBUST STABILIZATION OF NONLINEAR SYSTEMS WITH MARKOVIAN JUMPING PARAMETERS

E.K. Boukas

Mechanical Engineering Department
École Polytechnique de Montréal
P.O. Box 6079, station "Center-ville"
Montréal, Québec, Canada H3C 3A7. *

boukas@anas.meca.polymtl.ca

H. YANG

Department of Statistics
The University of Hong Kong
Hong Kong.†

hlyang@hkusua.hku.hk

Abstract: This paper deals with the class of uncertain nonlinear systems with Markovian jumping parameters. The uncertainties are assumed to be nonlinear and state dependent. The stabilization problem of this class of systems is studied here and sufficient conditions for the robust stabilizability are established.

1 INTRODUCTION

The class of systems with Markovian jumping parameters represents an interesting class of systems that we can use to model a variety of physical systems. This class of systems has two components in the state vector. The first one which varies continuously is referred to be the continuous state of the system

*Research of this author was supported by the Natural Sciences and Engineering Research Council of Canada under grants OGP0036444

†Research of this author was supported by the Research grant C337/017/0019, Tne University of Hong Kong.

and the second one which varies discretly is referred to be the mode of the system. This class of process was used in the past to model, for instance, manufacturing systems (see Sethi and Zhang (1994) or Gershwin (1994) and the references therein). Other applications using this class of systems can be found in Martion (1990) and the references therein.

In this paper, we will deal with the stochastic stabilizability and its robustness for the class of the nonlinear Markovian jumping parameters and try to establish conditions under which the optimal control of the nominal nonlinear system have a robust stabilizing control law for the uncertain nonlinear system. The uncertainty we will consider in this paper is a nonlinear and state dependent one. We will also consider the nonlinear changes in the control gain. The two nonlinearities will be assumed to be memoryless.

2 PROBLEM STATEMENT

Let us assume that the class of systems we consider in this paper be described by the following nonlinear differential equations:

$$\begin{aligned} \dot{x}(t) &= A(x(t), r(t)) + B(x(t), r(t))\left[d(x(t), r(t)) + p(u(t))\right], \\ & x(0) = x_0, r(0) = r_0 \end{aligned} \tag{2.1}$$

where $x(t) \in R^n$ is the state vector of the system at time t, $u(t) \in R^m$ is the control input of the system at time t ($p(u(t))$ describes the nonlinear changes in the control gain) and $r(t)$ is a continuous-time Markov process taking values in a finite state space denoted by $\mathcal{S} = \{1, 2, \ldots, n_s\}$; $d(x(t), r(t)) \in R^m$ is the system disturbance; $A(x(t), r(t))$ and $B(x(t), r(t))$ are matrix functions with appropriate dimensions; $x(0) = x_0$ and $r(0) = r_0$ are respectively the initial values of the state and the mode at time $t = 0$.

The evolution of the stochastic process $\{r(t), t \geq 0\}$ that determines the mode of the system is assumed to be described by the following probability transitions:

$$P[r(t+h) = \beta | r(t) = \alpha] = \begin{cases} q_{\alpha\beta} h + o(h), & \text{if } \alpha \neq \beta \\ 1 + q_{\alpha\alpha} h + o(h), & \text{otherwise} \end{cases} \tag{2.2}$$

with $q_{\alpha\beta} \geq 0$ for all $\alpha \neq \beta$ and $q_{\alpha\alpha} = -\sum_{\beta \in \mathcal{S}} q_{\alpha\beta}$ for all $\alpha \in \mathcal{S}$, and $\lim_{h \to 0} \frac{o(h)}{h} = 0$.

The control variable is assumed to be not constrained. Our objective in this paper is to seek an admissible feedback control law that minimises the following cost function:

$$J_u(x_0, r_0) = \mathsf{E}\left[\int_0^\infty \left[g(x(t)) + h(u(t))\right] dt | x(0) = x_0, r(0) = r_0\right] \tag{2.3}$$

where E stands for the mathematical expectation operator.

The question we will address in this paper is how can we design a control that stabilizes the system and guarantees its robustness.

Let $x(t, x_0, \alpha)$ denote the trajectory of the state $x(t)$ from the initial state (x_0, r_0). We introduce the following stochastic stability and stochastic stabilizability concepts for continuous-time jump nonlinear systems.

Definition 1 For system (2.1) with $u(t) \equiv 0$ (i.e. $p(u(t)) \equiv 0$) and $d(x(t), r(t)) \equiv 0$, for all $r(t) \in \mathcal{S}$, the equilibrium point 0 is *stochastically stable*, if for every initial state (x_0, r_0) the following:

$$\int_0^\infty \mathsf{E}\left\{\|x(t, x_0, r_0)\|^2\right\} dt < \infty$$

holds

Definition 2 We say that the system (2.1) is *stochastically stabilizable*, if for every initial state $(x_0, r(0))$, there exists a feedback control law $u(t) = -k(x(t), r(t))$, such that the closed-loop system

$$\dot{x}(t) = A(x(t), r(t)) + B(x(t), r(t))\left[d(x(t), r(t)) + p(-k(r(t))\right]$$

is stochastically stable for admissible uncertainty $d(x(t), r(t))$.

We will assume in this paper that the data of this optimization control problem are smooth enough to satisfy the existence of the control law of the nominal optimization control problem. The details of the assumptions will be given in the appropriate section. We will also assume that the mode of the system is completly accessible for control. The functions in the instantaneous cost will be chosen as follows:

$$g(x(t), r(t)) = g_0(x(t), r(t)) + g_1(x(t), r(t)) \quad (2.4)$$

$$h(x(t), r(t)) = \frac{1}{2} u^\top(t) R(r(t)) u(t) \quad (2.5)$$

where the function $g_1(x(t), r(t))$ is chosen to compensate the effect of the nonlinear state dependant disturbance $d(x(t), r(t))$ while the function $g_0(x(t), r(t))$ is chosen to accelerate the convergence rate under the worst possible disturbance $d(x(t), r(t))$. The expression of $h(u(t), r(t))$ is chosen to simplify the expression of the feedback control law of the nominal nonlinear optimization problem.

3 JUMP NONLINEAR OPTIMIZATION PROBLEM

To solve the optimization problem of the previous section, let us consider the more general case and then based on its solution get the one we are looking for. Let us consider that the system dynamics is described by the following differential nonlinear system of equations:

$$\dot{x}(t) = f(x(t), u(t), r(t)), x(0) = x_0, r(0) = r_0 \quad (3.1)$$

where the variables above keep the same definitions as in previous section, with $f(0, 0, \alpha) = 0$ for all $\alpha \in \mathcal{S}$.

Let the cost function be defined by:

$$J_u(x_0, r_0) = \mathsf{E}\left[\int_0^\infty \left[g(x(t), r(t)) + h(u(t), r(t))\right] dt | x(0) = x_0, r(0) = r_0\right] \quad (3.2)$$

where for each $\alpha \in \mathcal{S}$ we have $g(0,\alpha) = 0$ and $g(x(t),\alpha) > 0$ for all $x \neq 0$ for $g(x(t),r(t))$; and $h(0,\alpha) = 0$ and $h(u(t),\alpha) > 0$ for all $u \neq 0$ for $h(u(t),r(t))$.

Definition 3 A control $u(\cdot) = \{u(t) : t \geq 0\}$ is said to be *admissible* if: (i) $u(\cdot)$ is adapted to the σ-algebra generated by the random process $r(\cdot)$, denoted as $\sigma\{r(s) : 0 \leq s \leq t\}$, and (ii) $u(t) \in R$ for all $t \geq 0$.

Let $\mathcal{U}$ denote the set of all admissible controls of our control problem.

Definition 4 A measurable function $u(x(t),r(t))$ is an admissible *feedback* control, if (i) for any given initial x and α of the continuous state and the mode, the following equations have an unique solution $x(\cdot)$:

$$\dot{x}(t) = f(x(t),u(t),r(t)), x(0) = x \tag{3.3}$$

and (ii) $u(\cdot) = u(x(\cdot),r(\cdot)) \in \mathcal{U}$.

Let us assume that the functions $f(x(t),u(t),r(t))$, $g(x(t),r(t))$ and $h(u(t),r(t))$ satisfy all the required assumptions that guarantee the existence of the optimal feedback control law in the form:

$$k : R^n \times \mathcal{S} \to R^m \tag{3.4}$$

$$(x(t),r(t)) \to k(x(t),r(t)) \tag{3.5}$$

Let the value function $v(x(t),r(t))$ be defined by:

$$v(x(t),\alpha) = \min_{u(t)} J_u(x(t),\alpha), \text{when } r(t) = \alpha \tag{3.6}$$

Using the dynamic programming principle (see, Rishel (1975) or Boukas (1993)), and the expression of the value function, i.e.:

$$v(x(t),\alpha) = \min_{u(t)} \mathsf{E}\left[\int_0^\infty [g(x(s),r(s)) + h(u(s),r(s))]\, ds | x(t),\alpha\right] \tag{3.7}$$

it can be shown that the Hamilton-Jacobi-Bellman equation is given by the following:

$$\min_{u(t)} [(\mathcal{A}_u v)(x(t),r(t)) + g(x(t),r(t)) + h(u(t),r(t))] = 0 \tag{3.8}$$

where $(\mathcal{A}_u v)(x(t),r(t))$ is defined as follows:

$$\begin{aligned}(\mathcal{A}_u v)(x(t),r(t)) &= v_x^{\mathsf{T}}(x(t),r(t)) f(x(t),u(t),r(t)) \\ &\quad + \sum_{\beta \in \mathcal{S}} q_{\alpha\beta} v(x(t),\beta)\end{aligned} \tag{3.9}$$

In the next section we will return to the problem of section 2 and try to find the optimal control law of the feedback type of the nominal nonlinear system Markovian jumping parameters.

4 OPTIMALITY CONDITIONS OF THE NOMINAL NONLINEAR SYSTEM

The dynamics of the nominal nonlinear system of the one considered in section 2 is described by the following ordinary nonlinear differential equations:

$$\dot{x}(t) = A(x(t), r(t)) + B(x(t), r(t))u(t), x(0) = x_0, r(0) = r_0$$

with $A(0, \alpha) = 0$ for all $\alpha \in \mathcal{S}$

If we choose the functions $g(x(t), r(t))$ and $h(u(t), r(t))$ of the instantaneous cost as follows:

$$g(x(t), r(t)) = g_0(x(t), r(t)) + g_1(x(t), r(t)) \quad (4.1)$$

$$h(x(t), r(t)) = \frac{1}{2}u^\top(t)R(r(t))u(t) \quad (4.2)$$

with for all $\alpha \in \mathcal{S}$, $g_0(0, \alpha) = 0$ and $g_0(x(t), \alpha) > 0$ for all $x \neq 0$ and $g_1(0, \alpha) = 0$; and $R(\alpha)$ is positive-definite matrix.

We will assume that the following:

$$g_0(x(t), \alpha) \geq \rho(\alpha)||x(t)||^2, \quad (4.3)$$

hold for each $\alpha \in \mathcal{S}$.

Theorem 1 If the data of the previous nonlinear system is sufficiently smooth, therefore the optimality conditions of the nominal problem are given by:

$$\min_{u(t)}[(\mathcal{A}_u v)(x(t), r(t)) + g_0(x(t), r(t)) + g_1(x(t), r(t)) + \frac{1}{2}u^\top(t)R(r(t))u(t)] = 0 \quad (4.4)$$

where $(\mathcal{A}_u v)(x(t), r(t))$ is defined as follows:

$$(\mathcal{A}_u v)(x(t), r(t)) = v_x^\top(x(t), r(t))\,[A(x(t), r(t)) + B(x(t), r(t))u(t)] + \sum_{\beta \in \mathcal{S}} q_{\alpha\beta} v(x(t), \beta) \quad (4.5)$$

The corresponding optimal control is given by:

$$k(x(t), r(t)) = -R^{-1}(r(t))B^\top(r(t))v_x^\top(x(t), r(t)) \quad (4.6)$$

Proof: The proof of this theorem can be adapted from the one in Boukas (1993).

5 ROBUST STABILIZATION

Let us now return to the optimization problem we formulated in section 2 in which we have an additive nonlinear state dependent disturbance $d(x(t), r(t))$ to the changes in the control gain which in turn are assumed to be also nonlinear uncertainties. Our goal in this section is to design a stabilizing control law for the corresponding uncertain system (2.1)-(2.2).

Theorem 2 If for every $\alpha \in \mathcal{S}$, there exist a positive number $\gamma(\alpha)$ such that the following conditions

$$2p^{\top}(u(t))R(\alpha)u(t) - [1+\gamma(\alpha)]\, u^{\top}(t)R(\alpha)u(t) \geq 0, \tag{5.1}$$
$$2\gamma(\alpha)g_1(x(t),\alpha) - d^{\top}(x(t),\alpha)R(\alpha)d(x(t),\alpha) \geq 0, \tag{5.2}$$

hold. Then the optimal control law given by Eq. (4.6) is robustly stabilizes the uncertain system (2.1)-(2.2).

Proof: Let the value function be the candidate Lyapunov function. The infenitesimal operator of this function is given by:

$$\begin{aligned}\tilde{A}v(x,r(t)) &= v_x^{\top}(x,r(t))[A(x(t),\alpha) + B(x(t),r(t))d(x(t),r(t)) \\ &\quad + B(x(t),r(t))p(k(x(t),r(t))] + \sum_{\beta\in\mathcal{S}} q_{\alpha\beta}v(x(t),\beta)\end{aligned} \tag{5.3}$$

After some calculation, using the conditions above, we can prove that

$$\tilde{A}v(x,r(t)) \leq -\rho\|x(t)\|^2 \tag{5.4}$$

where $\rho > 0$.

The rest of the proof can be done as in Boukas and Yang (1997).

Remark The conditions (5.1)-(5.2) are conservative and the design approach requires the resolution of the optimization problem of the nominal nonlinear with Markovian jumping parameters.

The next theorem gives a less conservative conditions.

Theorem 3 If the condition (5.1) is satisfied and in some neighborhood $\mathcal{X}$ of $x = 0$, for $x \neq 0$ the following:

$$\begin{aligned}2k^{\top}(x(t),\alpha)R(\alpha)d(x(t),\alpha) &+ \gamma(\alpha)k^{\top}(x(t),\alpha)R(\alpha)k(x(t),\alpha) \\ &+ 2g_1(x(t),\alpha) \geq 0, \forall\alpha\in\mathcal{S}\end{aligned} \tag{5.5}$$

holds. Then $x = 0$ is asymptotically stochatically stable equilibrium point of system (2.1)-(2.2) under the control law given by (4.6).

Proof: The proof of this theorem follows the same steps of the one of Theorem 2. The detail is omitted.

Remark Notice that the design of the control law requires the resolution on the optimization problem of the nonlinear nominal system with Markovian jumping parameters which is in general not an easy task. A design procedure that overcome this will be preferable.

Theorem 4 If the condition (5.1) is satisfied and that there exist a positive definite function $v(x(t),r(t))$ and positive number $\gamma_1(r(t))$ and $\gamma_2(r(t))$, such that in a neighborhood $\mathcal{X}$ of $x = 0$, for $x \neq 0$ the following:

$$2\gamma_2(r(t))\,(1-\gamma_1(r(t))\,\Big[-v_x^{\top}(x(t),\alpha)A(x(t),r(t))$$

$$
\begin{aligned}
&-d^\top(x(t),r(t))R(r(t))d(x(t),r(t)) \\
&+\frac{1}{2}v_x^\top(x(t),r(t))B(x(t),r(t))R^{-1}(r(t))B^\top(x(t),r(t))v_x^\top(x(t),r(t))\Big] \\
&\geq 0
\end{aligned}
\tag{5.6}
$$

holds. Then under the feedback control law (4.6), $x = 0$ is asymptotically stochatically stable equilibrium point of system (2.1)-(2.2).

Proof: The proof of this theorem follows the same steps of the one of Theorem 2. The detail is omitted.

6 NUMERICAL EXAMPLES

Consider a production system consisting of one machine producing one item. Let the Markov process $r(t)$ has two modes, i.e., $\mathcal{S} = \{1, 2\}$, and let its dynamics be described by the following transition matrix:

$$
Q = \begin{bmatrix} -0.1 & 0.1 \\ 0.5 & -0.5 \end{bmatrix} \tag{6.1}
$$

- mode 1:

$$
\begin{aligned}
\dot{x}_1(t) &= -x_1(t) \\
\dot{x}_2(t) &= -x_2(t) + u - \frac{a_1(t) sin x_1(t)}{l_1}
\end{aligned}
$$

where l_1 is a positive constant and $a_1(t)$ is the uncertainty.

- mode 2:

$$
\begin{aligned}
\dot{x}_1(t) &= -x_1(t) \\
\dot{x}_2(t) &= -x_2(t) + u - \frac{a_2(t) sin x_1(t)}{l_2}
\end{aligned}
$$

where l_2 is a positive constant and $a_2(t)$ is the uncertainty.

Let

$$
R(r(t)) = \begin{bmatrix} 1 & 0 \\ 0 & 1 \end{bmatrix}
$$

for all $r(t) \in \mathcal{S}$. In this example, for both mode 1 and mode 2,

$$
\begin{aligned}
A &= \begin{bmatrix} -x_1(t) \\ -x_2(t) \end{bmatrix} \\
B &= \begin{bmatrix} 0 & 0 \\ 0 & 1 \end{bmatrix} \\
p(u(t)) &= \begin{bmatrix} 0 \\ u(t) \end{bmatrix}
\end{aligned}
$$

for mode 1,

$$d(x(t)) = \begin{bmatrix} 0 \\ -\frac{a_1(t) sin x_1(t)}{l_1} \end{bmatrix} \quad (6.2)$$

for mode 2,

$$d(x(t)) = \begin{bmatrix} 0 \\ -\frac{a_2(t) sin x_1(t)}{l_2} \end{bmatrix} \quad (6.3)$$

Assume that $g(x(t), r(t)) = g_0(x(t), r(t)) + g_1(x(t), r(t))$ satisfies that $g_0(x(t), r(t)) \geq \frac{1}{2}||x(t)||^2$ for all $r(t)$, and $g_1(x(t), r(t)) \geq L > 0$. If we assume that the uncertainties satisfy the following condition:

$$\frac{|a_1(t)a_2(t)|}{l_1 l_2} \leq L \quad (6.4)$$

Then conditions (5.1)-(5.2) of Theorem 2 are satisfied, and the system is robustly stabilizable.

References

[1] Boukas, E. K., (1993). "Control of systems with controlled jump Markov disturbances," Control Theory and Advanced Technology, **9**(2), 577–595.

[2] Boukas, E. K., (1995). "Robust stability of linear piecewise deterministic systems under matching conditions," Control Theory and Advanced Technology, **10**(4), 1541–1549.

[3] Boukas, E. K. and H. Yang, (1997). "Robust stabilization by dynamic combined state and output feedback compensator for nonlinear systems with jumps," Journal of Optimization Theory and Applications, **92**(1), 63–75.

[4] Gershwin, S. B. (1994) Manufacturing Systems Engineering, Prentice Hall.

[5] Ji, Y. and H. J. Chizeck, (1990). "Controllability, stabilizability, and continuous-time Markovian jump linear quadratic control," IEEE Trans. Automat. Control, **35**(7), 777–788.

[6] Mariton, M., (1990). "Jump linear systems in Automatic control," Marcel Dekker, New York.

[7] Rishel, R., (1975). "Control of systems with jump Markov disturbances," IEEE Trans. Automat. Control, **24**(4), 241–244.

[8] Sethi, S. P. and Q. Zhang, (1994). Hierarchical decision making in stochastic manufacturing systems. Birkhauser, Boston-Basel, Berlin.

LINEAR QUADRATIC OPTIMAL CONTROL: FROM DETERMINISTIC TO STOCHASTIC CASES

Shuping Chen*

Department of Applied Mathematics
Zhejiang University, Hangzhou 310027, China.

Xinhui Yu

Institute of Mathematics
Fudan University, Shanghai 200433, China

1 DETERMINISTIC LQ PROBLEM: A BRIEF HISTORICAL OVERVIEW

Given the following linear system

$$\dot{x}(t) = Ax(t) + Bu(t), \quad x(0) = x_0. \tag{1.1}$$

and the cost functional

$$J(x_0, u(\cdot)) = \langle Gx(T), x(T)\rangle + \int_0^T [\langle Qx(t), x(t)\rangle + \langle Ru(t), u(t)\rangle]\, dt \tag{1.2}$$

where $x_0 \in \mathbb{R}^n$, $x(t) \in \mathbb{R}^n$ and $u(t) \in \mathbb{R}^m$ for all $t \geq 0$, A, B, G, Q, R are matrices of correspondent dimension, and G, Q, R are symmetric. Our problem is:

Problem (DLQ). Minimize (1.2), subject to (1.1).

In the optimal control theory, the above problem is referred to as deterministic linear quadratic optimal control problem (DLQ problem, for short). Following are two definitions associated with it. Problem (DLQ) is said to be *finite* if

$$\forall x_0 \in \mathbb{R}^n, \quad V(x_0) \triangleq \inf_{u(\cdot)\in L^2(0,T;\mathbf{R}^m)} J(x_0, u(\cdot)) > -\infty.$$

*This work is supported in part by the NSFC, under grant 79790130.

Problem (DLQ) is said to be (*uniquely*) *solvable* if

$$\forall x_0 \in \mathbb{R}^n,\ \exists \bar{u}(\cdot) \in L^2(0,T;\mathbb{R}^m),\ \text{ s.t. } V(x_0) = J(x_0,\ \bar{u}(\cdot)).$$

LQ problem has roots going back to the very beginning of optimal control theory, the fundamental work was due to R. E. Kalman [5] in 1960. Since then, a lot of works have been done on DLQ problem, among them, we mention A. M. Letov [7], J. C. Willems [11], B. P. Molinari [10] and V. A. Yakubovic [13] etc. for finite dimensional cases, and I. Lasiecka and R. Triggiani [6], X. Li and J. Yong [8], S. Chen [2] and J. L. Lions [9] for infinite dimensional cases.

By now, it is very well known that

• If Problem (DLQ) is finite, by which, we mean that the infimum of the cost functional is finite, then $R \geq 0$.

• If $R \geq 0$, then Problem (DLQ) is (uniquely) solvable if and only if

$$\begin{cases} \dot{x}(t) = Ax(t) + Bu(t), & x(0) = x_0, \\ \dot{p}(t) = -Qx(t) - A^T p(t), & p(T) = Gx(T), \\ Ru(t) + B^T p(t) = 0, \end{cases} \tag{1.3}$$

admits a (unique) solution (x, u, p).

• If $R > 0$ and Riccati equation

$$\dot{P} = -(PA + A^T P + Q) + PB^T R^{-1} BP, \quad P(T) = G. \tag{1.4}$$

admits a solution $P(\cdot)$, then Problem (DLQ) is uniquely solvable with the optimal control

$$\bar{u}(t) = -R^{-1} B^T P \bar{x}(t). \tag{1.5}$$

• If $Q \geq 0$, $G \geq 0$, $R > 0$, then Problem (DLQ) is uniquely solvable.

2 STOCHASTIC LQ PROBLEM: A NEW PHENOMENON

Let $(\Omega,\ \mathcal{F},\ \mathcal{P},\ \{\mathcal{F}_t\}_{t\geq 0})$ be a complete filtered probability space on which defined a one dimensional standard Brownian motion $w(\cdot)$ such that $\{\mathcal{F}_t\}_{t\geq 0}$ is the natural filteration generated by $w(\cdot)$, augmented by all the $\mathcal{P}$-null sets in $\mathcal{F}$. Consider the following linear controlled stochastic diffenrential equation:

$$dx(t) = [Ax(t) + Bu(t)]dt + [Cx(t) + Du(t)]dw(t), \quad x(\tau) = \xi \tag{2.1}$$

and the cost functional

$$J(\tau, \xi; u(\cdot)) = E\Big\{\langle Gx(T), x(T)\rangle + \int_\tau^T \{\langle Qx(t), x(t)\rangle + \langle Ru(t), u(t)\rangle\}dt\Big\}, \tag{2.2}$$

where $\tau \in \mathcal{T}[0,T]$, the set of all $\{\mathcal{F}_t\}_{t\geq 0}$-stopping times taking values in $[0,T]$, $\xi \in \chi_\tau \triangleq L^2_{\mathcal{F}_\tau}(\Omega,\ \mathbb{R}^n)$, the set of all $\mathbb{R}^n$-valued $\mathcal{F}_\tau$-measurable square-integrable random variables; A, B, C, D are matrix-valued $\{\mathcal{F}_t\}_{t\geq 0}$-adapted bounded processes, Q, R are symmetric matrix-valued $\{\mathcal{F}_t\}_{t\geq 0}$-adapted bounded processes, G is symmetric matrix-valued $\mathcal{F}_T$-measurable bounded random variable, $u(\cdot)$ is a control process and $x(\cdot)$ is the corresponding state process. Our problem is:

Problem (SLQ). Minimize (2.2) subject to (2.1).
Here, our control process $u(\cdot)$ is taken from $\mathcal{U}[\tau,T] = L^2_{\mathcal{F}}(\tau,T;\mathbb{R}^m)$, the set of all $\mathbb{R}^m$-valued $\{\mathcal{F}_t\}_{t\geq 0}$-adapted square-integrable processes defined on the random interval $[\tau,T]$ with $\tau \in \mathcal{T}[0,T]$. Let $\Delta[0,T] = \cup_{r\in\mathcal{T}[0,T]}[\{r\}\times\chi_r]$, $V(\tau,\xi) = \inf_{u(\cdot)\in\mathcal{U}[\tau,T]} J(\tau,\xi,u(\cdot))$. Then one has

$$V : \Delta[0,t] \to \mathbb{R}.$$

Problem (SLQ) has recently been extensively studied in [3, 4], which is referred to as the stochastic linear quadratic optimal control problem (SLQ problem, for short) with random coefficient. Problem (SLQ) is said to be *finite* at $(\tau,\xi)\in\Delta[0,T]$, if

$$V(\tau,\xi) > -\infty$$

Problem (SLQ) is said to be *solvable* at $(\tau,\xi)\in\Delta[0,T]$, if there exists a $\bar{u}(\cdot)$, such that

$$J(\tau,\xi;\bar{u}(\cdot)) = V(\tau,\xi).$$

Comparing with its deterministic counterpart, Problem (SLQ) have some new interesting features. Firstly, in the deterministic case, $R\geq 0$ is a necessary condition for the finiteness; but in the stochastic case, $R\geq 0$ is no longer necessary for the finiteness. Example 2.1 shows that SLQ problem with $R<0$ may uniquely solvable. Secondly, $R>0$ can assure the unique solvability of Problem (DLQ), but can't even assure the finiteness of Problem (SLQ) sometimes. Example 2.2 shows this fact.

Example 2.1 Consider:

$$\dot{x}(t) = u(t), \quad x(0) = x_0 \in \mathbb{R},$$

with the cost functional

$$J(u(\cdot)) = x(T)^2 - \int_0^T u(t)^2 dt.$$

Thus, $R=-1<0$.

$$\inf_{u(\cdot)} J(u(\cdot)) = -\infty.$$

Consider the stochastic version:

$$dx(t) = u(t)dt + \delta u(t)dw(t), \quad x(0) = x_0,$$

with the cost functional

$$J(u(\cdot)) = E\Big\{x(T)^2 - \int_0^T u(t)^2 dt\Big\}.$$

It can be proved that, If $|\delta| > 0$ and

$$\delta^2(2\ln|\delta| - 1) > T - 1,$$

then Problem (SLQ) is solvable on $[0, T]$.

Example 2.2 Consider: $(0 < T < 1)$

$$\dot{x}(t) = u(t), \quad x(0) = x_0 \in \mathbb{R}.$$

with the cost functional

$$J(u(\cdot)) = -x(T)^2 + \int_0^T u(t)^2 dt.$$

Thus, $R = 1 > 0$. We can show that Problem (DLQ) is uniquely solvable with optimal control:

$$\bar{u}(t) = \frac{x(t)}{t + 1 - T}.$$

Now, consider the stochastic version:

$$dx(t) = u(t)dt + \delta u(t)dw(t), \quad x(0) = x_0.$$

with the cost functional

$$J(u(\cdot)) = E\Big\{ - x(T)^2 + \int_0^T u(t)^2 dt\Big\}.$$

It can be proved: If $|\delta| > 1$, then Problem (SLQ) is not finite.

Then one will ask: *what's the necessary condition for the finiteness of Problem (SLQ)?* The following theorem answer this.

Theorem 2.1 ([4]) *If Problem (SLQ) is finite at some $(\tau, \xi) \in \Delta[0, T]$, then*

$$R(T) + D^T(T)GD(T) \geq 0, \quad \text{a.s. } \omega \in \Omega. \tag{2.3}$$

From Theorem 2.1 and the arguments above, we can easy see that D plays a very interesting role in the finiteness and solvability of problem (SLQ). When A, B, C, D, G, Q, R are matrices, (2.3) becomes

$$R + D^T GD \geq 0. \tag{2.4}$$

In this case, if $D = 0$, i.e., control does not appear in the diffusion, (2.4) becomes $R \geq 0$, the same as the DLQ problem. In Example 2.2,

$$R + D^T GD = 1 - \delta^2.$$

Thus, $\delta > 1$ violates the necessary condition.

3 STOCHASTIC MAXIMUM PRINCIPLE AND STOCHASTIC RICCATI EQUATION

The following forward-backward stochastic differential equation (FBSDE, for short)

$$\begin{cases} dx(t) = [Ax(t) + Bu(t)]dt + [Cx(t) + Du(t)]dw(t), \\ dp(t) = -[A^T p(t) + C^T q(t) + Qx(t)]dt + q(t)dw(t), \\ x(\tau) = \xi, \qquad p(T) = Gx(T). \end{cases} \tag{3.1}$$

will play a very important role in the *stochastic maximun principle*:

Theorem 3.1 ([4]) *Problem (SLQ) is solvable at $(\tau, \xi) \in \Delta[0, T]$, with an optimal pair $(\bar{x}(\cdot), \bar{u}(\cdot))$ if and only if the following forward-backward stochastic differential equation (FBSDE)*

$$\begin{cases} d\bar{x}(t) = [A\bar{x}(t) + B\bar{u}(t)]dt + [C\bar{x}(t) + D\bar{u}(t)]dw(t), \\ d\bar{p}(t) = -[A^T \bar{p}(t) + C^T \bar{q}(t) + Q\bar{x}(t)]dt + \bar{q}(t)dw(t), \\ \bar{x}(\tau) = \xi, \qquad \bar{p}(T) = G\bar{x}(T). \end{cases} \tag{3.2}$$

admits an adapted solution $(\bar{x}(\cdot), \bar{p}(\cdot), \bar{q}(\cdot))$ such that

$$I_{\Omega(\tau,\xi)} \Big[R\bar{u}(t) + B^T \bar{p}(t) + D^T \bar{q}(t) \Big] = 0, \quad in\ L^2_{\mathcal{F}}(\tau, T; \mathbb{R}^m), \tag{3.3}$$

and for any $u(\cdot) \in \mathcal{U}[\tau, T]$, the unique adapted solution $(x(\cdot), p(\cdot), q(\cdot))$ of (3.1) with $\xi = 0$ satisfies

$$E \int_\tau^T \langle Ru(t) + B^T p(t) + D^T q(t),\, u(t) \rangle\, dt \geq 0. \tag{3.4}$$

Another important thing associated with the solvability of Problem (SLQ) is the following *stochastic backward Riccati equation* for symmetric matrix-valued process $(P(\cdot), \Lambda(\cdot))$ in a random time interval:

$$\begin{cases} dP = -\{PA + A^T P + C^T PC + \Lambda C + C^T \Lambda + Q \\ \qquad -(PB^T + C^T PD + \Lambda D)(R + D^T PD)^{-1} \\ \qquad\qquad \cdot(B^T P + D^T PC + D^T \Lambda)\}dt + \Lambda dw(t), \quad t \in [\tau, T], \\ P(T) = G. \\ \det[R(t) + D(t)^T P(t) D(t)] \neq 0, \quad t \in [\tau, T], \text{ a.s. } \omega \in \Omega, \end{cases} \tag{3.5}$$

where $\tau \in \mathcal{T}[0.T]$.

Theorem 3.2 ([3, 4]) *If (3.5) admits an adapted solution (P, Λ), then Problem (SLQ) is uniquely solvable at (τ, ξ), with the optimal control*

$$u(t) = -(R + D^T PD)^{-1}(B^T P + D^T PC + D^T \Lambda)x(t).$$

4 SOLVABILITY OF RICCATI EQUATION WITH DETERMINISTIC COEFFICIENT

Now, we consider Problem (SLQ) with deterministic coefficient, i.e., $\tau \in [0, T]$, $\xi \in \mathbb{R}^n$, $A, C, Q, G \in \mathbb{R}^{n\times n}$, $B, D \in \mathbb{R}^{n\times m}$, $R \in \mathbb{R}^{m\times m}$. In this case, the stochastic maximum principle is as following

Theorem 4.1 ([4]) *Let Problem (SLQ) be uniquely solvable. Then (3.1) admits a solution satisfies*

$$Ru(t) + B^T p(t) + D^T q(t) = 0. \tag{4.1}$$

Conversely, if (3.1)–(3.2) admits a (unique) solution $(\bar{x}, \bar{u}, \bar{p}, \bar{q})$, and any adapted solution (x, u, p, q) of (3.1) with $\xi = 0$ satisfies

$$E \int_0^T \langle Ru(t) + B^T p(t) + D^T q(t), u(t)\rangle dt \geq 0, \tag{4.2}$$

then Problem (SDLQ) is (uniquely) solvable; in addition, if R^{-1} exists, then

$$\bar{u}(t) = -R^{-1}[B^T \bar{p}(t) + D^T \bar{q}(t)], \tag{4.3}$$

is the optimal control.

The stochastic backward Riccati equation (3.5) becomes:

$$\begin{cases} P = -(PA + A^T P + C^T PC + Q) \\ \qquad +(PB^T + C^T PD)(R + D^T PD)^{-1}(B^T P + D^T PC), \quad t \in [\tau, T], \\ P(T) = G. \end{cases} \tag{4.4}$$

For Riccati equation (4.4), we have (see [3, 4])

(1) If $R > 0, Q \geq 0, G \geq 0$, then (4.4) is globally solvable.

(2) If $C = 0$. Then Riccati equation (4.4) is globally solvable if and only if there exists $K^{\pm}$, such that

$$K^+ \geq R + D^T \Psi(K^+) \geq R + D^T \Psi(K^-) D \geq K^-, \tag{4.5}$$

where $\Psi(K)$ is the solution of

$$\begin{cases} \dot{P} = -(PA + A^T P + C^T PC + Q) \\ \qquad +(PB^T + C^T PD)K^{-1}(B^T P + D^T PC), \\ P(T) = G. \end{cases} \tag{4.6}$$

References

[1] A. Bensoussan, *Lecture on Stochastic Control, part 1,* Lecture Notes in Math., 972(1983), 1-39.

[2] S. Chen, *Riccai Equations Arising in Infinite Dimensional Optimal Control Problems,* Control Theory Appl. , 2(1985), 64-72.

[3] S. Chen, X. Li and X. Y. Zhou, *Stochastic Linear Quadratic Regulators with Indefinite Control Weight Costs*, SIAM J. Control and Optim., toappear.

[4] S. Chen and J. Yong, *Stochastic Linear Quadratic Optimal Control Problems, I,II,* preprint.

[5] R. E. Kalman, *Contributions to the Theory of Optimal Control*, Bol. Soc. Mat. Mex. , 5(1960), 102–199.

[6] I. Lasiecka and R.Triggiani, *Differential and Algebraic Riccati Equations with Applications to Boundary/Point Control Problems: Continuous Theory and Approximation Theory,* Lecture Notes in Control & Inform. Sci., Vol. 164, Springer-Verlag, Berlin, 1991.

[7] A. M. Letov, *Analytical Design of Regulators,* Avtomat. i Telemekh. 4(1960), 436-446; 5(1960), 561-571; 6(1960), 661-669. English transl. in Automat. Remote Control, 21(1960).

[8] X. Li and J.Yong, *Optimal Control Theory for Infinite Dimensional System,* Birkhäuser, Boston, 1995.

[9] J. L. Lions, *Optimal Control of Systems Governed by Partial Differential Equations,* Springer-Verlag, New York, 1971.

[10] B. P. Molinari, *The Time-invariant Linear-Quadratic Optimal Control Problem,* Automatica, 13(1977), 347-357.

[11] J. C. Willems, *Least Squares Stationary Optimal Control and the Algebraic Riccati Equation*, IEEE Trans. Auto. Control, 16(1971), 621-634.

[12] W. M. Wonham, *On A Matrix Riccati Equation of Stochastic Control,* SIAM J. Control, 6(1968), 312-326.

[13] V. A. Yakubovic, *Singular Stationary Nonhomogeneous Linear Quadratic Optimal Control,* Amer. Math. Transl., 155(1993), 129-167.

OPTIMAL PORTFOLIO SELECTION WITH TRANSACTION COSTS *

Phillip Collings and Ulrich G. Haussmann

Department of Mathematics, University of British Columbia
1984 Mathematics Road, Vancouver, B.C., Canada, V6T 1Z2.

uhaus@math.ubc.ca

1 INTRODUCTION

Consider an investor who has the following instruments available to him: a bank account paying a fixed rate of interest r and n risky assets ("stocks") whose prices are modeled as geometric Brownian motions. The investor is allowed to consume at a rate $c(t)$ from the bank account and is subject to the constraint that he remain solvent at all times. Any trading in the stocks must be self-financing, and incurs a transaction cost which is proportional to the amount being traded. The investor's objective is to maximize his expected discounted utility of lifetime consumption.

Similar problems were first studied by Constantinides [2] and Magill and Constantinides [9] for $n = 1$, and solved by Davis and Norman [3]. Technically speaking, the optimal process is a reflecting diffusion inside the no trade region (a wedge) and the buying and selling strategies $L_1(t)$ and $M_1(t)$, cf below, are local times at the boundaries.

Akian, Menaldi and Sulem [1] consider the multi-asset case described above ($n > 1$), assuming that the noise terms are uncorrelated. Using dynamic programming methods, the value function is shown to be the unique viscosity solution of a variational inequality. This inequality is then discretized and solved numerically to provide the optimal strategy and indicate the shape of the trading boundaries, but convergence is not proved.

We present a different numerical scheme based on some ideas of Kushner and Martin [8], Kushner and Dupuis [7] and Fleming and Fitzpatrick [4], and we show convergence of the value functions.

Section 2 deals with the formulation and theoretical study of the portfolio selection problem. We introduce the model employed by Akian et al. [1], but

*This work was supported by the Natural Sciences and Engineering Research Council of Canada under Grant 88051.

with correlated noise, and we formulate the control problem associated with the model. The analytical methods of [1] carry over without much difficulty to show that the value function is the unique viscosity solution of a variational inequality. Then we transform the singular control problem into a new problem involving only absolutely continuous controls, using a random time change. It is shown that the value functions for the two problems are closely related and that the value function for the transformed problem is the unique viscosity solution of the corresponding Hamilton-Jacobi-Bellman (H.J.B) equation.

The Markov chain approximation to the transformed problem is constructed in Section 3, and convergence of the the value functions of the transformed problem is established as the discretization parameter goes to zero. Although the discretization method follows that of Kushner et al, [8], [7], the convergence arguments do not. They rely on the viscosity solution techniques used by Fitzpatrick and Fleming, [4].

Details will be given in the forthcoming thesis of the first author.

2 PROBLEM FORMULATION AND ANALYSIS

Let $(\Omega, \mathcal{F}, P)$ be a probability space with a given filtration $(\mathcal{F}_t)_{t\geq 0}$. We let the processes $S_0(t)$ and $S_i(t)$ be the amount of money invested in the bank account and the ith risky asset at time t respectively. Then

$$
\begin{aligned}
dS_0(t) &= [rS_0(t) - c(t)]dt + \sum_{i=1}^{n}[-(1+\lambda_i)dL_i(t) + (1-\mu_i)dM_i(t)], \\
dS_i(t) &= b_i S_i(t)dt + \sum_{j=1}^{n} \sigma_{ij} S_i(t) dW_j(t) + dL_i(t) - dM_i(t), \qquad (2.1) \\
S_i(0) &= x_i, \quad i = 0, \cdots, n
\end{aligned}
$$

where $W_j(t)$, $j = 1, \cdots, n$, are independent Brownian motions and $L_i(t)$ and $M_i(t)$ represent the cumulative purchase and sale of the ith risky asset on $[0,t]$ respectively. The constants r and b_i represent the interest rate and the mean return for stock i respectively and $\sigma \equiv (\sigma_{i,j})_{i,j=1,\ldots,n}$ is the volatility matrix which measures the level of "noisiness" of the stock price processes. The coefficients λ_i and μ_i are the proportionality constants of the transaction costs associated with stock i, so that purchasing 1 dollar's worth of asset i will cost $(1+\lambda_i)$ dollars, which is transferred from the bank account. Conversely selling 1 dollar's worth of stock i will result in a payment of $(1-\mu_i)$ dollars into the bank account.

Definition 2.1 *A policy for investment and consumption is a vector* $(c(t), (L_i(t), M_i(t))_{i=1,\cdots,n})$ *of measurable adapted processes such that*

1. $c(t) \geq 0$ *and* $\int_0^t c(\theta)\, d\theta < \infty$ *for a.e* (t, ω),

2. $L_i(t)$ *and* $M_i(t)$ *are non decreasing, left continuous processes with right limits and* $L_i(0) = M_i(0) = 0$.

For a policy $\mathcal{P} := (c, (L_i, M_i)_{i=1,\cdots,n})$, the corresponding processes $S_0(t)$ and $S_i(t)$ are left continuous with right hand limits and (2.1) is equivalent to

$$\begin{aligned} S_0(t) &= x_0 + \int_0^t [rS_0(\theta) - c(\theta)]d\theta + \sum_{i=1}^n [-(1+\lambda_i)L_i(t) + (1-\mu_i)M_i(t)], \\ S_i(t) &= x_i + \int_0^t b_i S_i(\theta)d\theta + \sum_{j=1}^n \int_0^t \sigma_{ij} S_i(\theta) dW_j(\theta) + L_i(t) - M_i(t), \quad (2.2) \end{aligned}$$

for $t \geq 0$. We define the solvency region to be

$$\mathcal{S} = \{x = (x_0, x_1, \cdots, x_n) \, : \, \mathcal{W}(x) > 0\} \tag{2.3}$$

where

$$\mathcal{W}(x) = x_0 + \sum_{i=1}^n \min\{(1-\mu_i)x_i, (1+\lambda_i)x_i)\}. \tag{2.4}$$

Note that $\mathcal{W}(x)$ represents the the net wealth, that is the amount of money in the bank account after the investor has liquidated all her assets.

Definition 2.2 *A policy $\mathcal{P} \equiv (c(t), (L_i(t), M_i(t))_{i=1,\cdots,n})$ is admissible if $S(t) \in \overline{\mathcal{S}}$ a.s. $t \geq 0$, i.e.*

$$\mathcal{W}(S(t)) \geq 0 \qquad a.s. \ t \geq 0.$$

Denote by $\mathcal{U}(x)$ the set of all admissible polices. Observe that when $S(t) \in \partial\mathcal{S}$, the only admissible action is to trade to zero, i.e. $S(t+) = 0$ with no consumption. Thereafter $S = 0$. For $x \in \overline{\mathcal{S}}$ and $\mathcal{P} \in \mathcal{U}(x)$, define

$$J(x, \mathcal{P}) = E_x \int_0^\infty e^{-\delta t} \frac{c(t)^\gamma}{\gamma} dt, \tag{2.5}$$

where E_x is the expectation operator given an initial endowment x, δ is a positive discount factor and $0 < \gamma < 1$. Here (2.5) represents the investor's expected discounted utility of lifetime consumption. The problem is to find

$$V(x) := \sup_{\mathcal{P} \in \mathcal{U}(x)} J(x, \mathcal{P}). \tag{2.6}$$

We now make the following assumptions:

- [A.1] σ is invertible and

$$\delta > \gamma\left(r + \frac{1}{2(1-\gamma)}(\mathbf{b} - r\mathbf{1})^T \mathbf{a}^{-1}(\mathbf{b} - r\mathbf{1})\right),$$

  where $\mathbf{b} = (b_1, \cdots, b_n)^T$, $\mathbf{1} = (1, \cdots, 1)^T$ and $\mathbf{a} = \sigma\sigma^T$.

- [A.2] $\mu_i,\ \lambda_i \in [0,1),\ \lambda_i + \mu_i > 0 \quad \forall i = 1, \cdots, n.$

Remark 2.1 *Assumption [A.1] reduces to the one stated in Akian, Menaldi and Sulem [1] when* **a** *is diagonal. When transaction costs are equal to zero (hence here also), assumption [A.1] implies that the value function for the problem is finite, cf. Karatzas, Lechoczky, Sethi and Shreve [6].*

Let

$$\mathcal{G} = \left\{ f \in C(\mathbb{R}^n) \,:\, \sup_x \frac{|f(x|}{1 + ||x||^\gamma} < \infty \right\}.$$

The following result can be proved as in [1].

Theorem 2.1 *Suppose that assumptions [A.1] and [A.2] hold. Then (i) the value function V is concave and γ-Hölder continuous on $\overline{\mathcal{S}}$, and (ii) V is the unique viscosity solution in $\mathcal{G}$ of the variational inequality*

$$\max\left\{ AV + G\left(\frac{\partial V}{\partial x_0}\right),\ \max_{1 \le i \le n} \mathcal{L}_i V,\ \max_{1 \le i \le n} \mathcal{M}_i V \right\} = 0 \quad \text{in } \mathcal{S}$$
$$V = 0 \quad \text{in } \partial\mathcal{S}, \tag{2.7}$$

where

$$AV = \frac{1}{2}\sum_{i=1}^{n}\sum_{i=1}^{n} a_{ij} x_i x_j \frac{\partial^2 V}{\partial x_i x_j} + \sum_{i=1}^{n} b_i x_i \frac{\partial V}{\partial x_i} + r x_0 \frac{\partial V}{\partial x_0} - \delta V, \tag{2.8}$$

$$\mathcal{L}_i V = -(1+\lambda_i)\frac{\partial V}{\partial x_0} + \frac{\partial V}{\partial x_i}, \tag{2.9}$$

$$\mathcal{M}_i V = (1-\mu_i)\frac{\partial V}{\partial x_0} - \frac{\partial V}{\partial x_i}, \tag{2.10}$$

$$G(p) = \max_{c \ge 0}(-cp + \frac{c^\gamma}{\gamma})$$
$$= (\frac{1}{\gamma} - 1)p^{\gamma/(\gamma-1)}, \tag{2.11}$$

and a_{ij} is the $(i,j)^{th}$ entry of the matrix **a**.

Using a time change (stretching time when trades occur) and adding another state and control variable, we can transform the above singular problem into a problem involving only absolutely continuous controls, cf. Kushner and Martin

[8], Kushner and Dupuis [7] and Haussmann and Suo [5]. Let

$$\hat{S}_0(t) = x_0 + \int_0^t \Big\{ [r\hat{S}_0(\theta) - \hat{c}(\theta)]\hat{u}(\theta)$$
$$+ \sum_{i=1}^{n} [-(1+\lambda_i)\hat{l}_i(\theta) + (1-\mu_i)\hat{m}_i(\theta)](1-\hat{u}(\theta)) \Big\} d\theta$$
$$\hat{S}_i(t) = x_i + \int_0^t \Big\{ b_i \hat{S}_i(\theta)\hat{u}(\theta) + (\hat{l}_i(\theta) - \hat{m}_i(\theta))(1-\hat{u}(\theta)) \Big\} d\theta$$
$$+ \sum_{j=1}^{n} \int_0^t \sigma_{ij} \hat{S}_i(\theta)\sqrt{\hat{u}(\theta)}\, d\hat{W}_j(\theta), \quad i = 1, \cdots, n, \tag{2.12}$$
$$\hat{S}_{n+1}(t) = x_{n+1} + \int_0^t \hat{u}(\theta) d\theta.$$

The control variable for this problem is given by a vector of measurable, adapted processes, $\hat{\mathcal{P}} \equiv (\hat{c}(t), (\hat{l}_i(t), \hat{m}_i(t))_{i=1,\cdots,n}, \hat{u}(t))$, taking values in $\mathbb{R}_+ \times \mathcal{Z} \times [0,1]$, where $\mathcal{Z} = \{z \in \mathbb{R}_+^{2n} : \sum_{i=1}^{2n} z_i = 1\}$. Note that $\hat{l}_i$ ($\hat{m}_i$) is the rate at which stock i is bought (sold), and $\hat{u} = 0$ when trading occurs. Define

$$\hat{\mathcal{S}} = \mathcal{S} \times \mathbb{R},$$

where $\mathcal{S}$ is defined by (2.3).

Definition 2.3 *A control $\hat{\mathcal{P}} \equiv (\hat{c}(\cdot), (\hat{l}_i(\cdot), \hat{m}_i(\cdot))_{i=1,\cdots,n}, \hat{u}(\cdot))$ is said to be admissible for the transformed problem if*

$$\hat{S}(t) \in \overline{\hat{\mathcal{S}}} \quad a.s.\ t \geq 0.$$

Let $\hat{\mathcal{U}}(x, x_{n+1})$ be the set of admissible polices for the transformed problem. Observe that it is independent of x_{n+1}, and as before, if $\hat{S}(t) \in \partial\hat{\mathcal{S}}$, then the only admissible action is to take $\hat{u} = 0$ and to trade (continuously) to zero. Given the initial endowment $\hat{x} := (x, x_{n+1}) \in \hat{\mathcal{S}}$ and $\hat{\mathcal{P}} \in \hat{\mathcal{U}}(\hat{x})$, define

$$\hat{J}(\hat{x}, \hat{\mathcal{P}}) = \hat{E}_{\hat{x}} \int_0^\infty e^{-\delta \hat{S}_{n+1}(t)} \frac{\hat{c}(t)^\gamma}{\gamma} \hat{u}(t) dt, \tag{2.13}$$

where $\hat{E}_{\hat{x}}$ is the expectation operator with respect to the new probability measure $\hat{P}$ given the initial endowment $\hat{x}$. The transformed problem is to find

$$\hat{V}(\hat{x}) := \sup_{\hat{\mathcal{P}} \in \hat{\mathcal{U}}(\hat{x})} \hat{J}(\hat{x}, \hat{\mathcal{P}}). \tag{2.14}$$

It can be shown as in [5] that the two control problems are equivalent, that is

Proposition 2.1

$$\hat{V}(\hat{x}) = e^{-\delta x_{n+1}} V(x) \equiv e^{-\delta x_{n+1}} V(x_0, \cdots, x_n).$$

The Hamilton-Jacobi-Bellman equation corresponding to problem (2.14) is

$$\max_{(\hat{l},\hat{m},\hat{u})\in\mathcal{Z}\times[0,1]}\left\{\hat{u}\hat{A}\hat{V}(x)+\hat{u}\exp\left(\frac{\delta x_{n+1}}{\gamma-1}\right)G\left(\frac{\partial\hat{V}(x)}{\partial x_0}\right)\right.$$
$$\left.+(1-\hat{u})\sum_{i=1}^{n}[(\mathcal{L}_i\hat{V}(x))\hat{l}_i+(\mathcal{M}_i\hat{V}(x))\hat{m}_i]\right\}=0 \quad x\in\hat{\mathcal{S}} \qquad (2.15)$$
$$\hat{V}(x)=0 \quad x\in\partial\hat{\mathcal{S}}$$

where

$$\hat{A}\hat{V}=\frac{1}{2}\sum_{i=1}^{n}\sum_{j=1}^{n}a_{ij}x_ix_j\frac{\partial^2\hat{V}}{\partial x_ix_j}+\sum_{i=1}^{n}b_ix_i\frac{\partial\hat{V}}{\partial x_i}+rx_0\frac{\partial\hat{V}}{\partial x_0}+\frac{\partial\hat{V}}{\partial x_{n+1}}. \qquad (2.16)$$

Let $\hat{\mathcal{G}}:=\{e^{-\delta x_{n+1}}f\ :\ f\in\mathcal{G}\}\subset C(\mathbb{R}^{n+1})$. Then we have

Theorem 2.2 *Assume [A.1], [A.2]. The value function $\hat{V}$ is the unique viscosity solution in $\hat{\mathcal{G}}$ of (2.15).*

3 MARKOV CHAIN APPROXIMATION

Let $h>0$ be an approximation parameter and define Σ^h to be the lattice

$$\Sigma^h=\{\hat{x}\in\mathbb{R}^{n+2}\ :\ \hat{x}=h\sum_{j=0}^{n+1}e_jK_j,\quad K_j=0,\pm1,\pm2,\cdots,\ j=0,\cdots,n,$$
$$K_{n+1}=0,1,\cdots\},$$

and $\Sigma^h_N=\{\hat{x}\in\Sigma^h\ :\ x_i\le Nh\}$,where N is some positive integer which will be chosen so that $Nh\to\infty$ as $h\to0$. Define the set $\hat{\mathcal{S}}^h_N$, the boundary $\partial\hat{\mathcal{S}}^h_N$ and the outer boundary $\partial\hat{\mathcal{S}}^{h+}_N$ to be

$$\begin{aligned}\hat{\mathcal{S}}^h_N &= \hat{\mathcal{S}}\cap\Sigma^h_N\\ \partial\hat{\mathcal{S}}^h_N &= \Sigma^h_N\cap([\partial\hat{\mathcal{S}}\pm e_ih\pm e_jh]\setminus\hat{\mathcal{S}}\quad i=0,\cdots,n).\\ \partial\hat{\mathcal{S}}^{h+}_N &= \{x\in\hat{\mathcal{S}}^h_{N+1}\ :\ x_i=(N+1)h\quad\text{for at least one } i=0,\cdots,n+1\}.\end{aligned}$$

We will now construct a Markov chain defined on $\hat{\mathcal{H}}^h_N:=\hat{\mathcal{S}}^h_N\cup\partial\hat{\mathcal{S}}^h_N\cup\partial\hat{\mathcal{S}}^{h+}_N$ whose behaviour "closely matches" that of the continuous process $\hat{S}$.

Define $\hat{U}^h_N:=[0,KNh]\times\mathcal{Z}\times[0,1]$ where K is an artificial bound, which will disappear in the limit as $h\to0$ and $Nh\to\infty$. For given $\hat{S}^h(0)$, define the Markov chain $\{\hat{S}^h(k):k=0,1,\cdots\}$ recursively by

$$\hat{S}^h(k+1)=\hat{S}^h(k)+\hat{W}^h(k)\quad k=0,1,\cdots,$$

where, for each k, given state $\hat{S}^h(k)=\hat{x}=(x_0,\cdots,x_{n+1})\in\hat{\mathcal{H}}^h_N$ and control $\alpha:=(\hat{c},\hat{l},\hat{m},\hat{u})\in\hat{U}^h_N$, the (conditional) distribution of $\hat{W}^h(k)$ is denoted by $\hat{P}(\cdot\,|\,\hat{x},\alpha)$, and is defined as follows.

For $\hat{x} \in \hat{\mathcal{S}}_N^h$,

$$\hat{P}^h(e_i h \mid \hat{x}, \alpha) = \frac{\hat{u} a_{ii} x_i^2/2 - \hat{u} \sum_{j \neq i} |a_{ij} x_i x_j|/2 + h(b_i x_i^+ \hat{u} + (1-\hat{u})\hat{l}_i)}{Q^h(\hat{x})},$$

$$\hat{P}^h(-e_i h \mid \hat{x}, \alpha) = \frac{\hat{u} a_{ii} x_i^2/2 - \hat{u} \sum_{j \neq i} |a_{ij} x_i x_j|/2 + h(b_i x_i^- \hat{u} + (1-\hat{u})\hat{m}_i)}{Q^h(\hat{x})},$$

$$\hat{P}^h(e_0 h \mid \hat{x}, \alpha) = \frac{h(r x_0^+ \hat{u} + (1-\hat{u}) \sum_{i=1}^n (1-\mu_i)\hat{m}_i)}{Q^h(x)},$$

$$\hat{P}^h(-e_0 h \mid \hat{x}, \alpha) = \frac{h(r x_0^- \hat{u} + (1-\hat{u}) \sum_{i=1}^n (1+\lambda_i)\hat{l}_i + \hat{c}\hat{u})}{Q^h(\hat{x})},$$

$$\hat{P}^h(e_{n+1} h \mid \hat{x}, \alpha) = \frac{h\hat{u}}{Q^h(x)}, \quad i = 1, \cdots, n,$$

$$\hat{P}^h(e_i h \pm e_j h \mid \hat{x}, \alpha) = \frac{\hat{u}(a_{ij} x_i x_j)^{\pm}/2}{Q^h(\hat{x})} \quad i, j = 1, \cdots, n \quad i \neq j,$$

$$\hat{P}^h(-e_i h \mp e_j h \mid \hat{x}, \alpha) = \frac{\hat{u}(a_{ij} x_i x_j)^{\pm}/2}{Q^h(\hat{x})} \quad i, j = 1, \cdots, n \quad i \neq j,$$

$$\hat{P}(0 \mid \hat{x}, \alpha) = 1 - \sum_{w \neq 0} \hat{P}(w \mid \hat{x}, \alpha). \tag{3.1}$$

The normalizing constant $Q^h(\hat{x})$ is taken so that for all $\alpha \in \hat{U}_N^h$

$$Q^h(\hat{x}) \geq \left[\sum_{i=1}^n a_{ii} x_i^2 \hat{u} - \sum_i \sum_j |a_{ij} x_i x_j| \hat{u}/2 + h \left\{ \sum_{i=1}^n [b_i |x_i| \hat{u} + (\hat{l}_i + \hat{m}_i)(1-\hat{u})] \right.\right.$$
$$\left.\left. + r|x_0|\hat{u} + \sum_{i=1}^n [(1+\lambda_i)\hat{l}_i + (1-\mu_i)\hat{m}_i](1-\hat{u}) + \hat{c}\hat{u} + \hat{u} \right\} \right].$$

For $\hat{x} \in \partial\hat{\mathcal{S}}_N^h$,

$$\hat{P}(-\hat{x} \mid \hat{x}, \alpha) = 1. \tag{3.2}$$

Hence $\partial\hat{\mathcal{S}}_N^h$ is an absorbing set for the chain, in fact $\{0\}$ is an absorbing state. This ensures that we obtain the correct boundary value for our approximation.

For $\hat{x} \in \partial\hat{\mathcal{S}}_N^{h+}$,

$$\hat{P}(\hat{z} - \hat{x} \mid \hat{x}, \alpha) = 1, \tag{3.3}$$

where $\hat{z}$ is the point in $\hat{\mathcal{S}}_N^h$ nearest to $\hat{x}$. This makes $\partial\hat{\mathcal{S}}_N^{h+}$ reflecting.

We need to ensure that the above expressions are transition probabilities, i.e. we want

$$\hat{u} \left[a_{ii} x_i^2 - \sum_{j: j \neq i} |a_{ij} x_i x_j| \right] \geq 0, \; i = 1, \cdots, n, \; x \in \mathcal{S}.$$

To this end, define $\rho_i = a_{ii} / \sum_{j:j\neq i} |a_{ij}|$, $i = 1, \cdots, n$, and let

$$\mathcal{K} := \{x \in \mathbb{R}^{n+2} : \rho_i^{-1} \leq \left|\frac{x_i}{x_j}\right| \leq \rho_j,\ i,j = 1, \cdots, n\}.$$

Here the ρ's may assume the value ∞. Now restrict $\hat{u}(\hat{x}) = 0$ for $\hat{x} \notin \mathcal{K}$. We are in fact forcing the system to trade into $\mathcal{K}$. From a heuristic analysis, cf [1], we expect that the no trade region is a cone in the first orthant. Our method requires that this cone be contained in $\mathcal{K}$. In practice, one proceeds under this assumption, and then checks at the end whether the boundary of $\mathcal{K}$ intersects the no trade region. If not, then all is well, but if it does, then another method must be found. We formalize this assumption as

- [A.3] $\mathcal{K}$ contains the no trade region.

Of course, if a is diagonal then [A.3] holds.

Definition 3.1 *A control* $\hat{\mathcal{P}}^h = (\hat{\mathcal{P}}^h(k))_{k=0}^{\infty} := ((\hat{c}(k), \hat{l}(k), \hat{m}(k), \hat{u}(k)))_{k=0}^{\infty}$ *is said to be admissible for the Markov chain if*

$$\hat{S}(k) \in \hat{\mathcal{H}}_N^h \quad a.s.\ for\ k = 0, 1, \cdots$$

The set of admissible controls is denoted by $\hat{\mathcal{U}}_N^h$.

For $\hat{x} \in \hat{\mathcal{H}}_N^h$ and $\hat{\mathcal{P}}^h \in \hat{\mathcal{U}}_N^h$, define

$$\hat{J}^h(\hat{x}, \hat{\mathcal{P}}^h) = \hat{E}_{\hat{x},\hat{\mathcal{P}}^h}\left[\sum_{k=0}^{\infty} e^{-\delta \hat{S}_{n+1}^h(k)} \frac{\hat{c}(k)^\gamma}{\gamma} \hat{u}(k) \Delta t^h(\hat{S}^h(k))\right],$$

where

$$\Delta t^h(\hat{x}) = \begin{cases} h^2/Q^h(\hat{x}) & \text{if } \hat{x} \in \hat{S}_N^h, \\ 0 & \text{if } \hat{x} \in \partial\hat{S}_N^{h+} \cup \partial\hat{S}_N^h \end{cases}$$

and $\hat{E}_{\hat{x},\hat{\mathcal{P}}^h}$ is the expectation operator given $\hat{S}^h(0) = \hat{x}$ and control $\hat{\mathcal{P}}^h$. The value function is then defined to be

$$\hat{V}^h(\hat{x}) = \sup_{\hat{\mathcal{P}}^h \in \hat{\mathcal{U}}_N^h} \hat{J}^h(\hat{x}, \hat{\mathcal{P}}^h). \tag{3.4}$$

The problem is to find a control $\tilde{\mathcal{P}}^h \in \hat{\mathcal{U}}_N^h$ such that

$$\hat{V}^h(\hat{x}) = \hat{J}^h(\hat{x}, \tilde{\mathcal{P}}^h) \quad \forall \hat{x} \in \hat{S}_N^h \cup \partial\hat{S}_N^{h+}.$$

Remark 3.1 *If* $\hat{x} \in \partial\hat{S}_N^h$, *then it remains there, so* $\Delta t^h(\hat{S}(k)) = 0$, $k = 0, \cdots$. *This implies that* $\hat{J}^h(\hat{x}, \hat{\mathcal{P}}^h) = 0$ *for any admissible control sequence* $\hat{\mathcal{P}}^h$, *which in turn implies that* $\hat{V}^h(\hat{x}) = 0$.

The following convergence result can now be established.

Theorem 3.1 *Assume [A.1]-[A.3]. Then for* $\hat{x} \in \overline{\hat{S}}$

$$\lim_{\substack{h \downarrow 0 \\ \hat{y}^h \to \hat{x} \\ y^h \in \hat{S}^h_N}} \hat{V}^h(\hat{y}^h) = \hat{V}(\hat{x})$$

where $\hat{V}$ *is the unique viscosity solution of (2.15).*

References

[1] M. AKIAN, J. L. MENALDI, AND A. SULEM, *On an investment-consumption model with transaction costs,* SIAM J. Control. Optim, 1 (1996) pp. 1-35.

[2] G. M. CONSTANTINIDES, *Capital market equilibrium with transaction costs,* J. of Political Economy, 94 (1986), pp. 842-864.

[3] M. H. A. DAVIS AND A. R. NORMAN, *Portfolio selection with transaction costs,* Math. Oper. Res, 15 (1990), pp. 676-713.

[4] B. G. FITZPATRICK AND W. H. FLEMING, *Numerical methods for an optimal investment-consumption model,* Math. Oper. Res, 16 (1991) pp. 823-841.

[5] U. G. HAUSSMANN AND W. SUO, *Existence of singular optimal control laws for stochastic differential equations,* Stochastics and Stochastics Reports, 48 (1994) pp. 249-272.

[6] I. KARATZAS, J. P. LEHOCZKY, S. P. SETHI AND S. E. SHREVE, *Explicit solution of a general consumption-investment problem,* Math. Oper. Res, 11 (1986), pp. 261-294.

[7] H. J. KUSHNER AND P. G. DUPUIS, *Numerical methods for stochastic control problems in continuous time* , Springer Verlang, New York, (1992).

[8] H. J. KUSHNER AND L. F. MARTIN, *Routing and singular control for queuing networks in heavy traffic,* SIAM J. Control, Optim, 28 (1990), pp. 1209-1233.

[9] M. J. MAGILL AND G. M. CONSTANTINIDES, *Portfolio selection with transaction costs,* J. Economic Theory, 13 (1976), pp. 245-263.

SOME APPROACHES TO ERGODIC AND ADAPTIVE CONTROL OF STOCHASTIC SEMILINEAR SYSTEMS

T. E. Duncan

Department of Mathematics, University of Kansas
Lawrence, KS 66045
U.S.A.*

duncan@math.ukans.edu

Abstract: Some ergodic and adaptive control problems for stochastic semilinear systems are described. The stochastic systems have cylindrical noise processes and the control is either distributed or boundary/point. The development of adaptive control for stochastic distributed parameter systems is described commencing with adaptive control problems for linear stochastic systems with quadratic cost functionals.

1 INTRODUCTION

The study of adaptive control of stochastic distributed systems has developed primarily in this decade. For adaptive control, the systems are only partially known and they must be controlled so there are the problems of identification and control. For the control part it is clear that some results for the control of the completely known systems should be available, so it is natural that there is some lag time between the results for optimal control and adaptive control of these systems. If it is desired that an adaptive control achieves the optimal cost for the control of the known system then an ergodic cost criterion is natural. If an adaptive control achieves this optimal cost then it is called self optimal. The semigroup approach has been an important method for the analysis of distributed parameter systems. The initial work on adaptive control using the semigroup approach is done in [2] where an adaptive control problem of a partially known linear stochastic system, for example, a stochastic partial differential equation, with a distributed control and a quadratic cost functional is solved. Since it seems to be more natural to use boundary or point control,

*Research supported in part by NSF Grant DMS 9623439.

instead of distributed control, some adaptive control problems for partially known linear stochastic distributed parameter systems with boundary/point control and quadratic cost functionals are solved in [3, 4]. These adaptive control problems use the optimal controls for deterministic, linear distributed parameter systems with infinite time horizon, quadratic cost functionals. It is well known, at least for finite dimensional systems, that the optimal control for the deterministic system gives the optimal control of the stochastic system with the corresponding ergodic cost. However this relation assumes that the systems are linear and the cost functional is quadratic.

For stochastic semilinear systems the situation is significantly more complicated. There are questions of existence and uniqueness of the solutions of the stochastic differential equations that describe the stochastic semilinear systems. The transition measures and semigroup for the solution of a stochastic semilinear system with a Markov-type control have to be shown to be strongly Feller, tight and irreducible so that invariant measures exist and are unique. These results are important for the ergodic control problem. For the adaptive control problem it is important to verify the continuity of the optimal cost and the optimal control with respect to parameters. These properties are important because typically a strongly consistent family of estimates of the unknown parameters is constructed and it is desired to show that a certainty equivalence adaptive control, that is obtained by using the estimate of the unknown parameter in place of the true parameter in the optimal control, converges in some sense to the optimal control and the "running" costs for this adaptive control converge to the optimal ergodic cost for the known system.

For modelling of the controlled stochastic semilinear system there are distributed and boundary/point control. For boundary/point control some results have been obtained for ergodic control and continuity properties for adaptive control in [6]. While these results provide important information for both ergodic and adaptive control, an optimal control is not given explicitly so that the continuity of an optimal control with respect to parameters is not known. Thus, if a certainty equivalence adaptive control is used with a strongly consistent family of estimates of the unknown parameter, then it is not clear that these adaptive controls converge to an optimal control for the true system. For distributed control of parameter dependent semilinear stochastic systems, some results have been obtained for ergodic control and continuity properties for adaptive control in [5]. However, an optimal control is not given explicitly, so it is difficult to verify that a family of adaptive controls converges to an optimal control. Another approach to ergodic cost, distributed control is given in [8]. While these results have potential application to adaptive control, it seems that more information is required about an optimal control.

For some ergodic, distributed control problems of stochastic semilinear systems, a Hamilton-Jacobi-Bellman equation is solved and an optimal control is given in [9]. In [7] the results in [9, 10] are used as a starting point for the verification of continuity properties of the optimal cost and an optimal control with respect to parameters. These results are used to solve an adaptive control problem. Some of these results are described explicitly in this paper. The proofs of the results that are given here are contained in [7]. The noise process that occurs in these semilinear stochastic systems is cylindrical, that is, it does not exist in the Hilbert space where the state of the system is described. Thus the semigroup associated with the linear part of the semilinear system is required to regularize this cylindrical noise.

2 PRELIMINARIES

A partially known stochastic semilinear system is described with a distributed control.

Let $(X(t),\, t \geq 0)$ be an H-valued, parameter dependent, controlled process that satisfies the (formal) stochastic differential equation

$$\begin{aligned} dX(t) &= (AX(t) + f(\alpha, X(t)) - u(t))dt + Q^{1/2}dW(t) \\ X(0) &= x \end{aligned} \tag{2.1}$$

where H is a real, separable Hilbert space with inner product $\langle\cdot,\cdot\rangle$ and norm $|\cdot|$, $A : \mathrm{Dom}(A) \to H$ is a densely defined, unbounded linear operator on H, $f(\alpha,\cdot) : H \to H$ for each $\alpha \in \mathcal{A} \subset \mathbb{R}^d$ that is a compact set of parameters, $(W(t),\, t \geq 0)$ is a standard, cylindrical H-valued Wiener process defined on a filtered probability space $(\Omega, \mathcal{F}, (\mathcal{F}_t), \mathbb{P})$ and $Q \in \mathcal{L}(H)$. The family of admissible controls is

$$\mathcal{U} = \{u : \mathbb{R}_+ \times \Omega \to B_R \mid u \text{ is measurable and } (\mathcal{F}_t) \text{ adapted}\} \tag{2.2}$$

where $B_R = \{y \in H \mid |y| \leq R\}$ and $R > 0$ is fixed. A family of Markov controls, e.g., $u(t) = \tilde{u}(X(t))$, is also considered where $\tilde{u} \in \tilde{\mathcal{U}}$ and

$$\tilde{\mathcal{U}} = \{\tilde{u} : H \to B_R \mid \tilde{u} \text{ is Borel measurable}\}. \tag{2.3}$$

The cost functionals $J(x,\lambda,u)$ and $\tilde{J}(x,u)$ are given as

$$J(x,\lambda,u) = \mathbb{E}_{x,u} \int_0^\infty e^{-\lambda t}(\psi(X(t)) + h(u(t)))dt \tag{2.4}$$

and

$$\tilde{J}(x,u) = \liminf_{T\to\infty} \mathbb{E}_{x,u} \frac{1}{T} \int_0^T (\psi(X(t)) + h(u(t)))dt \tag{2.5}$$

where $\lambda > 0$ and $h : B_R \to \mathbb{R}_+$ and $\psi : H \to \mathbb{R}$ that describe a discounted and an ergodic control problem, respectively.

The following assumptions are selectively used to solve an adaptive control problem.

(A1) The linear operator $Q = Q^{1/2^*}Q^{1/2}$ is invertible, $Q^{-1} \in \mathcal{L}(H)$ and $(S(t),\, t \geq 0)$, where $S(t) = e^{tA}$, is an exponentially stable semigroup of contractions, that is, $\|S(t)\|_{\mathcal{L}(H)} \leq e^{-\omega t}$ for all $t \geq 0$ and some $\omega > 0$. Furthermore, the semigroup is Hilbert-Schmidt and there is a $\gamma > 0$ such that $\int_0^T t^{-\gamma}\|S(t)\|_{\mathrm{HS}}^2 dt < \infty$ for some $T > 0$ where $\|\cdot\|_{\mathrm{HS}}$ is the Hilbert-Schmidt norm.

(A2) The function $f(\alpha, \cdot) : H \to H$ is Lipschitz continuous and Gateaux differentiable. The Gateaux derivative $Df(\alpha, \cdot)h$ is continuous on H for each $h \in H$ and $\alpha \in \mathcal{A}$ and there is a $\beta \in \mathbb{R}$ such that $\langle Df(\alpha, x)h, h\rangle \leq \beta |h|^2$ for all $x \in H$, $h \in H$ and $\alpha \in \mathcal{A}$.

(A3) There are constants $p > 0$, $\theta > 0$ and $C > 0$ such that $|f(\alpha_1, x) - f(\alpha_2, x)| \leq C|\alpha_1 - \alpha_2|^\theta (1 + |x|^p)$ for all $\alpha_1, \alpha_2 \in \mathcal{A}$ and $x \in H$.

(A4) $\psi \in C_b(H)$.

(A5) The function $\psi : H \to \mathbb{R}$ is convex and bounded on bounded sets and continuous. The function $\tilde{H} : H \to \mathbb{R}$ given by $\tilde{H}(x) = \sup_{|y| \leq R}[\langle y, x\rangle - h(y)]$ is Lipschitz continuous and Gateaux differentiable. Each directional derivative of $\tilde{H}$ is continuous on H.

Some implications of the assumptions (A1)–(A5) are described now. For the linear stochastic differential equation associated with (2.1), that is, $f \equiv u \equiv 0$, there are a unique mild solution and a unique invariant Gaussian probability measure, $\mu = N(0, Q_\infty)$ where Q_∞ is a trace class operator on H. If (A2) is satisfied then (2.1) has a unique mild solution for each $u \in \mathcal{U}$ and $\alpha \in \mathcal{A}$. If the control in (2.1) has the feedback form $u(t) = \tilde{u}(X(t))$ where $\tilde{u} \in \tilde{\mathcal{U}}$ then the solution of (2.1) is obtained by an absolute continuity of measures as a weak solution in the probabilistic sense.

The assumption (A3) is used to verify a suitable continuous dependence of the solutions of the ergodic Hamilton-Jacobi-Bellman equations on the parameter which is important to verify the self-optimality of a certainty equivalence adaptive control. The assumptions (A4) and (A5) are standard conditions on a cost functional in the stochastic control of semilinear systems (e.g., [9, 10]). Note that (A5) is satisfied in the case where $h(x) = |x|^2$ so that $\tilde{H}(x) = \hat{H}(\|x\|)$ where

$$\hat{H}(r) = \begin{cases} \frac{r^2}{4} & \text{if } |r| \leq 2R \\ R|r| - R^2 & \text{if } |r| > 2R \end{cases}$$

To provide some additional perspective for the adaptive control problem, an example of a stochastic partial differential equation that satisfies the assumptions is given. Consider the stochastic PDE

$$\frac{\partial y}{\partial t}(t, \xi) = \gamma \frac{\partial^2 y}{\partial \xi^2}(t, \xi) + F(\alpha, y(t, \xi)) - u(t, \xi) + \eta(t, \xi) \tag{2.6}$$

for $(t, \xi) \in \mathbb{R}_+ \times (0, 1)$ with the initial condition $y(0, \xi) = y_0(\xi)$. $\xi \in (0, 1)$ and the Dirichlet boundary conditions

$$y(t, 0) = y(t, 1) = 0$$

for $t \geq 0$ and $\gamma > 0$ is a constant, $F : \mathcal{A} \times \mathbb{R} \to \mathbb{R}$ satisfies $F(\alpha, \cdot) \in C^1(\mathbb{R})$ for each $\alpha \in \mathcal{A}$, $F'(\alpha, y) \leq \beta$ and $|F(\alpha_1, y) - F(\alpha_2, y)| \leq C|\alpha_1 - \alpha_2|^\theta(1 + |y|)$ for some constants $\beta \in \mathbb{R}_+$, $C > 0$ and all $\alpha, \alpha_1, \alpha_2 \in \mathcal{A}$, $y \in \mathbb{R}$ and η formally denotes a space time white noise. The control $(u(t), t \geq 0)$ is assumed to be adapted to the noise process and to take values in a ball, B_R, in $L^2(0,1)$. The formal equation (2.6) can be rigorously described in a standard way as an equation of the form (2.1) in the Hilbert space $H = L^2(0,1)$, $A = \gamma(\partial^2/\partial\xi^2)$, $\mathrm{Dom}(A) = \{\varphi \in L^2(0,1) \mid \varphi, \varphi'$ are absolutely continuous, $\varphi'' \in L^2(0,1)$, $\varphi(0) = \varphi(1) = 0\}$, $f(\alpha, x)(\xi) = F(\alpha, x(\xi))$ for $x \in H$, $\alpha \in \mathcal{A}$, $\xi \in (0,1)$ and a cylindrical Wiener process with $Q = \delta I$ where $\delta > 0$ is a constant and I is the identity on I. For ψ and h in the cost functionals (2.4, 2.5) arbitrary $\psi \in C_b(L^2(0,1))$ and $h : L^2(0,1) \to \mathbb{R}_+$ satisfying (A5) can be chosen, e.g., $h(u) = |u|^2$. It is well known that all of the assumptions (A1)–(A5) are satisfied where $\gamma \in (0, 1/4)$ in (A1).

3 MAIN RESULTS

The formal Hamilton-Jacobi-Bellman equations corresponding to the control problems (2.1, 2.4) and (2.1, 2.5) are respectively

$$\begin{aligned}\frac{1}{2}\mathrm{Tr}\, QD^2 v_\alpha^\lambda(x) + \langle Ax, Dv_\alpha^\lambda(x)\rangle + \langle f(\alpha, x), Dv_\alpha^\lambda(x)\rangle \\ - \tilde{H}(Dv_\alpha^\lambda(x)) + \psi(x) = \lambda v_\alpha^\lambda(x) \qquad (3.1)\end{aligned}$$

$$\begin{aligned}\frac{1}{2}\mathrm{Tr}\, QD^2 v_\alpha(x) + \langle Ax, Dv_\alpha(x)\rangle + \langle f(\alpha, x), Dv_\alpha(x)\rangle \\ - \tilde{H}(Dv_\alpha(x)) + \psi(x) = \rho(\alpha). \qquad (3.2)\end{aligned}$$

In (3.2) it is necessary to solve for the pair $(v_\alpha, \rho(\alpha))$, $\rho(\alpha) \in \mathbb{R}$ for each $\alpha \in \mathcal{A}$.

It is clear that the existence of strong solutions to (3.1) and (3.2) cannot be expected because of the first two terms on the left hand side of these equations, that is, Q is not trace class and A is only densely defined in H. The approach in [9] is to replace the first two terms in (3.1) and (3.2) by the generator of an Ornstein-Uhlenbeck semigroup in a suitable function space. The results of [9, 10] are used but for simplicity the solutions of (3.1) and (3.2) are defined in a weaker sense which is suitable for the applications to adaptive control.

Let $\mu = N(0, Q_\infty)$ be the invariant measure and $(R_t\varphi)(x) = \mathbb{E}_x\varphi(Z(t))$ be the Markov transition semigroup for the Ornstein-Uhlenbeck process $(Z(t), t \geq 0)$ that is the solution of (2.1) with $f \equiv u \equiv 0$. It is well known that $(R_t, t \geq 0)$ is a strongly continuous semigroup on the Hilbert space $\mathcal{H} = L^2(H, \mu)$. Let $\mathcal{L}$ be the infinitesimal generator of the semigroup $(R_t, t \geq 0)$ in $\mathcal{H}$. Furthermore, let $\mathcal{L}_0$ be given by

$$\mathcal{L}_0\varphi(x) = \frac{1}{2}\mathrm{Tr}\, QD^2\varphi(x) + \langle x, A^*D\varphi(x)\rangle \qquad (3.3)$$

for $x \in H$ and $\varphi \in \mathrm{Dom}(\mathcal{L}_0)$ where $\mathrm{Dom}(\mathcal{L}_0) = \{\varphi \in C_b^2(H) \mid (1/2)\mathrm{Tr}\, QD^2\varphi(\cdot) \in C_b(H),\ \langle \cdot, A^*D\varphi(\cdot)\rangle \in C_b(H)\}$.

Let $\varphi \in \mathrm{Dom}(\mathcal{L}_0)$ and use the Itô formula to obtain

$$\begin{aligned}(\mathcal{L}\varphi)(x) &= \lim_{t\downarrow 0} \frac{1}{t}(R_t\varphi(x) - \varphi(x)) \\ &= \lim_{t\downarrow 0} \frac{1}{t}(\mathbb{E}_x\varphi(Z(t)) - \varphi(x)) \\ &= (\mathcal{L}_0\varphi)(x)\end{aligned} \tag{3.4}$$

so $\mathcal{L}$ is a closed extension of the operator $\mathcal{L}_0$. This equality, (3.4), motivates the following definition of solution of (3.1) and (3.2).

Definition 3.1 A function $v_\alpha^\lambda \in \mathrm{Dom}(\mathcal{L})$ and a pair $(v_\alpha, \rho(\alpha)) \in \mathrm{Dom}(\mathcal{L}) \times \mathbb{R}$ are solutions to (3.1) and (3.2) respectively if

$$\mathcal{L}v_\alpha^\lambda + \langle f(\alpha,\cdot), Dv_\alpha^\lambda\rangle - \tilde{H}(Dv_\alpha^\lambda) + \psi = \lambda v_\alpha^\lambda \tag{3.5}$$

and

$$\mathcal{L}v_\alpha + \langle f(\alpha,\cdot), Dv_\alpha\rangle - \tilde{H}(Dv_\alpha) + \psi = \rho(\alpha) \tag{3.6}$$

are satisfied.

This definition of the solutions to (3.1) and (3.2) requires only that the solutions be in $\mathrm{Dom}(\mathcal{L}) \subset L^2(H,\mu)$ so the equations (3.5) and (3.6) are understood in an $L^2(H,\mu)$ sense. This relatively weak notion of solution is used to avoid some technical complications. Some results on the solutions to (3.1) and (3.2) are given in [9] and [10]. It is shown that the solutions are more regular than that required in the Definition 1.3.1. For the following two propositions the parameter $\alpha \in \mathcal{A}$ is fixed.

Proposition 3.1 *If* (A1), (A2), (A4) *and* (A5) *are satisfied, then the equation* (3.1) *has one and only solution* v_α^λ *in* $\mathrm{Dom}(\mathcal{L}) \cap C_b^1(H)$. *Furthermore,*

$$v_\alpha^\lambda(x) = \inf_{u\in\mathcal{U}} J(x,\lambda,u) \tag{3.7}$$

so that v_α^λ *gives the optimal cost and an optimal control in feedback form is* $\hat{u}_\alpha^\lambda(x) = D\tilde{H}(Dv_\alpha^\lambda(x))$ *for the discounted cost control problem* (2.1, 2.4).

This proposition has been basically proven by Gozzi and Rouy [10] when $f(\alpha,\cdot)$ is bounded. The generalization in Proposition 1.3.1 has been done by Goldys and Maslowski [9].

The ergodic control problem is usually considered to be more difficult than the discounted control problem because the Hamilton-Jacobi-Bellman equation (3.2) has an intrinsic degeneracy, that is, there is no uniqueness of the solution to (3.2) because if $(v_\alpha,\ \rho(\alpha))$ is a solution of (3.2) then $(v_\alpha + c,\ \rho(\alpha))$ for $c \in \mathbb{R}$ is also a solution.

Let $R > 0$ in (A5), satisfy

$$R < \frac{\sqrt{\omega_1}}{|Q^{-1/2}|_{\mathcal{L}(H)}k(\omega_1)\sqrt{\pi}} \tag{3.8}$$

where $\omega - \beta > 0$, $\omega_1 \in (0, \omega - \beta)$ is fixed and $k(\omega_1) > 0$ is a constant depending only on ω_1 that is obtained from an upper bound on the Fréchet derivative of the Markov transition semigroup induced by the solution of (2.1) with $u \equiv 0$.

The following proposition is a "$W^{1,2}(H,\mu)$-version" of a result in [9] on the existence and the uniqueness of a solution to the ergodic Hamilton-Jacobi-Bellman equation. While the existence result is weaker than the one in [9], the family of solutions for uniqueness is enlarged. The parameter $\alpha \in \mathcal{A}$ in the following proposition is fixed and

$$\begin{aligned}\mathcal{W} = \{\varphi \in W^{1,2}(H,\mu) \,|\, &\varphi \in \mathrm{Dom}(\mathcal{L}),\ ||D\varphi|| < \infty \\ &|\varphi(x)| + |\mathcal{L}\varphi(x)| \leq k(1+|x|^q) \text{ for all } x \in H \\ &\text{and some positive real numbers } k \text{ and } q\}.\end{aligned}$$

Proposition 3.2 *If* (A1), (A2), (A4), (A5) *with* $\omega - \beta > 0$ *and* (3.8) *are satisfied then there is a unique solution* $(v_\alpha, \rho(\alpha)) \in (\mathcal{W} \times C(H)) \times \mathbb{R}$ *of* (3.2) *such that* $v_\alpha(0) = 0$. *Furthermore,*

$$\rho(\alpha) = \inf_{u \in \mathcal{U}} \tilde{J}(x,u)$$

so that $\rho(\alpha)$ *is the optimal cost and an optimal control in feedback form is* $\hat{u}_\alpha(x) = D\tilde{H}(Dv_\alpha(x))$ *for the ergodic control problem* (2.1, 2.5).

Consider the equation (2.1) with the true parameter value $\alpha_0 \in \mathcal{A}$, that is,

$$\begin{aligned}dX(t) &= (AX(t) + f(\alpha_0, X(t)) - \tilde{u}(t))dt + Q^{1/2}dW(t) \\ X(0) &= x\end{aligned} \tag{3.9}$$

where

$$\tilde{u}(t) = D\tilde{H}(Dv_{\alpha(t)}(X(t))) \tag{3.10}$$

where $(\alpha(t),\, t \geq 0)$ is an adapted, measurable, $\mathcal{A}$-valued process satisfying

$$\lim_{t \to \infty} \alpha(t) = \alpha_0 \qquad \text{a.s. } \mathbb{P}$$

and v_α is given in Proposition 1.3.2. For notational simplicity, let $v = v_{\alpha_0}$ and $\rho = \rho(\alpha_0)$.

Proposition 3.3 *If* (A1)–(A5) *are satisfied,* $\omega - \beta > 0$ *and the inequality* (3.8) *are satisfied, then*

$$\tilde{J}(x, \tilde{u}) = \rho(\alpha_0)$$

for each $x \in H$, *that is, the adaptive control* $\tilde{u}$ *given by* (3.10) *is self-optimizing for the ergodic control problem* (2.1, 2.5).

References

[1] G. Da Prato and J. Zabczyk, Stochastic Equations in Infinite Dimensions, Cambridge Univ. Press, Cambridge, 1992.

[2] T. E. Duncan, B. Goldys and B. Pasik-Duncan, Adaptive control of linear stochastic evolution systems, Stochastics Stochastic Rep. 36(1991), 71–90.

[3] T. E. Duncan, B. Maslowski, and B. Pasik-Duncan, Adaptive boundary and point control of linear stochastic distributed parameter systems, SIAM J. Control Optim. 32(1994), 648–672.

[4] T. E. Duncan, B. Maslowski, and B. Pasik-Duncan, Adaptive boundary control of linear distributed parameter systems described by analytic semigroups, Appl. Math. Optim. 33(1996), 107–138.

[5] T. E. Duncan, B. Maslowski, and B. Pasik-Duncan, Ergodic distributed control for parameter dependent stochastic semilinear systems in Hilbert space, Proc. Conf. Stochastic Differential and Differentia Equations, Birkhäuser, 1997.

[6] T. E. Duncan, B. Maslowski, and B. Pasik-Duncan, Ergodic boundary/point control of stochastic semilinear systems, SIAM J. Control Optim. 36(1998), 1020–1047.

[7] T. E. Duncan, B. Maslowski, and B. Pasik-Duncan, Adaptive control for semilinear stochastic systems, preprint.

[8] T. E. Duncan, B. Pasik-Duncan and L. Stettner, On ergodic control of stochastic evolution equations, Stochastic Process. Appl. 15(1997), 723–750.

[9] B. Goldys and B. Maslowski, Ergodic control of semilinear stochastic equations and Hamilton-Jacobi equations, Statistics Dept., Univ. New South Wales, Rept. S97-13.

[10] F. Gozzi and E. Rouy, Regular solutions to second-order stationary Hamilton-Jacobi equations, J. Differential Equations 130(1996), 201–234.

A ONE-DIMENSIONAL RATIO ERGODIC CONTROL PROBLEM

Yasuhiro Fujita*
Department of Mathematics
Faculty of Science
Toyama University
Toyama 930-0887, Japan
yfujita@sci.toyama-u.ac.jp

Abstract: In this paper we consider a one-dimensional ratio ergodic control problem. We obtain both the optimal control and the minimum value by solving the corresponding Bellman equation rigidly in C^2- class.

1. INTRODUCTION

Let $\kappa \leq a(\cdot) \in C_b(\mathbb{R})$, $b(\cdot,\cdot) \in C_b(\mathbb{R} \times \Gamma)$, $f(\cdot), g(\cdot) \in C(\mathbb{R})$ be given functions, where $\kappa > 0$ is a positive constant and Γ is a compact set in a separable metric space. We consider the ergodic control problem to minimize the cost :

$$K(v) = \liminf_{T\to\infty} \frac{\displaystyle\int_0^T f(x_t)\, dt}{\displaystyle\int_0^T g(x_t)\, dt} \quad \text{a.s.} \tag{1.1}$$

subject to one–dimensional stochastic differential equations

$$dx_t = b(x_t, v(x_t))\, dt + \sqrt{a(x_t)}\, dw_t \ , \qquad x_0 = \xi \ , \tag{1.2}$$

over a class of $\Gamma-$ valued Borel measurable functions $v = v(x)$ on $\mathbb{R}$, where ξ is a constant in $\mathbb{R}$ and $w = (w_t)$ is a one–dimensional standard Brownian motion on a complete probability space $(\Omega, \mathcal{F}, \mathbb{P}; \{\mathcal{F}_t\})$. Such a function $v(\cdot)$ is called a Markov control with control set Γ.

When g is a constant function, the problem above is called a classical ergodic control problem. Classical ergodic control problems have been investigated by

*The author was partially supported by the Grant-in-Aid for Encouragement of Young Scientists (No. 09740140) by the Ministry of Education Science, and Culture of Japan.

many authors [1; 2; 3; 5] etc. We are interested in the case for non–constant function g, which we call a ratio ergodic control problem. The Bellman equation corresponding to the ratio ergodic control problem above is given by

$$\lambda g(x) = \frac{1}{2} a(x) \phi''(x) + \inf_{\gamma \in \Gamma} \left[\phi'(x) b(x, \gamma) \right] + f(x) , \qquad x \in \mathbb{R} . \quad (1.3)$$

If the function g is bounded below and bounded above by some positive constants, then the ratio ergodic control problem above is reduced to a classical ergodic control problem. Indeed, in this case, dividing the both sides of (3) by $g(x)$, we can consider the control problem above as the classical ergodic control problem to minimize the cost

$$L(v) = \liminf_{T \to \infty} \frac{1}{T} \int_0^T f(y_t)/g(y_t) \, dt \quad \text{a.s.}$$

subject to one–dimensional stochastic differential equations

$$dy_t = \frac{b(y_t, v(y_t))}{g(y_t)} dt + \sqrt{\frac{a(y_t)}{g(y_t)}} d\tilde{w}_t , \qquad y_0 = \xi ,$$

over the same class of Markov controls as above. Here $(\tilde{w}_t)$ is a one–dimensional standard Brownian motion. change method.

Our aim of the present paper is to obtain the optimal control and the minimum value of the ratio ergodic control problem (1.1), (1.2) for functions g which may vanish at some points or may be unbounded. For example, we can treat the case that $g(x) = |x|^\gamma$ for some constant $\gamma > 0$. For this aim, we have two keys. The first key is the pathwise asymptotic behavior of (x_t) as $t \to \infty$. This is necessary to treat the cost $K(v)$ defined almost surely. The second key is to solve corresponding Bellman equation (1.3).

The present paper is organized as follows : In § 2, we solve Bellman equation (1.3) rigidly in C^2 - class. In §3 we consider ratio ergodic control problem (1.1), (1.2).

2. BELLMAN EQUATION

Throughout this paper we assume :

(A1) $0 \le f, g \in C(\mathbb{R})$, $0 < g(x)$ $(x \neq 0)$, $\lim_{R \to \infty} \inf_{|x|>R} g(x) > 0$. There exist constants $\beta \ge 0$ and $C > 0$ such that

$$f(x) + g(x) \le C(1 + |x|^\beta) , \qquad x \in \mathbb{R} .$$

(A2) $f/g \in C^1(\mathbb{R} \setminus \{0\})$. f/g is strictly decreasing on $(-\infty, 0)$ and is strictly increasing on $(0, \infty)$. In addition

$$\lim_{x \downarrow 0} \frac{f(x)}{g(x)} = \lim_{x \uparrow 0} \frac{f(x)}{g(x)} ,$$

$$\lim_{x\to\infty}\frac{f(x)}{g(x)}=\lim_{x\to-\infty}\frac{f(x)}{g(x)}\,.$$

(A3) Γ is a compact set in a separable metric space.

(A4) $a \in C_b(\mathbb{R})$ satisfies $\kappa \le a(\cdot) \le \kappa^{-1}$ on $\mathbb{R}$ for some constant $0<\kappa<1$ and

$$\left|\sqrt{a(x)}-\sqrt{a(y)}\right| \le K|x-y|\,, \qquad x,y\in\mathbb{R}$$

for some constant $K>0$. $b\in C_b(\mathbb{R}\times\Gamma)$ satisfies

$$\sup_{\gamma\in\Gamma}\left|\, b(x,\gamma)-b(y,\gamma)\,\right| \to 0 \qquad (\;|x-y|\to 0\;)\,.$$

(A5)

$$\limsup_{x\to\infty}\frac{x}{a(x)}\inf_{\gamma\in\Gamma} b(x,\gamma) < -\frac{\beta+1}{2}\,, \tag{2.1}$$

$$\liminf_{x\to-\infty}\frac{x}{a(x)}\inf_{\gamma\in\Gamma} b(x,\gamma) > 0\,, \tag{2.2}$$

$$\liminf_{x\to\infty}\frac{x}{a(x)}\sup_{\gamma\in\Gamma} b(x,\gamma) > 0\,, \tag{2.3}$$

$$\limsup_{x\to-\infty}\frac{x}{a(x)}\sup_{\gamma\in\Gamma} b(x,\gamma) < -\frac{\beta+1}{2}\,. \tag{2.4}$$

Example 1 below shows that conditions (2.1) and (2.4) can not be weakened in general.

Example 1 *Let* $\Gamma=[-1,1]$, $f(x)=|x|^\beta$ $(\beta>0)$, $g(x)\equiv 1$, $a(x)\equiv 1$ *and* $b(x,\gamma)=\dfrac{\gamma(\beta+1)}{2(1+|x|)}$. *Then, except (2.1) and (2.4), (A1) ~ (A5) hold and we have*

$$\limsup_{x\to\infty}\frac{x}{a(x)}\inf_{\gamma\in\Gamma} b(x,\gamma)=\limsup_{x\to-\infty}\frac{x}{a(x)}\sup_{\gamma\in\Gamma} b(x,\gamma)=-\frac{\beta+1}{2}\,.$$

In this case we can consider that $K(v)=\infty$ *a.s. for all* Γ *- valued Borel measurable functions* v *on* $\mathbb{R}$.

Indeed, fix an arbitrary Γ – valued Borel measurable function v on $\mathbb{R}$, and let (x_t) be the response to v. Furthermore, let (x_t^*) be the response to $v^*(x)=-\mathrm{sgn}(x)$. By Theorem 2.1 in p.357 of [6], we can consider that $|x_t|\ge|x_t^*|$ $(t\ge 0)$ a.s. Thus $K(v)\ge K(v^*)$ a.s. On the other hand, by Theorem 16 in p.46 of [7], the invariant measure $p^*(x)\,dx$ of (x_t^*) is given by $p^*(x)=(1+|x|)^{-(\beta+1)}$ for

$x \in \mathbb{R}$. Thus p^* belongs to $L^1(\mathbb{R}, dx)$, because $\beta > 0$. Let $N > 0$ be arbitrary. By equality (80) in p.52 of [7], we have with probability 1

$$\lim_{T\to\infty} \frac{1}{T} \int_0^T |x_t^*|^\beta \, \chi_{\{|x_t^*|\le N\}} \, dt = \frac{\int_{\mathbb{R}} |x|^\beta \, \chi_{\{|x|\le N\}} \, p^*(x) \, dx}{\int_{\mathbb{R}} p^*(x) dx}$$

$$= \frac{\beta}{2} \int_{-N}^{N} |x|^\beta \, (1+|x|)^{-(\beta+1)} \, dx \, .$$

Since $|x|^\beta \, (1+|x|)^{-(\beta+1)}$ does not belong to $L^1(\mathbb{R}, dx)$, we have

$$\lim_{T\to\infty} \frac{1}{T} \int_0^T |x_t^*|^\beta \, dt \ge \lim_{T\to\infty} \frac{1}{T} \int_0^T |x_t^*|^\beta \, \chi_{\{|x_t^*|\le N\}} \, dt \to \infty \qquad (N \to \infty) \, .$$

Thus $K(v^*) = \infty$ a.s., and we can consider that $K(v) = \infty$ a.s. for all Γ – valued Borel measurable functions v on $\mathbb{R}$. Thus we have shown the assertion of Example 1.

Let

$$\begin{aligned} Q(x) &= \exp\left\{ 2 \int_0^x \frac{1}{a(t)} \sup_{\gamma\in\Gamma} b(t,\gamma) \, dt \right\}, \qquad x \in \mathbb{R} \, , \\ R(x) &= \exp\left\{ 2 \int_0^x \frac{1}{a(t)} \inf_{\gamma\in\Gamma} b(t,\gamma) \, dt \right\}, \qquad x \in \mathbb{R} \, . \end{aligned}$$

For $x \in \mathbb{R}$, let

$$F(x) = \frac{\int_{-\infty}^x \frac{f(t)}{a(t)} Q(t) \, dt}{\int_{-\infty}^x \frac{g(t)}{a(t)} Q(t) \, dt} \, , \qquad G(x) = \frac{\int_x^\infty \frac{f(t)}{a(t)} R(t) \, dt}{\int_x^\infty \frac{g(t)}{a(t)} R(t) \, dt} \, .$$

The functions F and G are well–defined by (A1) $\sim$ (A5).

Lemma 1 *We assume (A1) $\sim$ (A5). Then there exists a constant $\alpha \in \mathbb{R}$ such that $F(\alpha) = G(\alpha)$ and*

$$\begin{aligned} F(x) &< G(x) \, , \qquad \alpha < x \, , \\ G(x) &< F(x) \, , \qquad x < \alpha \, . \end{aligned}$$

Furthermore, $F(x)$ is strictly decreasing on $(-\infty, \alpha]$ and $G(x)$ is strictly increasing on $[\alpha, \infty)$.

Using the differential calculus and describing the graphs of F and G, we can show Lemma 1.

We introduce the class in which a unique solution of (1.3) is constructed.

Definition We define the class $\mathcal{H}$ by all pairs (μ, ϕ) such that
(1) $\mu \in \mathbb{R}$ and $\phi \in C^2(\mathbb{R})$,
(2) there exists a constant $C > 0$ such that

$$|\phi'(x)| \le C(1 + |x|^{\beta+1}), \qquad x \in \mathbb{R} .$$

(3) there exists a piecewise continuous function $L(x) > 0$ on $\mathbb{R}$ such that $\dfrac{\phi'(x)}{L(x)}$ is strictly increasing function on $\mathbb{R}$ and

$$\lim_{x\to-\infty} \frac{\phi'(x)}{L(x)} < 0 < \lim_{x\to\infty} \frac{\phi'(x)}{L(x)} .$$

Theorem 1 *Under (A1) ~ (A5), Bellman equation (1.3) has a unique solution (λ, ϕ) in $\mathcal{H}$, and it is given by*

$$\lambda = F(\alpha) = G(\alpha) , \tag{2.5}$$

$$\phi'(x) = \begin{cases} 2\displaystyle\int_x^\infty \left[\frac{f(t)}{a(t)} - \lambda\,\frac{g(t)}{a(t)}\right] \frac{R(t)}{R(x)}\,dt , & \alpha \le x , \\ -2\displaystyle\int_{-\infty}^x \left[\frac{f(t)}{a(t)} - \lambda\,\frac{g(t)}{a(t)}\right] \frac{Q(t)}{Q(x)}\,dt , & x < \alpha . \end{cases} \tag{2.6}$$

In addition, $\operatorname{sgn}(\phi'(x)) = \operatorname{sgn}(x - \alpha)$.

The proof of this theorem is similar to that of the case for the classical ergodic control (cf. [5]). So we omit it.

Example 2 . *Let $\Gamma = [-1, 1]$, $f(x) = |x|^2$, $g(x) = |x|$, $a(x) \equiv 2$ and $b(x, \gamma) = \gamma$. In this case, Bellman equation (1.3) is reduced to*

$$\lambda\,|x| = \phi''(x) - |\phi'(x)| + |x|^2 , \qquad x \in \mathbb{R} .$$

Its solution (λ, ϕ) is given by $\lambda = 2$, $\phi'(x) = x\,|x|$.

3. RATIO ERGODIC CONTROL

In this section we consider the ratio ergodic control problem (1.1), (1.2). Define $\mathcal{V}$ by the class of all $\Gamma-$ valued Borel measurable functions v on $\mathbb{R}$ such that

$$\lim_{R\to\infty} \sup_{|x|>R} \frac{x}{a(x)} b(x, v(x)) < -\frac{\beta+1}{2} . \tag{3.1}$$

For each $v \in \mathcal{V}$, (1.2) has a unique strong solution (x_t) by [8]. We investigate several properties of the response (x_t) to $v \in \mathcal{V}$. For $v \in \mathcal{V}$, set

$$p_v(x) = \frac{1}{a(x)} \exp\left\{ 2 \int_0^x \frac{1}{a(t)} b(t, v(t))\, dt \right\}, \qquad x \in \mathbb{R} , \tag{3.2}$$

and choose $\delta = \delta(v) > 0$ so that

$$\lim_{R\to\infty} \sup_{|x|>R} \frac{x}{a(x)} b(x, v(x)) < -\delta < -\frac{\beta+1}{2} \tag{3.3}$$

Since there exists a constant $r_1 > 0$ such that

$$p_v(x) \le \kappa^{-2} \max\{ p_v(r_1) , p_v(-r_1) \} \, |x/r_1|^{-2\delta} , \quad r_1 < |x|$$

and since $\beta - 2\delta < -1$, we have

Lemma 2 *We assume (A1) $\sim$ (A5). Then, for every $v \in \mathcal{V}$,*

$$p_v^{-1} \notin L^1(0,\infty) , \quad p_v^{-1} \notin L^1(-\infty, 0) , \quad |x|^\beta p_v(x) \in L^1(\mathbb{R}) . \tag{3.4}$$

In particular,

$$\lim_{T\to\infty} \frac{1}{T} \int_0^T g(x_t)\, dt = \frac{\displaystyle\int_{\mathbb{R}} g(x)\, p_v(x)\, dx}{\displaystyle\int_{\mathbb{R}} p_v(x) dx} > 0 \quad \text{a.s.} \tag{3.5}$$

Equality (3.5) follows (80) in p.52 of [7]. Throughout Lemmas 3 $\sim$ 5 below, we denote by $C_1, C_2, \cdots$ constants which are independent of the time variable $T > 0$.

Lemma 3 *We assume (A1) $\sim$ (A5). Then*

$$\mathbb{E} \sup_{0\le t\le T} |x_t|^{2\delta+1} \le C_1 \left(|\xi|^{2\delta+1} + T \right) , \qquad T \ge 0 .$$

Lemma 4 *We assume (A1) ~ (A5). Then, there exists a constant* $T_0 > 0$ *such that*

$$\mathbb{E} \sup_{0 \le t \le T} |x_t|^{\beta+2} \le C_2 T^{\frac{\beta+2}{2\delta+1}} , \qquad T \ge T_0 ,$$

$$\mathbb{E} \sup_{0 \le t \le T} \left| \int_0^t \sqrt{a(x_s)}\ \phi^{'}(x_s)\, dw_s \right| \le C_3 T^{\frac{1}{2} + \frac{\beta+2}{2(2\delta+1)}} , \qquad T \ge T_0 .$$

Lemma 5 *Let* (X_t) *be a stochastic process satisfying*

$$\mathbb{E} \sup_{0 \le t \le T} |X_t| \le C_4 T^{\theta} , \qquad T \ge T_1$$

for some constants $0 < \theta < 1$ *and* $T_1 > 0$*. Then, with probability 1,*

$$\frac{1}{T} \sup_{0 \le t \le T} |X_t| \to 0 , \qquad T \to \infty .$$

Lemmas 3 and 4 follow from the stochastic calculus. Lemma 5 follows from Borel– Cantelli lemma. Combining Lemmas 4, 5 and remarking that $\dfrac{\beta+2}{2\delta+1} < 1$, we have

Lemma 6 *We assume (A1) ~ (A5). Then, for every* $v \in \mathcal{V}$*, we have with probability 1*

$$\frac{1}{T} \sup_{0 \le t \le T} |x_t|^{\beta+2} \to 0 , \qquad (T \to 0) ,$$

$$\frac{1}{T} \sup_{0 \le t \le T} \left| \int_0^t \sqrt{a(x_s)}\ \phi^{'}(x_s)\, dw_s \right| \to 0 , \qquad (T \to 0) .$$

By the measurable selection theorem (cf. Lemma 16.30 of [4]), we have

Lemma 7 *We assume (A1) ~ (A5). Then there exists* $\rho \in \mathcal{V}$ *satisfying*

$$b(x, \rho(x)) = \begin{cases} \inf_{\gamma \in \Gamma} b(x, \gamma) , & \alpha \le x , \\ \sup_{\gamma \in \Gamma} b(x, \gamma) , & x < \alpha . \end{cases}$$

Now we state the main results of this section.

Theorem 2 *We assume (A1) $\sim$ (A5). Then,*

(1) for each $v \in \mathcal{V}$, with probability 1, the cost $K(v)$ of (1.1) is equal to a constant which is independent of $\xi \in \mathbb{R}$ of (1.2).

(2) $\lambda = \min\{ K(v) \ : \ v \in \mathcal{V} \} = K(\rho)$.

Sketch of the proof. We obtain (1) by Lemma 2 and equality (80) in p.52 of [7]. Next, applying Ito formula to $\phi(x_t^*)$, where (x_t^*) is the response to ρ, and using the fact that (λ, ϕ) is a solution of (1.3), we obtain $\lambda = K(\rho)$ by (3.5) and Lemmas 6, 7. For $v \in \mathcal{V}$, we obtain $\lambda \leq K(\rho)$ similarly.

References

[1] A. Bensoussan *Perturbation Methods in Optimal Control.* John Wiley and Sons, Paris, 1988.

[2] A. Bensoussan and J. Frehse On Bellman equations of ergodic control in $\mathbb{R}^n$. *J. Reine angewandte Math.*, 429:125–160, 1992.

[3] V. S. Borkar *Optimal control of diffusion processes.* Pitman Research Notes in Mathematics Series No. 203, Longman Scientific and Technica, Harlow, 1989.

[4] R. J. Elliott *Stochastic Calculus and Applications.* Springer-Verlag, New York, 1982.

[5] Y. Fujita and H. Morimoto Ergodic control of stochastic differential systems with controller constraints . *Stochastics and Stochastics Reports*, 58:245–257, 1996.

[6] N. Ikeda and S. Watanabe *Stochastic Differential Equations and Diffusion Processes.* North-Holland, Amsterdam, 1981.

[7] A. V. Skorokhod *Asymptotic Methods in the Theory of Stochastic Differential Equations* . American Mathematical Society, Translations of Mathematical Monographs. Vol. 87, Providence, 1989.

[8] A.Ju. Veretennikov On strong solutions and explicit formulas for solutions of stochastic integral equations. *Math. USSR. Sbornik.*, 39:387–403, 1981.

NONLINEAR H_∞ CONTROL: A STOCHASTIC PERSPECTIVE*

M.R. James

Department of Engineering
Faculty of Engineering and Information Technology
Australian National University
Canberra, ACT 0200
Australia

E-mail: Matthew.James@anu.edu.au

Abstract: The last few years have seen a number of exciting developments in the area known as nonlinear H_∞ control. The aim of work in this area is to design robust controllers for nonlinear systems using generalizations of the highly successful H_∞ methods used in linear systems theory. An interesting and important feature of the H_∞ problem is that it has close connections with differential games, risk-sensitive stochastic control, and dissipative systems, and as a consequence these subjects have enjoyed renewed interest in recent years. Indeed, these topics have a lot to contribute to the understanding of robust control. In this paper I will discuss some recent developments in robust control from a "stochastic" perspective. Results to be presented include (i) singular information states and stability, and (ii) semigroups and generators.

1 INTRODUCTION

Consider the following standard generalized regulator arrangement with nonlinear plant $G : (w, u) \mapsto (z, y)$ and nonlinear controller $K : y \mapsto u$. The problem is to find a controller K for which the resulting closed-loop system $(G, K) : w \mapsto z$ enjoys the following two properties. (i) *Dissipation.* Given gain $\gamma > 0$ dissipation means that there exists a non-negative function β with $\beta(0) = 0$ for which the *dissipation inequality* holds:

$$\begin{cases} \frac{1}{2}\int_0^T |z(s)|^2\, ds \ \leq \ \gamma^2 \frac{1}{2}\int_0^T |w(s)|^2\, ds + \beta(x_0) \\ \text{for all} \ \ w \in L_{2,T} \text{ and all } T \geq 0. \end{cases} \tag{1.1}$$

*Research supported in part by the Australian Research Council.

(ii) *Stability.* By stability of the closed-loop system we mean that if G is initialized at any x_0, then if $w(\cdot) \in L_2[0,\infty)$, then in the closed-loop defined by $u = K(y)$ the signals $u(\cdot), y(\cdot), z(\cdot)$ belong to L_2 and the plant state $x(t)$ converges to 0 as $t \to \infty$.

The problem just specified is a generalization of basic problems from classical control engineering, and may at first sight have little to do with stochastic control, nor is any stochastic-like interpretation immediately apparent. Indeed, no probability was used in the problem definition.

Background to a consideration of stochastic interpretations are the following facts. (i) The H_∞ control problem was originally formulated in the frequency domain for linear systems. The problem can also be interpreted in the time domain (as above) as one involving L_2 gains using the *dissipative systems* framework. (ii) It was shown in [5] that the linear H_∞ problem is equivalent to a *stochastic risk-sensitive* optimal control problem, in the sense that the solutions are the same (via the same Riccati equation). Further, it has been shown [7] that the risk-sensitive solution is equivalent to a *deterministic differential game.* (iii) Except for the frequency domain formulation, all of the above control concepts and optimization problems have natural interpretations for nonlinear systems. Investigation of these concepts has been the subject of a considerable amount of research activity over the last decade. In particular, the H_∞ control problem posed above can be solved by viewing it as a *minimax* game problem, where the controller K attempts to minimize a worst-case cost function defined by maximizing over the disturbance w, [2]. (iv) The so-called *max-plus* algebra provides an elegant framework for treating deterministic optimal control problems, see [1], [12], etc. In the max-plus algebra, ordinary addition is replaced by the maximum binary operator, and so optimal control problems involving maximization over a control variable have natural stochastic analogs where the maximization is interpreted as an integration. There are also very close connections with the theory of *large deviations,* see [3], [11].

So since the nonlinear H_∞ control problem can be formulated as a minimax game, it lends itself to interpretation using the max-plus algebebra, with it's strong stochastic analogs. Moreover, the minimax formulation is directly related to risk-sensitive stochastic optimal control via small noise limits (large deviations), see [8], etc.

The *information state* approach to solving partially observed stochastic optimal control problems has been well known since at least the 1960's, [13]. Recently this approach has proved very fruitful for solving output feedback risk-sensitive optimal control problems and minimax differential games, [8], and from this an information state theory for nonlinear H_∞ control is being developed [9], [6]. Interestingly, the "stochastic" concept of information state is the key to a general solution of the nonlinear H_∞ control problem.

Development of the information state approach is nontrivial; indeed new and deep mathematical questions have arisen. Key issues include: (i) *Dynamic programming PDE.* The dynamic progrogramming PDE is defined on an infinite dimensional space. There is little mathematical theory available, and questions concerning the correct definition of solution, uniqueness theorems, etc, are unanswered, [10]. (ii) *Stability.* The stability or asymptotic behavior of the information state dynamical system is only just beginning to be understood, [6]. The max-plus framework appears to be essential for interpreting information state convergence in general.

2 INFORMATION STATE SOLUTION

We consider nonlinear plants G of the form

$$G: \begin{cases} \dot{x} = A(x) + B_1(x)w + B_2(x)u \\ z = C_1(x) + D_{12}(x)u \\ y = C_2(x) + D_{21}(x)w. \end{cases} \tag{1.2}$$

Here, $x(t) \in \mathbf{R}^n$ denotes the state of the system, and is not in general directly measurable; instead an output $y(t) \in \mathbf{R}^p$ is observed. The additional output quantity $z(t) \in \mathbf{R}^m$ is a performance measure, depending on the particular problem at hand. The control input is $u(t) \in \mathbf{R}^m$, while $w(t) \in \mathbf{R}^p$ is regarded as an opposing disturbance input. We assume that all of the functions appearing in (1.2) are as smooth and bounded as necessary, and that zero is an equilibrium: $A(0) = 0$, $C_1(0) = 0$ and $C_2(0) = 0$. In order to simplify the notation as much as possible, we take $D_{12}(x) = I_m$ and $D_{21}(x) = I_p$; this is known as the "one block problem"(this includes some important problems from classical control, such as the mixed sensitivity problem).

For $p : \mathbf{R}^n \to \mathbf{R} \cup \{-\infty\}$ (more will be said about p later) and a controller K define the cost functional

$$J_{p_0}(K) = \sup_{T \geq 0} \sup_{w \in L_{2,T}} \sup_{x_0 \in \mathbf{R}^n} \left\{ p_0(x(0)) + \tfrac{1}{2} \int_0^T \left[|z(s)|^2 - |w(s)|^2 \right] ds \right\}.$$

This can be interpreted as a requirement that a max-plus expectation be negative. Further, this captures the notion of dissipation, since (G, K) is dissipative if and only if $J_{-\beta}(K) \leq 0$, for some $\beta \geq 0$, $\beta(0) = 0$.

For fixed $u, y \in L_{2,loc}$, the *information state* p_t is defined by

$$p_t(x) = p_0(\xi(0)) + \tfrac{1}{2} \int_0^t \left[|C_1(\xi(s)) + u(s)|^2 - |y(s) - C_2(\xi(s))|^2 \right] ds, \tag{1.3}$$

where $\xi(\cdot)$ is the solution of

$$\dot{\xi} = A(\xi) + B_1(\xi)(y - C_2(\xi)) + B_2(\xi)u \tag{1.4}$$

with *terminal* condition $\xi(t) = x$. Using the information state, the dissipative property can be characterized as:

$$J_{p_0}(K) = \sup_{T \geq 0} \sup_{y \in L_{2,T}} \{ \langle p_t \rangle \} \leq 0, \tag{1.5}$$

for some p_0, where $\langle p \rangle = \langle p + 0 \rangle$ and $\langle p + q \rangle \triangleq \sup_{x \in \mathbf{R}^n}\{p(x) + q(x)\}$ is the *max-plus inner product* (called "sup-pairing" in [8]).

The dynamics for p_t when smooth is a partial differential equation: for fixed $u \in L_{2,loc}$ and $y \in L_{2,loc}$ we have

$$\dot{p}_t = F(p_t, u(t), y(t)), \tag{1.6}$$

where $F(p, u, y)$ is the differential operator $F(p, u, y) = -\nabla_x p \cdot (A + B_1(y - C_2) + B_2 u) + \frac{1}{2}|C_1 + u| - \frac{1}{2}|y - C_2|^2$.

The desired controller is obtained by finding the controller which minimizes $J(K)$. If the H_∞ problem for G is solvable, then the function $W(p) = \inf_K J_p(K)$ is finite on a certain domain dom$W$. By dynamic programming, W formally satisfies the (infinite dimensional) PDE

$$\inf_u \sup_y \{\nabla_p W(p)[F(p, u, y)]\} = 0. \tag{1.7}$$

Let $\mathbf{u}^*(p)$ and $\mathbf{y}^*(p)$ denote the values of u and y which attain the infimum and supremum in (1.7). Indeed, a direct calculation gives

$$\mathbf{u}^*(p) = \nabla_p W(p)[-C_1 + B_2' \nabla_x p] \tag{1.8}$$

These formulas define the *optimal information state controller* K^* by

$$K^*[y](t) = \mathbf{u}^*(p_t[y]), \tag{1.9}$$

where p_0 is suitably chosen in dom$W$ (the particular choice of p_0 is a very important issue.) In [9] it is shown for discrete-time problems that under suitable hypotheses, *K^* solves the H_∞ control problem for G if and only if the H_∞ control problem for G is solvable.* These and more detailed results have been obtained for continuous time problems in [6].

Regarding this solution, we make the following remarks. (i) Information states need not be everywhere finite (these are called *singular information states* in [6]). (ii) A framework is needed which will permit interpretation of the dynamic programming PDE (1.7), and the formulas (1.8) for the optimal control and observation as well as for convergence of information states (and hence stability). These points are discussed in detail in [6].

3 SINGULAR INFORMATION STATES AND STABILITY

A crucial issue in the information state solution is initialization of the controller $K^*_{p_0}$, and this in turn is very closely related to equilibrium solutions p_e of the information state equation: $0 = F(p_e, 0, 0)$. By stability of the information state system (1.6) we will mean convergence, in a sense to be described, to a particular stable equilibrium (modulo additive constants): $p_t \to p_e + c$ as $t \to \infty$, where $c \in \mathbf{R}$ and the equilibrium p_e is defined by

$$p_e(x) = \lim_{T \to \infty} \{\tfrac{1}{2} \int_{-T}^{0} [|C_1(\xi(s))|^2 - |C_2(\xi(s))|^2] \, ds\}, \tag{1.10}$$

where $\xi(\cdot)$ is the solution of (1.4) with $u = 0$, $y = 0$, and $\xi(0) = x$.

The nature of the equilibrium p_e depends on the stability properties of $A^\times = A - B_1 C_2$. If $A^\times$ is L_2 exponentially stable, then $p_e = \delta_0$ where δ_0 is a *singular* information state of the form

$$\delta_M(x) = \begin{cases} 0 & \text{if } x \in M, \\ -\infty & \text{if } x \notin M, \end{cases} \tag{1.11}$$

where $M \subset \mathbf{R}^n$. On the other hand, if $-A^\times$ is L_2 exponentially stable, then p_e is a function which is everywhere finite. If $A^\times$ is hyperbolic, with stable and antistable directions, then $p_e(x) = \delta_{M_{as}} + \breve{p}$ where M_{as} is the antistable manifold for $A^\times$, and $\breve{p}$ is a finite function on M_{as}. In [6], it is shown that the choice $p_0 = p_e$ is a natural and correct initialization of the information state controller. Thus it is important to make use of singular information states in the theory of nonlinear H_∞ control.

Let's consider the convergence $p_t \to \delta_0 + c$ in the case where $A^\times$ is L_2 exponentially stable, with initialization $p_0 = \delta_0$ and inputs $u, y \in L_2$. By [6], p_t is given by $p_t = \delta_{\xi(t)} + c(t)$, where $\xi(\cdot)$ is the trajectory of the system (1.4) with initial condition $\xi(0) = 0$ and inputs u, y, and $c(t)$ is the integral accummulated long this trajectory: $c(t) = \frac{1}{2}\int_0^t [|C_1(\xi(s)) + u(s)|^2 - |y(s) - C_2(\xi(s))|^2]\, ds$. The stability assumption ensures that $\xi(t) \to 0$ and $c(t) \to c(u, y)$ as $t \to \infty$, with $c(u, y) \in \mathbf{R}$. However, $\delta_{\xi(t)}$ does not converge **pointwise** to δ_0 as $t \to \infty$. To see this, consider the generic situation so that $\xi(t) \neq 0$ for $t > 0$ (recall $\xi(0) = 0$). Then $\delta_{\xi(t)}(0) = -\infty < \delta_0(0) = 0$ for all $t > 0$ implies that $\lim_{t\to\infty} \delta_{\xi(t)}(0) = -\infty \neq \delta_0(0) = 0$. Therefore it is necessary to relax the mode of convergence. We do this by making use of weak convergence in the max-plus sense. Here, the information states are interpreted as *max-plus measures*.

Let $\mathcal{X}_e$ denote the space of u.s.c. functions $p : \mathbf{R}^n \to \mathbf{R} \cup \{-\infty\}$ which are bounded above: $\langle p \rangle < +\infty$. This is the natural state space for the information state system. The *max-plus integral of f with respect to p* is defined by

$$p(f) = \sup_{x \in \mathbf{R}^n} \{p(x) + f(x)\} = \langle p + f \rangle. \tag{1.12}$$

A sequence $\{p_n\} \subset \mathcal{X}_e$ is said to *converge weakly* to $p_\infty \in \mathcal{X}_e$ if and only if

$$p_n(f) \to p_\infty(f)$$

in $\mathbf{R}_e$ for all $f \in C_b(\mathbf{R}^n)$ (continuous, bounded).

In the next theorem we apply this concept of weak convergence to obtain a stability result for the information state.

Theorem 1 *Assume $A^\times$ is L_2 exponentially stable. Let $u, y \in L_2[0, \infty)$ and $p_0 = p_e = \delta_0$. Then*

$$p_t \Longrightarrow p_e + c(u, y) \text{ as } t \to \infty \tag{1.13}$$

where $c(u, y)$ is a real number depending on u, y.

Proof. Given the above discussion, it remains to prove $\delta_{\xi(t)} \Longrightarrow \delta_0$ as $t \to \infty$. Select a test function $f \in C_b(\mathbf{R}^n)$. Then for any $x \in \mathbf{R}^n$,

$$\delta_x(f) = \langle \delta_x + f \rangle = f(x),$$

and so

$$\delta_{\xi(t)}(f) = f(\xi(t)) \to f(0) = \delta_0(f) \text{ as } t \to \infty.$$

□

In summary, we see that information states have natural and useful interpretation as max-plus measures, and using max-plus weak convergence, stability of the information state system can be addressed.

4 SEMIGROUPS AND GENERATORS

We turn now to the dynamic programming PDE (1.7) and the formula for the optimal control (1.8) Both of these expressions present serious difficulties since one would like to interpret them for p in the space $\mathcal{X}_e$. The main difficulties are the interpretation of the gradient $\nabla_p W$ (since $\mathcal{X}_e$ is not a standard space of functional analysis such as a Banach or Hilbert space) if we are fortunate enough for a derivative to exist, the definition of weak solution since in general W will not be differentiable in a classical sense, and interpretation of the terms in (1.7) and (1.8) involving $\nabla_x p$ when p is not differentiable, and in particular, when p is singular.

We now present a theory of semigroups and generators for the nonlinear H_∞ control problem that has recently been developed in [6]. This theory was inspired by the Nisio semigroup as studied in the case of partially observed stochastic control in [4]. The idea is to define the semigroup associated with the H_∞ problem and describe it generator on a class of test functions.

Let $S_t^{u,y}$ denote the operator which maps an initial information state $p_0 = p$ to the information state at time t determined by the inputs u, y. By dynamic programming, $S_t^{u,y}$ satisfies the semigroup property. The semigroup $S_t^{u,y}$ induces a semigroup $\mathcal{S}_t^{u,y}$ in the space $\mathbf{F}(\mathcal{X}_e) = \{p \mid p : \mathcal{X}_e \to \mathbf{R}\}$ of real valued functions defined on $\mathcal{X}_e$. For $\psi \in \mathbf{F}(\mathcal{X}_e)$, write

$$\mathcal{S}_t^{u,y}\psi(p) = \psi(S_t^{u,y}p) \tag{1.14}$$

whenever the RHS is defined. In general there will be a domain of functions ψ and points p for which the RHS is defined. Using this semigroup, we can express the value function W as $W(p) = \inf_K \sup_{T \geq 0, y \in L_{2,T}} \mathcal{S}_T^{K(y),y}\langle\,\cdot\,\rangle(p)$ (the semigroup is applied to the function $\psi(p) = \langle p \rangle$).

We shall say that an operator $\mathcal{L}^{u,y}$ is a *"generator"* for the transition operator $\mathcal{S}^{u,y}$ if there exists a nonempty set $\mathrm{dom}\mathcal{L}^{u,y} \subset \mathbf{F}(\mathcal{X}_e)$ such that for $\psi \in \mathrm{dom}\mathcal{L}^{u,y}$, and each constant pair $(u, y) \in \mathbf{R}^m \times \mathbf{R}^p$ the limit

$$\mathcal{L}^{u,y}\psi(p) = \lim_{t \to 0} \frac{\mathcal{S}_t^{u,y}\psi(p) - \psi(p)}{t} \tag{1.15}$$

exists for p belonging to a nonempty set $\mathrm{dom}\mathcal{L}^{u,y}\psi \subset \mathcal{X}_e$. In general it will be difficult or impossible to evaluate $\mathcal{L}^{u,y}\psi$ for arbitrary ψ. However, it is possible to evaluate $\mathcal{L}^{u,y}$ for certain types of functions ψ. Let us define the following class of test functions:

$$\hat{\mathcal{G}}_b = \{\psi \in \mathcal{G}_b \ : \ \psi(p) = g(\langle p+f_1\rangle, \ldots, \langle p+f_k\rangle),$$
$$\text{for some } k \geq 1,\ g \in C_b^1(\mathbf{R}^k),\ f_1, \ldots, f_k \in C_b^1(\mathbf{R}^n)\}$$

The next theorem provides explicit evaluation of $\mathcal{L}^{u,y}\psi$ for $\psi \in \hat{\mathcal{G}}_b$. Write $[[p+f]] = \mathrm{argmax}_{x\in\mathbf{R}^n}\{p(x)+f(x)\}$.

Theorem 2 *([6]) Let $\psi \in \hat{G}_b$ and let $p \in \mathcal{X}_e$ be tight. Then for each constant pair $(u, y) \in \mathbf{R}^m \times \mathbf{R}^p$*

$$\begin{aligned}\mathcal{L}^{u,y}\psi(p) &= \textstyle\sum_{i=1}^k \partial_i g(\langle p+f_1\rangle, \ldots, \langle p+f_k\rangle)\cdot \\ &\quad \sup_{\bar{x}\in[[p+f_i]]} \{\nabla_x f_i(\bar{x})\cdot(A^\times(\bar{x}) + B_1(\bar{x})y + B_2(\bar{x})u) \\ &\qquad + \tfrac{1}{2}|C_1(\bar{x}) + D_{12}u|^2 - \gamma^2\tfrac{1}{2}|y - C_2(\bar{x})|^2]\} \\ &= \textstyle\sum_{i=1}^k \partial_i g(\langle p+f_1\rangle, \ldots, \langle p+f_k\rangle)\cdot \sup_{\bar{x}\in[[p+f_i]]}\{F(-f_i, u, y)(\bar{x})\}\end{aligned}$$

A remarkable feature of the expression in the theorem is that it does not involve any derivatives of p; the derivatives have been "transfered" to the test functions f_i. Thus the theorem provides a "weak-sense" interpretation of the formal expression $\nabla_p\psi(p)[F(p,u,y)]$ (recall from the definition that $F(p,u,y)$ involves the gradient $\nabla_x p$, which may not exist, whereas $F(-f_i,u,y)$ involves $\nabla_x f_i$, which does exist).

In terms of the operator $\mathcal{L}^{u,y}$, the dynamic programming PDE (1.7) takes the form

$$\inf_{u\in\mathbf{R}^m}\sup_{y\in\mathbf{R}^p}\{\mathcal{L}^{u,y}W(p)\} = 0. \tag{1.7$'$}$$

This is a "weak-sense" view of the dynamic programming PDE of nonlinear H_∞ control. It is analogous to simlar equations in stochastic control.

In the important special case of *certainty equivalence* [2], the value function turns out to be $W(p) = \langle p+V\rangle$ (for some p), where V is the state feedback H_∞ value function, and one assumes that $[[p+V]] = \bar{x}$ is unique. Then we can evaluate $\mathbf{u}^*(p) = \operatorname*{argmin}_u \sup_y \mathcal{L}^{u,y}\langle p+V\rangle = -C_1(\bar{x}) - B_2(\bar{x})'\nabla_x V(\bar{x})'$ which makes sense for singular p (provided V is smooth). The p dependence is via $\bar{x} = \bar{x}(p)$.

References

[1] M. Akian, J.P. Quadrat and M. Viot, *Bellman Processes*, in: Lecture Notes in Control & Info. Science, 199, 1994, Springer Verlag, New York.

[2] T. Basar and P. Bernhard, "H^{∞}-Optimal Control and Related Minimax Design Problems: A Dynamic Game Approach", Second Edition, Birkhauser, Boston, 1995.

[3] P. Dupuis and R.S. Ellis, "A Weak Convergence Approach to the Theory of Large Deviations", Wiley, New York, 1997.

[4] W.H. Fleming, *Nonlinear Semigroup for Controlled Partially Observed Diffusions*, SIAM J. Control Optim., 20(2) 1982, 286-301.

[5] K. Glover and J.C. Doyle, *State Space Formulea for all Controllers that Satisfy an H_{∞} Norm Bound and Relations to Risk Sensitivity*, Systems and Control Letters, 11 (1988) 167-172.

[6] J.W. Helton and M.R. James, "Extending H_{∞} Control to Nonlinear Systems", monograph, to appear, SIAM, 1999.

[7] D.H. Jacobson, *Optimal Stochastic Linear Systems with Exponential Performance Criteria and Their Relation to Deterministic Differential Games*, IEEE Trans. Aut. Control, **AC-18** (2) (1973) 124-131.

[8] M.R. James, J.S. Baras and R.J. Elliott, *Risk-Sensitive Control and Dynamic Games for Partially Observed Discrete-Time Nonlinear Systems*, IEEE Trans. Automatic Control, 39 (1994), 780-792.

[9] M.R. James and J.S. Baras, *Robust H_{∞} Output Feedback Control for Nonlinear Discrete-Time Systems*, IEEE Trans. Aut. Control, 40 (1995), 1007-1017.

[10] M.R. James and J.S. Baras, *Partially Observed Differential Games, Infinite Dimensional HJI Equations, and Nonlinear H_{∞} Control*, SIAM J. Control Optim., 1996.

[11] T. Jiang, *The Metric of Large Deviation Convergence*, preprint, 1995.

[12] G.L. Litvinov and V.P. Maslov, *Correspondence Principle for Idempotent Calculus and Some Computer Applications*, preprint, 1995.

[13] M. Nisio, *On a Nonlinear Semigroup Attached to Optimal Stochastic Control*, Publ. RIMS Kyoto Univ. 13 (1976), 513-537.

REFLECTED FORWARD BACKWARD STOCHASTIC DIFFERENTIAL EQUATIONS AND CONTINGENT CLAIMS

Michael Kohlmann

Universitaet Konstanz, Germany

1 INTRODUCTION

In [8] we examined the properties of adjoint processes in stochastic control on the basis of the theory of stochastic flows. We extended earlier results by [2] and made extensive use of the important contributions to the theory by Pardoux, Peng , Zhou, and Renner, and of the results on stochastic partial differential equations by Antonelli , Pardoux, Peng , Zhou, which extend the pioneering works by Bensoussan, Bismut, Kunita, and many others.For references the reader is referred to [8]. We will here apply some of the results to the theory of pricing contingent claims. The valuation process will take the form of a possibly reflected BSDE (see e.g.[3]) which will be interpreted as the adjoint of a trivial, possibly singular control problem. Using the results in [1] we study sequential hedging problems.

2 A TRIVIAL CONTROL PROBLEM

A bond/stock price-asset is given in the usual notation by

$$dP_t^0 = r(P_t^0, \omega)dt \text{ (bond)}$$
$$dP_t^i = b_i(t, P_t, \omega)dt + \sigma_{ij}(t, P_t, \omega)dw_t^j \text{ (stock)}$$

on the time interval $[0, T]$. Let (P_t^{sx}) be the strong solution of the SDE with initials$(s, x) \in [0, T] \times \Re^n$, and let σ satisfy conditions such that a unique risk premium process

$$\theta(t, P_t, \omega) = \sigma^{-1}(t, P_t, \omega)\left[b(t, P_t, \omega) - r(t, P_t, \omega) \cdot 1_n\right]$$

exists. With this we are in an arbitrage free world (for details see e.g. [7]).We will assume that $n = d = 1$. In the first part of this article this is no real restriction. Most results can be generalized to higher dimensions. In the last section, however, this assumption becomes crucial, as there we make use of comparison theorems where processes are used which have similar properties as local times. As in [1] this makes an extension to higher dimensions impossible, at least at the moment. Also we should note that some of the results on spdes used below only hold when (P_t) is replaced by $(\log P_t)$.

Now consider the following trivial control problem

$$dz_{st} = -z_{st}\left[r(t, P_t^{sx}, \omega)dt + \theta(t, P_t^{sx}, \omega)dw_t\right]$$
$$z_{ss} = 1$$

with cost criterion $J = E\left[z_{sT}g(P_T^{sx})\right]$, where $g : \Re \times \Omega \to \Re^+$ is a nonnegative, bounded, non-anticipative process which is assumed to be once continuously differentiable in the first variable. We interpret this as a control problem with a trivial one-point control space. The formal adjoint process for this control problem is given by the backward equation (more exactly we had to call it a system of forward-backward sdes)

$y_t = g(P_T^{sx}) - \int_t^T \left[y_u r(u, P_u^{sx}) + Z_u\theta(u, P_u^{sx}, \omega)\right] du - \int_t^T Z_u dw,\ t \in [s,T]$.

Note that (z_{st}) is the deflator process and (y_t) is the price process for the claim $\xi_T := g(P_T^{sx})$, where the formal duality gives the interpretation

$$y_t = E\left[z_{tT}\xi_T \mid F_t\right], t \in [s,T]$$

Rewrite (y_t) as $y_t = \bar{E}\left[\exp(-\int_t^T r(s, P_s)ds)\xi_T \mid F_t\right]$,where $\bar{E}$ is the expectation with respect to the risk neutral measure associated with the Girsanov functional of θ. In this form we see that (y_t) corresponds to the risk neutral price of the claim in the classical notation. Also note that $z_{st}y_t$ is a P-martingale. So, in order to determine the price of the claim it is necessary to solve the BSDE.

Definition A solution of the BSDE

$$y_t = g(P_T^{sx}) - \int_t^T \left[y_u r(u, P_u^{sx}) + Z_u\theta(u, P_u^{sx}, \omega)\right] du - \int_t^T Z_u dw_u$$

is a pair (y, Z) such that (y_t) is a continuous, adapted process and (Z_t) is a predictable, square integrable process. The solution is unique if both processes are equal $P - a.s.$.

Conditions to guarantee a unique solution in the sense of 2.1 are found in [3].

Now we will try to characterize the solution in terms of a pricing system. This is defined to be a mechanism to bring (y_t) and (P_t) into a relation.

3 THE PRICING SYSTEM

Definition A stochastic pricing system for the claim ξ_T is a function $u : [0,T] \times \Re \times \Omega \to \Re$ which satisfies u is progressively measurable and $u(t, P_t^{sx}, \omega) = y_t(\omega)$ $P - a.s., t \in [s,T]$.

At this stage we do not impose conditions in the second variable, as such conditions must comply with the real world requirements.

Definition The pricing system will be called *convenient*, if $u(t,.,\omega) = u(t,.,\alpha_t)$
where α is a given diffusion process. It will be called *differentiable*, if $u(t,.,\omega) \in C^{12}$ $P-a.s.$,and it will be called *deterministic* if $u(t,x,\omega) = u(t,x)$.

a) the most general case

Let $E\left[z_{tT}g(P_T^{tx}) \mid F_t\right] = u(t,x,\omega)$, where $x = P_t^{sx}$. As we may assume that $E\left[z_{tT}g(P_T^{tx}) \mid F_t\right]$ is a special semimartingale (under appropriate conditions) we write $u(t,x,\omega)$ as an integral equation between random fields (see [9] for semimartingales with spatial parameters):

$$u(t,x,\omega) = u(0,x,\omega) + \int_0^t p(s,x,\omega)ds + \int_0^t k(s,x,\omega)dw_s.$$

Now apply the Itô-Ventcell formula as generalized in [9] to find

$$\begin{aligned}
u(t,P_t) &= u(T,P_T) - \int_t^T \left[p(s,P_s) + 1/2\sigma^2(s,P_s)u_{xx}(s,P_s)\right] ds \\
&\quad - \int_t^T \left[b(s,P_s)u_x(s,P_s + \sigma(s,P_s)k_x(s,P_s)\right] \\
&\quad - \int_t^T \left[k(s,P_s) + \sigma(s,P_s)u_x(s,P_s)\right] dw_s
\end{aligned}$$

Compare this to the backward s.d.e for (y_t)

$$y_t = g(P_T^{sx}) - \int_t^T \left[y_u r(u,P_u^{sx}) + Z_u\theta(u,P_u^{sx},\omega)\right] du - Z_u dw_u$$

to find

$$\begin{aligned}
u(T,P_T) &= g(P_T) = \xi_T \\
p &= -1/2\sigma^2 u_{xx} - bu_x - \sigma k_x + ru + (k + \sigma u_x)\theta \\
&= -1/2\sigma^2 u_{xx} - (b - \sigma\theta)u_x + ru + k\theta - \sigma k_x,
\end{aligned}$$

where $Z = k + \sigma u_x$.This means that the solution of the spde

$$u_t = \left[1/2\sigma^2 u_{xx} + (b - \sigma\theta)u_x - ru - k\theta + \sigma k_x\right] ds - kdw$$

with final condition $u(T,x) = g(x)$ is a stochastic price system in the sense of definition 3.1.

Remark (i) For conditions to ensure existence and uniqueness of this spde the reader is referred to [10].

(ii) When we take the hedging point of view as in [7] it is staightforward that the optimal hedging strategy is given by $\pi_t = \sigma^{-1}Z_t = u_x(t,P_t) + \sigma^{-1}(t,P_t)k(t,P_t)$.

(iii) Many arguments above become notationally more transparent when we use notations and results from Malliavin's calculus. However, in order to make the results easily comparable to the application below, we refrained from doing so.

(iv) The result $k = Z - \sigma u_x$ is closely related to an equation which appears in the maximum principle of a stochastic control problem, where both drift and diffusion are controlled. There the term corresponds to the second adjoint equation . An important question then arises, namely when equality holds between Z and σu_x, i.e. $k_t = 0$.

(v) The BSDE as a tool to model evaluations of claims has turned out to be extremely powerful. So it is easy to model the Foellmer-Schweizer hedging within this model: just subtract a martingale orthogonal to the Brownian motion from the original BSDE:

$$y_t = g(P_T^{sx}) - \int_t^T [y_u r(u, P_u^{sx}) + Z_u \theta(u, P_u^{sx}, \omega)]\, du - Z_u dw_u - M_t.$$

Or just as another example for the power of this tool: Recently we could derive the price and portfolio of an informed agent, i.e. an agent with anticipative knowledge about part of the market, by applying the BSDE-techniques to Protter´s [11]result on the connection between the enlargement of a filtration and Girsanov´s theorem. This simplifies the proof [6], and extends the result to include the Foellmer-Schweizer model.

The BSDE technique appears to be tailor made for finance purposes.

To compute the price of the claim and the optimal hedging strategy we have to
(i)solve the FBSDE and compute $\sigma^{-1} Z_t$
or equivalently
(ii) solve the spde (u, k) and compute $u_x(t, P_t) + \sigma^{-1} k(t, P_t)$
or
(iii) find cases where $k = 0$ and then compute $u_x(t, P_t)$ with different means.

b) the convenience rate case

To solve the problem of 3.3 (iii) we use a result in [12] where the relation between first and second adjoint was considered in the framework of stochastic control. From the point of view of finance our setting will be more general than necessary for the convenience rate problem, where only the drift coefficient would depend on a further forward SDE. However we hope to treat more general problems as e.g. a problem of pricing an asset which depends on an index of some kind in the coefficients:

Let the index be described by

$$d\alpha_t = a(t, \alpha_t)dt + c(t, \alpha_t)dw_t, a_0 = \alpha$$

and let all randomness in g, r, b, c come from (P_t, α_t), i.e.

$$dP_t = b(t, P_t, \alpha_t)dt + \sigma(t, P_t, \alpha_t)dw_t.$$

In this case a quite lengthy technical computation leads to the result

$$k_s = Z_s - \sigma u_x - \sigma u_\alpha = 0,$$

so that here we find a deterministic price system in the form $u(t, x, \alpha)$. The influence of the index on the portfolio is similar to the influence of the volatility of the stock price.

c) direct computations

Finally let us apply some results from the theory of stochastic flows to generate an explicit representation of (y_t). We here assume that we are in a Markovian world, that all coefficients are sufficiently differentiable, and that $r = 0$. The last assumption is made at the beginning to make the results from [2] applicable without change. The general case then is a simple obvious generalization. We are now going to compute the representation of the martingale

$$y_t = \bar{E}\left[g(P_T(x_0)) \mid F_t\right] = \bar{E}(g(P_T(x_0))) + \int_0^t \gamma d\widetilde{w},$$

where $\widetilde{w}$ is a Brownian motion under the risk neutral measure $\widetilde{P}$. From the Markov property we have

$$\begin{aligned} y_t &= \bar{E}\left[g(P_T(x_0)) \mid F_t\right] = E\left[z_{tT}(x) g(P_{tT}(x)) \mid F_t\right] \\ &= E\left[z_{tT}(x) g(P_{tT}(x))\right] = u(t, x). \end{aligned}$$

Arguing as in [4] the integrand (γ_s) must be equal to

$$\tfrac{\partial u}{\partial x}(s, P_{0s}(x_0))\sigma(s, P_{0s}(x_0)).$$

It is immediate from [4] that then for the general case $r = r(t, P_t) \not\equiv 0$

$$\begin{aligned} \frac{\partial u}{\partial x} &= \\ & E[z_{tT}(x)\{\int_t^T (\theta_\xi(r, P_{tr})D_{tr}d\widetilde{w}r + \int_t^T r_\xi(r, P_{tr})D_{tr}dr) \cdot g(P_{0T}(x_0)) \\ & + g_\xi(P_{tT}(x))D_{tT}(x)\}], \end{aligned}$$

and with this we are back in the classical case.

Remark (i) As in [4]we can derive a bpde for γ.

(ii) By repeating the representation in c) over and over we find a chaos decomposition as in [5]. From this we can compute the ratios of the claim.

4 THE AMERICAN CONTINGENT CLAIM

An American option allows to choose the exercise time at any time within the horizon. In order to hedge the risk of early exercise, so-called super-strategies have to be introduced. The price of the option then takes the form of a reflected backward sde as studied in [3].

Definition The reflected backward sde is described by

$$y_t = \xi_T - \int_t^T [ry_s + \theta Z_s]\, ds - \int_t^T Z_s dw_s + \int_t^T dC_t$$

subject to the constraints (i) $y_t \geq \xi_t$ and (ii) $\int_0^T (y_t - \xi_t) dC_t = 0$.

A solution is a triple $(y^{sxC}, Z^{sxC}, C^{sx})$ where (y, Z) has the usual properties and C is an adapted, continuously increasing process with $C_0^{sx} = 0$, such that (i) and (ii) holds for $\xi_t = g(P_t^{sx})$.

The coefficients here have the same general properties as in the first section. The price (y_t) is then given by

$$y_t = ess - \sup_{\tau \in [t,T]} E\left[z_{t\tau} \xi_\tau \mid F_t\right], \tau \text{ stopping time. } (*)$$

This is obvious from the following control-theoretical considerations:

(y_t) is the first adjoint of a singular control problem. On the other hand this first derivative coincides with the value function of a stopping problem, and this is the intuitive meaning of $(*)$.

In [1] we extended the above mentioned results to general diffusions and established the relation between singular control and optimal stopping for this generalized case. Then it was shown that the singularly influenced process corresponds to a process conctructed from a monotone sequence of stopping times.

Before we go into this problem we state the following variational result for the stochastic pricing system of the American claim.

Theorem 1 : *Let $u(t,x)$ be a random field which solves the obstacle problem*

$$\{(u(t,x) - g(x)) \wedge$$
$$(u(t,x) - g(x) + \int_t^T \tfrac{1}{2}\sigma^2 u_{xx} + ru_x - ru + \sigma k_x - k\theta ds + \int_t^T k dw_s)\}$$
$$= 0$$

with final condition $u(T,x) = g(x)$.

This system of variational equalities (in an appropriate space) gives the stochastic price of the American contingent claim $u(t, P_t) = y_t$.

From this it is clear that the role of the increasing process (C_t) is to keep (y_t) away from the obstacle (or the forbidden region) ξ_t ($y_t < \xi_t$, respectively). This will allow the seller of the option to fulfill the requirements of the option at any time in $[0, T]$. Now it is easy to guess, a bit more difficult to prove, but standard, that the optimal stopping time is given by

$$\tau_t = \inf\{T \geqslant s \geq t : y_s = \xi_s\} = \inf\{T \geq s \geq t : y_s = g(P_s)\}.$$

Thus τ_t is the first time of the first move of (C_t), and it is immediate that the American price is the European price with (random) exercise time τ_t.

In [1] we considered the mathematics behind a problem of installment options and related this to a problem of sequential stopping which turned out to be the limit of a family of impulse control problems. We will apply these results to describe sequential hedging. The idea is easily described: We consider the price of an American claim as decribed above by a RBSDE. At the first exercise time τ_t the seller offers a new option starting in (τ_t, P_{τ_t}) and we compute the price of this new claim with these new initials, to get a second stopping time

$$\tau_{\tau_t} = \inf\{T \geq s > \tau_t : y_s^{\tau_t \xi_{\tau_t}} = g(P_s^{\tau_t P_{\tau_t}})\}.$$

The price again coincides with the European price with starting parameters (τ_t, P_{τ_t}) and exercise time τ_{τ_t}. In this way we get an increasing sequence of stopping times. On the other hand let us work backwards to consider an increasing sequence of stopping times $(\tau_j)_{j=1,\dots,n+1}$ and a related obviously increasing family of deterministic states $0 = x_1 \leq x_2 \leq \dots \leq x_n \leq K$ such that

$$y_t^{sx(\tau_j x_j)} = g(P_T^{sx}) - \int_t^T (r(P_u^{sx})y_u + \theta(P_u^{sx})Z_u)du - \int_t^T Z_s dw_s + x_n - x_j$$

for $\tau_j \leq t < \tau_{j+1}$,where the x_j are minimal such that $y_t^{sx(\tau_j x_j)} \geq \xi_t$ for $\tau_j \leq t < \tau_{j+1}$. Define (ζ_t^n) by $\zeta_t^n = x_j$ for $\tau_j \leq t < \tau_{j+1}$, so that (ζ_t^n) is increasing and right continuous. The resulting impulsed process may thus be identified with a process which at random times τ_{j+1} jumps to a process with final condition $g(P_t^{sx}) + (x_n - x_j)$ and stays there until τ_j.

By choosing more and more support points it was then proved that finally there is a one-to-one correspondence between an exhaustive family of stopping times $\tau^* = (\tau_y)_{y\in J}$ derived from $\tau_j = \tau_{x_j}$ and an increasing continuous process (ζ_t) defined as the limit of the (ζ_t^n) constructed from (τ_j). (ζ_t) is independent of the approximating sequences so that we identify $(y^{sx\zeta}, Z^{sx\zeta}, \zeta) = (y^{sx\tau^*}, Z^{sx\tau^*})$.

Now let $(y^{sxC}, Z^{sxC}, C^{sx})$ be a solution of the RBSDE above, then $(y^{sxC}, Z^{sxC}, C^{sx}) = (y^{sx\tau^*}, Z^{sx\tau^*})$, so that we may summarize:

Theorem 2 *A self-financing superprice of the American claim is a solution of the RBSDE*

$$y_t = \xi_T - \int_t^T [ry_s + \theta Z_s]\, ds - \int_t^T Z_s dw_s + \int_t^T dC_t$$

with obstacle $y_t \geq \xi_t = g(P_t^{sx})$ such that $C_0 = 0$ and $\int_0^T (y_u - \xi_u)dC_u = 0$.

The solution is denoted by $(y^{sxC}, Z^{sxC}, C^{sx})$. Let the correspondence between C and τ^* be denoted by $\alpha(C) = \tau^*, \pi(\tau^*) = C$, then $(y^{sx\pi(\tau^*)}, Z^{sx\pi(\tau^*)}, \pi(\tau^*)) = (y^{sx\tau^*}, Z^{sx\tau^*})$ is a self-financing superprice.

For obvious reasons we call this price the price corresponding to the **rolling hedge** $(\sigma^{-1} Z^{sx\alpha(C)})$. In this way the superprice is characterized by the limit of European prices with random exercise times.

References

[1] Boetius, F.; Kohlmann, M.: Connections between optimal stopping and singular stochastic control; to appear in Stochastic Proc. Appl. (1998)

[2] Baras, J.M.; Elliott, R.J.; Kohlmann, M.: The partially observed stochastic minimum principle; SIAM Journal Control Optim. 27,6 (1989), 1279-1292

[3] El Karoui, N.; Peng, S.; Quenez, M.C.: Backward stochastic differential equations in finance; to appear in: Mathematical Finance

[4] Elliott, R.J.; Kohlmann, M.: The second order minimum principle and adjoint processes; Stochastics and Stochastics Rep. 46 (1994), 25-39

[5] Elliott, R.J.; Kohlmann, M.: Integration by parts, homogeneous chaos expansions and smooth densities; Annals of Prob. 17 (1989), 194-207

[6] Elliott,R.J.; Geman, H.; Korkie, B.M.: Portfolio optimization and contingent claim pricing with differential information; Sochastics and Stochastics Rep.60 (1997); 185-203

[7] Karatzas, I.; Shreve, S.E.: Methods of Mathematical Finance, to appear (1998)

[8] Kohlmann, M.: Adjoint equations in stochastic control, working paper (1998)

[9] Kunita, H.: Stochastic flows and stochastic differential equations, Cambridge University Press (1990)

[10] Ma, Y.; Yong, J.: Adapted solutions of a degenerate backward spde withapplications; Stochastic Proc. App. 70 (1997), 59-84

[11] Protter, P.: A connection between the expansion of filtrations and Girsanov's Theorem;in Stoch.Partial Diff. Eq. Appl. II, ed. Da Prato, G. Tubaro, L.; LNMath 1390, Springer Verlag, Berlin-Heidelberg-New York (1990)

[12] Renner, P.; Optimal control of diffusions, thesis Konstanz (1996)

SHORT TIME ASYMPTOTICS OF RANDOM HEAT KERNELS

Hiroshi KUNITA

Graduate School of Mahtematics
Kyushu University 36
Fukuoka 812-8581, Japan

kunita@math.kyushu-u.ac.jp

Abstract: We study the random fundamental solution of a linear stochastic parabolic partial differential equation of the second order. We are particularly interested in the short time asymptotics of the random fundamental solution. If the random differential operator is nondegenerate, it will be shown that the short time asymptotics is represented by a Gaussian kernel. However, if it is degenerate, the short time aymptotics will be represented by a joint distribution of a certain linear sum of a Brownian motion and its multiple Wiener-Stratonovich integrals.

Our approach is based on the asymptotic analysis of the stochastic flow which solves the stochastic partial differential equation. It is parallel to the study of the short time asymptotics of the fundamental solution of a second order (deterministic) partial differerntial equation in Kunita [3].

1 SHORT TIME ASYMPTOTICS OF HEAT KERNELS

Let us consider a second order linear partial differential operator L on a Euclidean space $\mathbf{R}^d$ represented by

$$L = \frac{1}{2}\sum_{j=1}^{r} X_j^2 + X_0, \tag{1.1}$$

where $X_0, X_1, ..., X_r$ are first order linear partial differential operators (=vector fields) with C^∞ coefficients. We assume that $X_0, X_1, ..., X_r$ are complete and generate an n dimensional Lie algebra $\mathcal{L}$, where $d \le n < \infty$. In [3], the author studied the short time asymptotics of the fundamental solution associated with the operator L, through the solution of a stochastic differential equation:

$$d\varphi_t = \sum_{j=1}^{r} X_j(\varphi_t) \circ dW^j(t) + X_0(\varphi_t)dt, \tag{1.2}$$

where $(W^1(t), ..., W^r(t))$ is a standard r-dimensional Brownian motion on the probability space $(\Omega, \mathcal{F}, P)$ and $\circ dW^j(t)$ are Stratonovich integrators. We shall summarize them in a different context. In the following discussions, Condition (A.2) in [3] will be not assumed, but results will be stated in a localized form.

We first consider the case where L is nondegenerate. Let $\varphi_t(x), t \geq 0$ be the solution of equation (2) starting from x at time 0. It is a diffusion process with the infinitesimal generator L. Let $P_t(x, E)$ be the distribution of the random variable $\varphi_t(x)$. It is not a Gaussian distribution unless $X_j, j = 1, ..., r$ are constant vectors (first order differential operators with constant coefficients). However, if $X_1, ..., X_r$ are commutative and $X_0 = \sum_{j=1}^r c^j X_j$, then we can choose a local coordinate system $\{(\psi_x, U_x), x \in \mathbf{R}^d\}$ such that for any $x \in \mathbf{R}^d$, the distribution $P_t(x, E)$ on the local coordinate (ψ_x, U_x) is Gaussian.

We will show this in the case where $r = d$ and $X_1, ..., X_d$ are linearly independent. Associated with $\{X_1, ..., X_d\}$ we consider a system of maps $\phi_x; \mathbf{R}^d \to \mathbf{R}^d$ with parameter x by

$$\phi_x(z) = \phi_x(z_1, ..., z_d) := \exp\{\sum_{j=1}^d z_j X_j\}(x), \tag{1.3}$$

where $\exp t\{\sum_j z_j X_j\}, t \in (-\infty, \infty)$ is the one parameter group of diffeomorpshisms generated by the vector field $\sum_j z_j X_j$. Then for each x there exists a neighborhood V_x of the origin such that $\phi_x : V_x \to U_x$ is a diffeomorphism. We define a *canonical local coordinate system* $\{(\psi_x, U_x), x \in \mathbf{R}^d\}$ by $\psi_x = \phi_x^{-1}; U_x \to V_x$. Then $\psi_x(x) = 0$ holds for any x. Now the solution $\varphi_t(x)$ of equation (2) is represented by

$$\varphi_t(x) = \exp\{\sum_{j=1}^d (W^j(t) + c^j t) X_j\}(x). \tag{1.4}$$

([1], p.210) Then $\psi_x(\varphi_t(x))$ is equal to $(W^1(t) + c^1 t, ..., W^d(t) + c^d t)$ in U_x. Therefore for any x, t, $P_t(x, E \cap U_x)$ is a Gaussian measure.

If $X_1, ..., X_d$ are not commutative, the process $\psi_x(\varphi_t(x))$ is not a Brownian motion. We will change the scale of the space and time, and consider the process $\varphi_t^{(r)}(x)$ represented with the canonical local coordinate (ψ_x, U_x) as

$$\psi_x(\varphi_t^{(r)}(x)) = \frac{1}{\sqrt{r}} \psi_x(\varphi_{rt \wedge \sigma_x}(x)), \quad t \geq 0, \tag{1.5}$$

where $\sigma_x = \inf\{t > 0; \varphi_t \notin U_x\}$. Then it converges weakly to a continuous stochastic process $\varphi_t^{(0)}(x), t \geq 0$ as $r \to 0$, which is represented by

$$\varphi_t^{(0)}(x) = \exp\{\sum_{j=1}^d W^j(t) X_j\}(x). \tag{1.6}$$

Then $\psi_x(\varphi_t^{(0)}(x))$ is equal to $(W^1(t), ..., W^d(t))$ in U_x, proving that for any x, t, $P_t^{(0)}(x, E) := P(\varphi_t^{(0)}(x) \in E)$ is a Gaussian measure if E is restricted to a Borel subset of U_x.

The fact enables us to determine the short time asymptotics of the fundamental solution associated with the operator (1). Let $p_t(x,y)$ and $p_t^{(0)}(x,y)$ be the density functions of the distributions $P_t(x,E)$ and $P_t^{(0)}(x,E)$. The former is the fundamental solution associated with the operator (1). The latter is represented by

$$p_t^{(0)}(x,y) = \frac{1}{(2\pi t)^{d/2}} \exp\left(-\frac{|\psi_x(y)|^2}{2t}\right) \rho_x(y), \quad y \in U_x \tag{1.7}$$

where

$$\rho_x(y) = \left|\det \frac{\partial \phi_x}{\partial z} \circ \psi_x(y)\right|^{-1} \tag{1.8}$$

Further, we showed in [3] that $p_t(x,y) \sim p_t^{(0)}(x,y)$ as $t \to 0$, provided that $y \in U_x$.

We shall discuss how the above argument will be extended to the case where L is degenerate. Let $J = \{j_1, ..., j_l\}$ be a multi-index of $j_k \in \{0, ..., r\}, k = 1, ..., l$. We define $X_J = [X_{j_1}, [X_{j_2}, [\cdots [X_{j_{l-1}}, X_{j_l}] \cdots]$. Set $V_i = \{X_J; \|J\| = i/2\}$ for $i \geq 1$, where $\|J\| = \sharp\{j \in J; j = 0\} + 2^{-1}\sharp\{j \in J; 1 \leq j \leq r\}$. Then it holds $\mathcal{L} = \cup_{i=1}^k V_i$ for some k. We will choose a basis $\{X_1, ..., X_n\}$ of $\mathcal{L}$ such that $V_1 = \{X_1, ..., X_{r_1}\}$, $V_2 - V_1 \cap V_2 = \{X_{r_1+1}, ..., X_{r_2}\}, \cdots, V_k - (\cup_{j=1}^{k-1} V_j) \cap V_k = \{X_{r_{k-1}+1}, ..., X_{r_k}\}$. Associated with the first d vector fields $\{X_1, ..., X_d\}$ in $\{X_1, ..., X_n\}$, we define a system of maps ϕ_x by (3) and introduce a canonical local coordinate system (ψ_x, U_x), similarly as the above. Now for $j = 1, ..., d$, set $\alpha_j = 2^{-1}l$ if $X_j \in V_l - \cup_{i=1}^{l-1} V_i$. Let R be the $d \times d$ diagonal matrix with diagonal elements $\alpha_j, j = 1, ..., d$. For $r > 0$ we set $r^R = \exp(\log r)R$. For each x, we introduce a family of local diffeomorphisms $\{\gamma_r^{(x)}, r > 0\}$ by $\gamma_r^{(x)}(y) = (\phi_x \circ r^R \psi_x)(y), \quad y \in U_x$. The point x is invariant by these local diffeomorphisms. It has the group property $\gamma_r^{(x)} \circ \gamma_s^{(x)} = \gamma_{rs}^{(x)}$ for $r, s > 0$. Further it holds $\gamma_r^{(x)}(y) \to x$ as $r \to 0$ for any $y \in U_x$ and $\gamma_r^{(x)}(y) \to \infty$ as $r \to \infty$ for any $y \in U_x$ such that $y \neq x$. We call $\{\gamma_r^{(x)}\}$ a *dilation* with invariant point x.

For $r > 0$, we define a stochastic process $\varphi_t^{(r)}(x), t \geq 0$ by

$$\varphi_t^{(r)}(x) = \gamma_{1/r}^{(x)}(\varphi_{rt \wedge \sigma_x}(x)), \tag{1.9}$$

where $\sigma_x = \inf\{t > 0, \varphi_t(x) \notin U_x\}$.

Lemma 1 *([3]) The family of stochastic processes $\{\varphi_t^{(r)}(x), t \geq 0\}$ converges weakly as $r \to 0$. The limit process is equivalent in law with the followings:*

$$\varphi_t^{(0)}(x) = \exp \eta_t^{(0)}(x), \quad t \geq 0. \tag{1.10}$$

Here

$$\eta_t^{(0)} = \sum_{j=1}^{d} c_t^j X_j, \quad t \geq 0, \tag{1.11}$$

where c_t^j is a linear sum of multiple Wiener-Startonovich integrals

$$W_t^J = \int_0^t \cdots \int_0^{t_2} \circ dW^{j_1}(t_1) \cdots \circ dW^{j_l}(t_l), \tag{1.12}$$

whose multi-indices J satisfies $\|J\| = \alpha_j$ and $P_{X_j} X^J \neq 0$.

The stochastic process $\{\eta_t^{(0)}, t \geq 0\}$ of (11) is no longer a Brownian motion, provided that $\alpha_j \neq 2^{-1}$ for some $1 \leq j \leq d$. However, it is self-similar with respect to $\{r^R, r > 0\}$, i.e., the law of the process $\{\eta_{rt}^{(0)}, t \geq 0\}$ is identical with the law of the process $\{r^R \eta_t^{(0)}, t \geq 0\}$ for any $r > 0$. Let $F_t, t > 0$ be distributions of the random variables $\eta_t^{(0)}, t > 0$. They have the self-similar property: $F_{rt}(A) = F_t(r^{-R}A)$ for any r, t, A.

Now we introduce Hormander's condition. Let $\mathcal{V}(X_1, ..., X_r)$ be the smallest class of vector fields such that $X_i \in \mathcal{V}(X_1, ..., X_r)$ for any $i = 1, ..., r$, and $[X, X_0] \in \mathcal{V}(X_1, ..., X_r)$ if $X \in \mathcal{V}(X_1, ..., X_r)$. We introduce

Condition (H). dim span$\{X(x); X \in \mathcal{V}(X_1, ..., X_r)\} = d$ for any x of $\mathbf{R}^d$.

Under Condition (H), distributions $F_t, t > 0$ have C^∞ density functions $f_t(z)$. They satisfy

$$f_{rt}(z) = \frac{1}{r^{\mathrm{tr}R}} f_t(r^{-R}z), \tag{1.13}$$

because of the self-similarity of distributions F_t.

Theorem 2 *([3]) Assume Condition (H). Then the distribution $P_t^{(0)}(x, E)$ of $\varphi_t^{(0)}(x)$ has a C^∞-density function $p_t^{(0)}(x, y)$. Further, it is represented by*

$$p_t^{(0)}(x, y) = \frac{1}{t^{\mathrm{tr}R}} f_1(t^{-R}\psi_x(y))\rho_x(y), \quad \forall y \in U_x, \tag{1.14}$$

where $\rho_x(y)$ is given by (8).

Theorem 3 *([3]) Let $p_t(x, y)$ be the fundamental solution associated with the operator L of (1). Assume Condition (H). Then, for any $y \in U_x$, it holds*

$$p_t(x, y) \sim p_t^{(0)}(x, y) \quad \text{as} \quad t \to 0 \tag{1.15}$$

2 SPDE AND ASSOCIATED RANDOM TRANSITION FUNCTIONS

We shall apply the above argument to the study of the short time asymptotics of the random fundamental solution associated with the stochastic partial differential operator

$$A(t) = \left(\frac{1}{2} \sum_{j=1}^r X_j^2 + X_0 \right) t + \sum_{k=1}^m Y_k W^k(t), \tag{1.16}$$

where $Y_1, ..., Y_m$ are linearly independent complete C^∞ vector fields on $\mathbf{R}^d$ and $(W^1(t), ..., W^m(t))$ is a standard m-dimensional Brownian motion on the probability space $(\Omega, \mathcal{F}, P)$. We assume that $\{X_0, X_1, ..., X_r, Y_1, ..., Y_m\}$ generates a finite dimensional Lie algebra $\tilde{\mathcal{L}}$.

We shall first construct random transition probabilities associated with the random infinitesimal generator $A(t)$ of (16). Let $\mathbf{C} = C([0,\infty), \mathbf{R}^r)$ be the spaces of continuous paths starting from the origin of the Euclidean spaces $\mathbf{R}^r$. Elements of $\mathbf{C}$ are denoted by w and their values at t are denoted by $B(t)$, i.e., $B(t)(w) = w(t)$. Let Q be a Wiener measure on $\mathbf{C}$. We consider the product probability space $(\Omega \times \mathbf{C}, P \otimes Q)$. Then $\{B(t) = (B^1(t), ..., B^r(t)), t \geq 0\}$ and $\{W(t) = (W^1(t), ..., W^m(t)), t \geq 0\}$ are independent standard Brownian motions on the product probability space.

Consider a stochastic differential equation on the probability space $P \times Q$.

$$\begin{aligned} d\xi_t &= \sum_{j=1}^r X_j(\xi_t) \circ dB^j(t) + X_0(\xi_t)dt + \sum_{k=1}^m Y_k(\xi_t) \circ dW^k(t), \quad t > s \\ \xi_s &= x, \quad x \in \mathbf{R}^d. \end{aligned} \tag{1.17}$$

The solution $\xi_{s,t}(x) = \xi_{s,t}(x, \omega, w), t \geq s$ of the above equation defines a stochastic flow of C^∞-diffeomorphisms of $\mathbf{R}^d$. We define random transition probabilities $\tilde{P}_{s,t}(x, E)$ by

$$\tilde{P}_{s,t}(x, E, \omega) = Q(\xi_{s,t}(x, \omega, \cdot) \in E). \tag{1.18}$$

and the random positive semigroup by

$$\tilde{T}_{s,t}f(x) = \int_{\mathbf{R}^d} \tilde{P}_{s,t}(x, dy)f(y). \tag{1.19}$$

Its random infinitesimal generator (in Stratonovich sense) is given by (17), that is to say, it satisfies

$$\tilde{T}_{s,t}f = f + \int_s^t \tilde{T}_{s,r}A(\circ dr)f \tag{1.20}$$

for any C^∞ function f with bounded derivatives.

Now let $\mathcal{V}(X_1, ..., X_r)$ be the smallest class of vector fields such that (1) $X_i \in \mathcal{V}(X_1, ..., X_r)$ for any $i = 1, ..., r$, (2) If $X \in \mathcal{V}(X_1, ..., X_r)$ then $[Y_j, X] \in \mathcal{V}(X_1, ..., X_r)$ for all $j = 0, 1, ..., m$. We introduce a condition similar to Hörmander's condition (H).

Condition ($\tilde{H}$). dim span$\{X(x); X \in \mathcal{V}(X_1, ..., X_r)\} = d$ for any x of $\mathbf{R}^d$.

Then it is shown in [2] that there exists a random function $\tilde{p}_{s,t}(x, y) = \tilde{p}_{s,t}(x, y, \omega)$, C^∞ with respect to x, y, such that

$$\tilde{P}_{s,t}(x, dy) = \tilde{p}_{s,t}(x, y)dy, \quad a.s. \ P. \tag{1.21}$$

We call $\tilde{p}_{s,t}(x, y)$ the *random fundamental solution or random heat kernel* associated with the operator $A(t)$. It satisfies the following properties.

(1) $\tilde{p}_{s,t}(x,y)$ is nonnegative and continuous in (s,x,t,y) a.s. P. Further, it is a $\mathbf{C}^\infty$-function of x and y a.s. P

(2) For any $s < t < u$,

$$\tilde{p}_{s,u}(x,z) = \int \tilde{p}_{s,t}(x,y)\tilde{p}_{t,u}(y,z)dy, \quad \forall x,z \in \mathbf{R} \quad a.s.\ P. \tag{1.22}$$

(3) Kolmogorov's backward equation holds true:

$$\tilde{p}_{s,t}(x,y) = \delta_{t,y} + \int_s^t A_x(\circ\hat{dr})\tilde{p}_{r,t}(x,y), \quad \forall s < t,\ \forall x,y \in \mathbf{R}^d \quad a.s.\ P. \tag{1.23}$$

(4) Kolmogorov's forward equation holds true:

$$\tilde{p}_{s,t}(x,y) = \delta_{s,x} + \int_s^t A'_y(\circ dr)\tilde{p}_{s,r}(x,y), \quad \forall s < t,\ \forall x,y \in \mathbf{R}^d \quad a.s.\ P. \tag{1.24}$$

Here $A'(t)$ is the formal adjoint of $A(t)$, i.e., $\int A(t)fgdx = \int fA'(t)gdx$ holds a.s. for any C^∞-functions f,g with compact supports.

3 ASYMPTOTICS OF RANDOM TRANSITION PROBABILITIES

We are interested in the asymptotics of random transition probabilities $\tilde{P}_{s,t}(x,E)$ and random fundamental solutions $\tilde{p}_{s,t}(x,y)$ as t tends to s. For this purpose we shall introduce a system of canonical local coordinates associated with the random differential operator $A(t)$.

For the convenience of notations, we set $X_{r+1} = X_0, X_{r+2} = Y_1, ..., X_{r+m+1} = Y_m$. Let $J = \{j_1,...,j_l\}$ be a multi-index of $j_i \in \{1,...,r+m+1\}$. Define $\tilde{V}_i = \{X_J; \langle J\rangle = i\}, i = 1,...,$ where $\langle J\rangle = \sharp\{j \in J; 1 \le j \le r\}$ It holds $\tilde{\mathcal{L}} = \cup_{i=1}^k \tilde{V}_i$ for some $k < \infty$. We will choose a basis $\{\tilde{X}_1,...,\tilde{X}_n\}$ of $\tilde{\mathcal{L}}$ such that $\tilde{V}_1 = \{\tilde{X}_1,...,\tilde{X}_{r_1}\}$, $\tilde{V}_i - (\cup_{j=1}^{i-1}\tilde{V}_j)\cap\tilde{V}_i = \{\tilde{X}_{r_{i-1}+1},...,\tilde{X}_{r_i}\}$. Associated with the first d vector fields $\{\tilde{X}_1,...,\tilde{X}_d\}$ in $\tilde{\mathcal{L}}$, we define a system of maps $\tilde{\phi}_x; \mathbf{R}^d \to \mathbf{R}^d$ with parameter x by $\tilde{\phi}_x(z) = \exp\sum_{j=1}^d z_j\tilde{X}_j(x)$. Then the canonical local coordinate system $\{(\tilde{\psi}_x, \tilde{U}_x), x \in \mathbf{R}^d\}$ is introduced similarly as before.

Let $\tilde{R}$ be a $d\times d$ diagonal matrix, whose diagonal elements are $\tilde{\alpha}_j = 2^{-1}i$, $j = 1,...,d$ if $\tilde{X}_j \in \tilde{V}_i - \cup_{j\le i-1}\tilde{V}_j$. We define a dilation $\{\tilde{\gamma}_r^{(x)}\}$ for each x as $\tilde{\gamma}_r^{(x)}(y) = (\tilde{\phi}_x \circ r^{\tilde{R}}\tilde{\psi}_x)(y), \quad y \in \tilde{U}_x$.

We shall rewrite equation (17) using new vector fields $\tilde{X}_1,\cdots,\tilde{X}_n$. Denote the projection of X_k (or Y_k) to the space $\{\tilde{X}_j\}$ by $c_{jk}\tilde{X}_j$ (or $d_{jk}\tilde{X}_j$) and set $\tilde{W}_j(t) = \sum_k d_{jk}W^k(t)$ (or $\tilde{B}_j(t) = \sum_k c_{jk}B^k(t)$). Then equation (17) is written as

$$d\xi_t = \sum_{j=1}^n \tilde{X}_j(\xi_t)\circ dZ^j(t), \quad \text{where} \quad Z^j(t) = \tilde{B}^j(t) + \tilde{W}^j(t). \tag{1.25}$$

Define

$$\xi^{(r)}_{s,s+t}(x) = \tilde{\gamma}^{(x)}_{1/r}(\xi_{s,(s+rt)\wedge\tilde{\sigma}_x}(x)), \tag{1.26}$$

where $\tilde{\sigma}_x = \inf\{t > s, \xi_{s,t}(x) \notin \tilde{U}_x\}$.

Lemma 4 *Let $r \to 0$. Then the family of stochastic processes $\{\xi^{(r)}_{s,s+t}(x), t \geq 0\}, r > 0$ converges weakly to $\{\xi^{(0)}_{s,s+t}(x), t \geq 0\}$. Further, it is represented by*

$$\xi^{(0)}_{s,s+t}(x) = \exp \eta^{(0)}_{s,s+t}(x). \tag{1.27}$$

Here

$$\eta^{(0)}_{s,s+t} = \sum_{j=1}^{d} c^j_{s,s+t} \tilde{X}_j, \tag{1.28}$$

where $c^j_{s,s+t}$ is a linear sum of multiple Wiener-Stratonovich integrals

$$Z^J_{s,s+t} = \int_s^{s+t} \cdots \int_s^{s+t_2} dZ^{j_1}(t_1) \cdots \circ dZ^{j_l}(t_l), \tag{1.29}$$

whose multi-indecis J satisfies $\|J\| = \tilde{\alpha}_j$ and $P_{\tilde{X}_j} \tilde{X}_J \neq 0$.

The stochastic process $\eta^{(0)}_{s,s+t}$ is self-similar with respect to $r^{\tilde{R}}$, i.e., the law of $(\eta^{(0)}_{s,s+rt}, P \times Q)$ is identical with that of $(r^{\tilde{R}}\eta^{(0)}_{s,s+t}, P \times Q)$ for any s, t, r. Let $F_{s,t} = F_{s,t}(\omega)$ be the conditional distribution given ω, i.e., $F_{s,t}(A, \omega) = Q(\eta^{(0)}_{s,t}(\omega, \cdot) \in A)$. It has the self-similar property $F_{s,s+rt}(A, \omega) = F_{s,s+t}(r^{-\tilde{R}}A, \omega^{(r)})$ holds for any s, t, r, A, where $\omega^{(r)}$ is the sample such that $W_t(\omega^{(r)}) = W_{rt}(\omega)$ holds for all $t > 0$. Assume Condition $(\tilde{H})$. Then $F_{s,t}$ has a C^∞ density function $f_{s,t}(z) = f_{s,t}(z, \omega)$. It satisfies

$$f_{s,s+rt}(z, \omega) = \frac{1}{r^{\mathrm{tr}\tilde{R}}} f_{s,s+t}(r^{-\tilde{R}}z, \omega^{(r)}), \quad a.s.\ P, \tag{1.30}$$

for any $r > 0$.

We can rewrite $\eta^{(0)}_{s,t}$ of (28) as: $\eta^{(0)}_{s,t} = \sum_{1 \leq j \leq d}(d^j_{s,t} + e^j_{s,t})\tilde{X}_j$. Here, $d^j_{s,t}$ is the collection of terms $Z^J(t)$ in $c^j_{s,t}$ such that $\langle J \rangle > 0$, and $e^j_{s,t}$ is the collection of terms in $c^j_{s,t}$ such that $\langle J \rangle = 0$. We define $\hat{\eta}^{(0)}_{s,t} = \sum_{1 \leq j \leq d} d^j_{s,t} X_j$. Let $\hat{F}_{s,t}(A, \omega) = Q(\hat{\eta}^{(0)}_{s,t}(\omega, \cdot) \in A)$. Under Condition $(\tilde{H})$, $\hat{F}_{s,t}$ has also a C^∞ density function $\hat{f}_{s,t}(z) = \hat{f}_{s,t}(z, \omega)$. It satisfies the equality similar to (30). We have further,

$$f_{s,t}(z) = \hat{f}_{s,t}(z - e_{s,t}). \tag{1.31}$$

Theorem 5 *Assume Condition $(\tilde{H})$. Let $\tilde{P}^{(0)}_{s,t}(x, E, \omega) = Q(\xi^{(0)}_{s,t}(x, \omega, \cdot) \in E)$. Then $\tilde{P}^{(0)}_{s,t}(x, \cdot, \omega)$ has a C^∞ density function $\tilde{p}^{(0)}_{s,t}(x, y, \omega)$ a.s. P. Further, it is represented by*

$$\begin{aligned} \tilde{p}^{(0)}_{s,s+t}(x, y, \omega) &= f_{s,s+t}(\tilde{\psi}_x(y), \omega)\tilde{\rho}_x(y), \quad a.s.\ P, \qquad (1.32) \\ &= \frac{1}{t^{\mathrm{tr}\tilde{R}}} \hat{f}_{s,s+1}(t^{-\tilde{R}}(\tilde{\psi}_x(y) - e_{s,t}), \omega^{(t)})\tilde{\rho}_x(y), \quad a.s.\ P \end{aligned}$$

if $y \in U_x$, where $\tilde{\rho}(x)$ is defined by (8) replacing ϕ_x and ψ_x by $\tilde{\phi}_x$ and $\tilde{\psi}_x$, respectively.

Finally, we consider the case where the operator $A(t)$ of (16) is nondegenerate. The diagonal matrix $\tilde{R}$ coincides with $2^{-1}I$, where I is the identity matrix. Further, in the expression of $\eta^{(0)}_{s.s+t}$, multiple Wiener-Stratonovich integrals are not involved so that $\eta^{(0)}_{s,s+t}$ is a linear sum of Brownian motions $Z^1(t), ..., Z^m(t)$. Then the conditional distribution of $\eta^{(0)}_{s,s+t}$ with respect to Q is Gaussian. Indeed, we have

Corollary 6 *Suppose that $A(t)$ is nondegenerate. Then, if $y \in U_x$,*

$$\tilde{p}^{(0)}_{s,s+t}(x,y) = \frac{1}{(2\pi t)^{d/2}} \exp\left(-\frac{|\tilde{\psi}_x(y) - (W(s+t) - W(s))|^2}{2t} \right) \tilde{\rho}_x(y), \quad a.s.\ P. \tag{1.33}$$

Theorem 7 *Assume Condition $(\tilde{H})$. Let $\tilde{p}_{s,s+t}(x,y)$ be the fundamental solution associated with the random differential operator $A(t)$. Then we have for $y \in U_x$,*

$$\tilde{p}_{s,s+t}(x,y) \sim \tilde{p}^{(0)}_{s,s+t}(x,y) \quad \text{as } t \to 0 \quad a.s.\ P. \tag{1.34}$$

The proof of theorems in this section will be given elsewhere.

References

[1] H. Kunita, *Stochastic flows and stochastic differential equations*, Cambridge Univ. Press, 1990.

[2] H. Kunita, Generalized solutions of stochastic partial differential equations, J. Theoretical Probability 7(1994), 279-308.

[3] H. Kunita, Asymptotic self-similarity and short time asympttics of stochastic flows, J. Univ. Tokyo

[4] S. Kusuoka, D.W. Stroock, Application of the Malliavin calculus, Part II, J. Fac. Sci. Univ. Tokyo 32(1985), 1-76.

[5] S. Kusuoka, D.W. Stroock, Application of the Malliavin calculus, Part III, J. Fac. Sci. Univ. Tokyo 32(1985), 103-147.

[6] S. Takanobu, Diagonal short time asymptotics fo Heat kernels for certain degenerate second order differential operators of Hörmander type, Pub. Res. Inst. Math. Sci. 24(1988), 169-203.

ROUGH ASYMPTOTICS OF FORWARD-BACKWARD STOCHASTIC DIFFERENTIAL EQUATIONS*

Jin Ma

Department of Mathematics
Purdue University
West Lafayette, IN 47907, USA
majin@math.purdue.edu

Tim Zajic

Department of Mathematics
University of Minnesota
Minneapolis, MN 55455, USA
zajic@math.umn.edu

Abstract: This note is concerned with the asymptotic behavior of adapted solutions to forward-backward SDEs when the forward diffusion contains small noises. We prove the sample path Large Deviation Principle (LDP) for the adapted solutions to the FBSDEs under appropriate conditions stated in terms of a certain type of convergence of the solutions to the associated quasilinear PDEs. As an application we apply the LDP results to a problem of rare event simulation. We provide a necessary condition for the existence of an asymptotically efficient estimator amongst the class of importance sampling estimators.

1 INTRODUCTION

Let $(\Omega, \mathcal{F}, P; \{\mathcal{F}_t\}_{t\geq 0})$ be a complete filtered probability space on which is defined a Brownian motion W such that $\{\mathcal{F}_t\}$ is the augmented natural filtration generated by W. We consider the following *forward-backward stochastic differ-*

*The first author is supported in part by the Office of Naval Research grant #N00014-96-1-0262; and the second author is supported in part by NSF Grant DMS-9508709.

ential equation (FBSDE) on a given time duration $[0,T]$:

$$\begin{cases} X_t = x + \int_0^t b(s,X_s,Y_s,Z_s)ds + \sqrt{\varepsilon}\int_0^t \sigma(s,X_s,Y_s)dW_s, \\ Y_t = g(X_T) + \int_t^T \hat{b}(s,X_s,Y_s,Z_s)ds + \int_t^T Z_s dW_s, \end{cases} \quad t \in [0,T], \quad (1.1)$$

where $0 < \varepsilon \ll 1$.

We refer the readers to El Karoui-Quenez-Peng (1997) [7] for an overview of backward SDEs, and to Ma-Protter-Yong (1994) [13], Hu-Peng (1995) [12], Duffie-Ma-Yong (1994) [6], and Cvitanić-Ma (1996) [2] for the theory and applications involving FBSDEs. Throughtout this note we denote the *adapted solution* of (1.1) by $\Theta^\varepsilon \triangleq (X^\varepsilon, Y^\varepsilon, Z^\varepsilon)$, and we assume that X^ε, Y^ε, and Z^ε are all one-dimensional. Also, we shall make use of the following assumptions:

(H1) The functions σ, b, $\hat{b}$ and g are continuous in (t,x,y,z), and continuously differentiable in (x,y,z) with uniformly bounded first order partial derivatives;

(H2) The function g belongs boundedly to $C^{2+\alpha}(\mathbf{R})$ for some $\alpha \in (0,1)$;

(H3) There exists positive numbers $\mu > 0$, and $K > 0$ such that

$$\begin{cases} \mu \le \sigma(t,x,y) \le K, & \forall (t,x,y); \\ |b(t,x,0,0)| + |\hat{b}(t,x,0,z)| \le K, & \forall (t,x,z). \end{cases}$$

The main purpose of this note is to study the asymptotic behavior of $\{\Theta^\varepsilon\}$ as $\varepsilon \to 0$. In particular we would like to show that the large deviations principle (LDP) holds for the distribution of $\{\Theta^\varepsilon\}$, as $\varepsilon \to 0$. It turns out that the following type of convergence of a family of functions will be essential in our future discussion.

Definition 1.1 (Hypothesis A) *Let $\varphi^\varepsilon \in C^{1,2}([0,T]\times \mathrm{R})$, $\varepsilon > 0$, be a family of functions, and $\varphi^0 \in C([0,T]\times \mathbf{R})$. We say that the family $\{\varphi^\varepsilon\}_{\varepsilon>0}$ satisfies "Hypothesis A" with limiting function φ^0, if the following conditions hold:*

(1) The family $\{\varphi^\varepsilon\}$ converges uniformly on compacts to φ^0, as $\varepsilon \to 0$; and, for each $M > 0$, there exists $K^M > 0$ such that

$$|\varphi^0(t,x) - \varphi^0(t,y)| \le K^M|x-y|, \qquad \forall (t,x) \in [0,T]\times[-M,M].$$

(2) $\lim_{\varepsilon\to 0} |\sqrt{\varepsilon}\varphi_x^\varepsilon| = 0$, *uniformly on compacts.*

Combining the LDP results with the simulation of the adapted solution of FBSDEs (cf. Douglas-Ma-Protter [5]), we then discuss the simulation of rare events for the paths of FBSDEs. Rare event simulation has been extensively studied in queuing and reliability models (cf. Heidelberg [11]). In the setting of diffusion processes, it can be used to asses the probability of exercising a deep out of the money option when the underlying stock possesses small volatility (see Fournié-Lasry-Touzi [9]). We consider *importance sampling* estimators and provide necessary conditions for the existence of an unbiased estimator that is *asymptotically efficient* (or *asymptotically optimal*) in the sense of Sadowsky-Bucklew [16].

2 LARGE DEVIATION FOR FBSDES

Let us recall some definitions concerning large deviations (cf. e.g. [4]). Let $\mathcal{X}$ be a topological space equipped with σ-field $\mathcal{B}$, and $I : \mathcal{X} \mapsto [0,\infty]$ be a lower semicontinuous function such that the level sets $\Psi_\alpha^I = \{x : I(x) \leq \alpha\}$ are compact for all $\alpha < \infty$. We say that a family of $\mathcal{X}$-valued random variables satisfies the LDP with *good rate function* I, if their distributions on $(\mathcal{X},\mathcal{B})$, denoted by $\{\mu_\varepsilon\}_{\varepsilon>0}$, satisfy

$$-\inf_{\gamma\in\Gamma^o} I(\gamma) \leq \liminf_{\varepsilon\to 0} \varepsilon \log \mu_\varepsilon(\Gamma) \leq \limsup_{\varepsilon\to 0} \varepsilon \log \mu_\varepsilon(\Gamma) \leq -\inf_{\gamma\in\overline{\Gamma}} I(\gamma)\,,$$

for all $\Gamma \in \mathcal{B}$, where Γ^o and $\overline{\Gamma}$ denote the interior and closure of Γ, respectively. In this note we consider the case when $\mathcal{X} = C^3([0,T];\mathbf{R})$ with sup-norm; denote

$$H_1(0,T) \triangleq \left\{ g_t = \int_0^t f_s ds,\ t\in[0,T] : \int_0^T |f_s|^2 ds < \infty \right\},$$

and endow H_1 with the norm $\|g\|_{H_1} \triangleq \sqrt{\int_0^T |f_s|^2 ds}$. Also, we denote, for functions $b,\sigma : [0,T]\times\mathbf{R}\to\mathbf{R}$ that satisfy (H1) and (H3),

$$F_{b,\sigma}(f) \triangleq \int_0^t \{[\dot{f}_s - b(s,f_s)]\sigma^{-1}(s,f_s)\}ds, \qquad \forall f\in H_1. \tag{2.1}$$

Following the *Four Step Scheme*, initiated by Ma-Protter-Yong in [13], we know that under the assumptions (H1)-(H3) (1.1) can be solved "explicitly" as follows:

$$\begin{cases} X_t^\varepsilon = x + \int_0^t \tilde{b}^\varepsilon(s,X_s^\varepsilon)ds + \int_0^t \tilde{\sigma}^\varepsilon(s,X_s^\varepsilon)dW_s, \\ Y_t^\varepsilon = \theta^\varepsilon(t,X_t^\varepsilon), \qquad Z_t^\varepsilon = -\sqrt{\varepsilon}\tilde{\sigma}^\varepsilon(t,X_t^\varepsilon)\theta_x^\varepsilon(t,X_t^\varepsilon), \end{cases} \tag{2.2}$$

where $\theta^\varepsilon \in C^{1,2}([0,T]\times\mathbf{R})$ is the classical solution to the following quasi-linear parabolic PDE:

$$\begin{cases} 0 = \theta_t^\varepsilon + \frac{1}{2}\varepsilon\tilde{\sigma}^\varepsilon(t,x)^2\theta_{xx}^\varepsilon + \tilde{b}^\varepsilon(t,x)\theta_x^\varepsilon + \hat{b}(t,x,\theta^\varepsilon,-\sqrt{\varepsilon}\sigma(t,x,\theta^\varepsilon)\theta_x^\varepsilon), \\ \theta^\varepsilon(T,x) = g(x), \end{cases} \tag{2.3}$$

and $\tilde{b}^\varepsilon(t,x) \triangleq b(t,x,\theta^\varepsilon(t,x),-\sqrt{\varepsilon}\tilde{\sigma}^\varepsilon(t,x)\theta_x^\varepsilon(t,x))$, $\tilde{\sigma}^\varepsilon(t,x) \triangleq \sigma(t,x,\theta^\varepsilon(t,x))$.

Our principle result is the following

Theorem 2.1 *Suppose that the solutions of (2.3), $\{\theta^\varepsilon\}_{\varepsilon>0}$, satisfy Hypothesis A with limiting function θ^0. Then the solution family $\{(X^\varepsilon,Y^\varepsilon,Z^\varepsilon)\}_{\varepsilon>0}$ satisfies the LDP in $C^3([0,T];\mathbf{R})$ with good rate function given by*

$$I_x(f,g,h) = \inf\left\{ \frac{1}{2}\int_0^T |\dot{F}_{\tilde{b},\tilde{\sigma}}(f)_s|^2 ds \,\middle|\, f\in H_1^x;\ g=\theta^0(\cdot,f),\ h=0 \right\}$$

where

$$\tilde{b}(t,x) = b(t,x,\theta^0(t,x),0), \qquad \tilde{\sigma}(t,x) = \sigma(t,x,\theta^0(t,x)); \tag{2.4}$$

and θ^0 *is the (unique) viscosity solution to the first order PDE*

$$\begin{cases} \theta^0_t + b(t,x,\theta^0,0)\theta^0_x + \hat{b}(t,x,\theta^0,0) = 0 \\ \theta^0(T,x) = g(x). \end{cases} \tag{2.5}$$

3 VERIFICATION OF HYPOTHESIS A

1. Small Duration Case.

We first show that if the duration T and the coefficients b, $\hat{b}$, σ, and g satisfy a certain "compatibility condition", then the family $\{\theta^\varepsilon\}$ will satisfy Hypothesis A. Such a condition contains in particular the case when the duration T is small.

To begin with, note that using a maximum-principle type of argument one can show that there exists a constant $M > 0$, independent of ε, such that

$$\sup_{\varepsilon>0} \sup_{(t,x)\in[0,T]\times\mathbf{R}^n} |\theta^\varepsilon(t,x)| \le M. \tag{3.1}$$

Next, with $\varepsilon > 0$, consider the FBSDE on $0 \le t \le s \le T$:

$$\begin{cases} X_s = x + \displaystyle\int_t^s b(r,X_r,Y_r,Z_r)dr + \sqrt{\varepsilon}\int_t^s \sigma(r,X_r,Y_r)dW_r, \\ Y_s = g(X_T) + \displaystyle\int_s^T \hat{b}(r,X_r,Y_r,Z_r)dr + \int_s^T Z_r dW_r, \quad s \in [t,T]. \end{cases} \tag{3.2}$$

Denote the adapted solution of (3.2) on $[t,T]$ by $(X^{t,x,\varepsilon}, Y^{t,x,\varepsilon}, Z^{t,x,\varepsilon})$. Since the coefficients are deterministic, it follows that $\theta^\varepsilon(t,x) \triangleq Y_t^{t,x,\varepsilon}$ is also deterministic, and is in fact the classical solution to the quasilinear PDE (2.3), thanks to the Four Step Scheme. The following result is more or less well understood in the theory of backward SDEs, and is a modification of those of Pardoux-Tang [15] and Cvitanic-Ma [3] in the FBSDE case.

Lemma 3.1 *Assume (H1)-(H3). Let* $T > 0$ *be given. Suppose that there exist positive constants* C_1, C_2, C_3, C_4 *and* $\lambda > 0$, *such that* $0 < 1 - KC_4 < 1$, *and the following "compatibility condition" holds:*

$$1 - \mu(T)K^2 > 0, \qquad \lambda_1 - \mu(T)KC_3 > 0, \tag{3.3}$$

where

$$\begin{cases} \lambda_1 = \lambda - 2K - K(C_1^{-1} + C_2^{-1}) - \varepsilon^2 K^2; \\ \lambda_2 = -\lambda - 2K - KC_3^{-1} - KC_4^{-1}, \\ \mu(T) = (KC_1 + \varepsilon K^2)B(\lambda_2,T) + \frac{1}{1-KC_4}A(\lambda_2,T)KC_2; \\ A(\lambda,t) = e^{-(\lambda\wedge 0)t}; \\ B(\lambda,t) = \dfrac{1-e^{-\lambda t}}{\lambda}. \end{cases} \tag{3.4}$$

Then there exists a constant $C > 0$, depending only on K and T, such that for all $(t_1, x_1), (t_2, x_2) \in [0, T] \times \mathbf{R}$, and $\varepsilon \in (0,1)$, it holds that

$$|\theta^\varepsilon(t_1, x_1) - \theta^\varepsilon(t_2, x_2)|^2 \leq C\{|x_1 - x_2|^2 + (1 + |x_1|^2 + |x_2|^2)|t_1 - t_2|\}. \quad (3.5)$$

Remark. It is easy to check that (3.3) is solvable at $T = 0$. A "continuous dependence" argument then shows that there exists a constant $T_0 > 0$, independent of ε, such that (3.3) holds for all $0 < T < T_0$. Consequently (3.5) holds when $T > 0$ is small. ∎

Lemma 3.1, together with (3.1), essentially says that $\{\theta^\varepsilon\}$ is uniformly bounded and (locally) equicontinuous. Thus in view of the Arzelà-Ascoli theorem we derive

Theorem 3.2 *Assume (H1)–(H3). Suppose that $T > 0$ is such that the compatibility conditions (3.3) and (3.4) are solvable. Then the family of solutions $\{\theta^\varepsilon\}_{0<\varepsilon\leq 1}$ of the PDE (2.3) satisfies Hypothesis A with limiting function θ^0, where θ^0 is a viscosity solution to the first order PDE (2.5). Moreover, the solution is unique among those that are uniform Lipschitz in the x variable. In particular, there exists a constant $T_0 > 0$, such that for all $0 < T < T_0$, the above statement holds true.*

2. Large Duration Case.

This case is more complicated, since the PDE theory tells us that it is quite remote to establish Hypothesis A for the solutions to nonlinear PDEs in general. Our purpose here is to find reasonable sufficient conditions.

To establish Hypothesis A-(1) we follow the so-called Barles-Perthame procedure (cf. Barles-Perthame [1], or Fleming-Soner [8]). Namely, we define two functions

$$\theta^*(t, x) = \limsup_{\substack{(s,y)\to(t,x) \\ \varepsilon\to 0}} \theta^\varepsilon(t, x) \quad \text{and} \quad \theta_*(t, x) = \liminf_{\substack{(s,y)\to(t,x) \\ \varepsilon\to 0}} \theta^\varepsilon(t, x).$$

Clearly, both θ^* and θ_* are well defined thanks to (3.1); $\theta_* \leq \theta^*$; and θ_* (resp. θ^*) is upper (resp. lower) semicontinuous. Next, we use standard arguments to show that, under (H1)–(H3), θ^* is a viscosity subsolution and θ_* is a viscosity supersolution to the first order PDE (2.5). A "comparison principle" is then established, showing $\theta^* \leq \theta_*$, and consequently that $\theta^* = \theta_* (\triangleq \theta^0)$ is continuous and the convergence must be uniform on compact sets. Thus Hypotheses A-(1) will follow, provided we can show that θ^0 is uniform Lipschitz in x.

Continuing, using gradient estimates for nonlinear PDEs, together with (H1)–(H3), as well as an extra condition that $b(t, x, y, 0) = 0$, we can show that there exists a constant $M > 0$, independent of ε, such that

$$\sup_{(t,x)} |\theta^\varepsilon_x(t, x)| \leq M(1 + \frac{1}{\varepsilon^r}), \qquad \forall \varepsilon \in (0,1). \quad (3.6)$$

which clearly leads to Hypothesis A-(2). Therefore we have the following

Theorem 3.3 *Assume (H1)–(H3) and that* $b(t,x,y,0)=0$, $\forall(t,x,y)\in[0,T]\times \mathbf{R}^2$. *Then the solution sequence* $\{\theta^\varepsilon\}$ *of (2.3) satisfies Hypothesis A with limiting function* θ^0 *which is the unique viscosity solution to (2.5).*

We remark here that the assumption $b(t,x,y,0)=0$ in Theorem 3.3 is almost necessary for (3.6) to hold, which is obviously stronger than Hypothesis A-(2). However, in the case when Hypothesis A-(2) can be verified independently, we can replace the assumption $b(t,x,y,0)=0$ by the requirement that $b(t,x,y,z)$ be independent of y, a slightly different situation.

4 RARE EVENT SIMULATION

In this section we turn to the problem of simulating the probability of certain events $\{\Theta^\varepsilon\in A\}\in\mathcal{F}$, where A is some Borel set of $C^3([0,T];\mathbf{R})$, such that $P\{\Theta^\varepsilon\in A\}\to 0$, as $\varepsilon\to 0$ (i.e., A is a "rare event"). Noting that (2.2) shows that Y^ε and Z^ε are functions of X^ε, we shall restrict our attention to estimating $\gamma_A(\varepsilon)\triangleq P\{X^\varepsilon\in A\}$, where A is now some Borel set of $C([0,T];\mathbf{R})$.

Consider the standard Monte Carlo estimator of $\gamma_A(\varepsilon)$: let $X^\varepsilon(1),\ \dots\ ,$ $X^\varepsilon(N)$ be N independent copies of sample paths X^ε, which can be simulated using the method of [5], and define $\hat\gamma^\varepsilon_N := \frac{1}{N}\sum_{n=1}^N 1_A(X^\varepsilon(n))$. Clearly, $\hat\gamma^\varepsilon_N$ is an unbiased estimator of $\gamma_A(\varepsilon)$ such that $\hat\gamma^\varepsilon_N\to\gamma_A(\varepsilon)$ almost surely as $N\to\infty$, thanks to the Strong Law of Large Numbers. A drawback of this estimator, however, is that as the event gets rarer (i.e., $\gamma_A(\varepsilon)\ll 1$) the *relative error* of the estimator (the quotient of the standard deviation of the estimator $\hat\gamma^\varepsilon_N$ and $\gamma_A(\varepsilon)$ itself), given by $\mathcal{E}_r(\hat\gamma^\varepsilon_N)\triangleq\sqrt{1-\gamma_A(\varepsilon)}/\sqrt{N\gamma_A(\varepsilon)}$, behaves like $1/\sqrt{N\gamma_A(\varepsilon)}$, which makes the formation of confidence intervals overly time consuming. In an attempt to overcome this difficulty we turn to the so-called *importance sampling estimators*, which we now describe.

For any $h\in H_1$ and $\varepsilon>0$, define a process

$$M^{h,\varepsilon}_t=\exp\left\{\frac{1}{\sqrt{\varepsilon}}\int_0^t\dot h_s dW_s+\frac{1}{2\varepsilon}\int_0^t|\dot h_s|^2ds\right\},\quad t\in[0,T].$$

It can be easily checked that $(M^{h,\varepsilon})^{-1}$ is a martingale. By the Girsanov theorem, under a new probability measure $P^{h,\varepsilon}$ defined by $dP^{h,\varepsilon}/dP=(M^{h,\varepsilon}_T)^{-1}$, $W^{h,\varepsilon}_t=W_t+\frac{1}{\sqrt{\varepsilon}}h_t$ will be a Brownian motion, and the pair $(X^\varepsilon, M^{h,\varepsilon})$ will satisfy the SDE:

$$\begin{cases} X^\varepsilon_t=x+\displaystyle\int_0^t[\tilde b^\varepsilon(s,X^\varepsilon_s)-\tilde\sigma^\varepsilon(s,X^\varepsilon_s)\dot h_s]ds+\int_0^t\sqrt{\varepsilon}\tilde\sigma^\varepsilon(s,X^\varepsilon_s)dW^{h,\varepsilon}_s\\ M^{h,\varepsilon}_t=1+\dfrac{1}{\sqrt{\varepsilon}}\displaystyle\int_0^t\dot h_s M^{h,\varepsilon}_s dW^{h,\varepsilon}_s. \end{cases}\tag{4.1}$$

Now if we simulate N independent copies of the sample paths of the solution to (4.1), $\{(\widetilde X^{h,\varepsilon}(n),\widetilde M^{h,\varepsilon}(n))\}_{n=1}^N$, on any probability space $(\widetilde\Omega,\widetilde{\mathcal{F}},\widetilde P)$, with $W^{h,\varepsilon}$

in (4.1) being replaced by any other $\widetilde{P}$-Brownian motion(!), and define

$$\widetilde{\gamma}_N^{h,\varepsilon} \triangleq \frac{1}{N}\sum_{n=1}^{N} 1_A(\widetilde{X}^{h,\varepsilon}(n))\widetilde{M}^{h,\varepsilon}(n),$$

then by the uniqueness in law of the SDE, it is seen that $\widetilde{\gamma}_N^{h,\varepsilon}$ is again an unbiased estimator of $\gamma_A(\varepsilon)$ such that $\widetilde{\gamma}_N^{h,\varepsilon} \to \gamma_A(\varepsilon)$, $\widetilde{P}$-almost surely, as $N \to \infty$. However, the relative error of the estimator $\widetilde{\gamma}_N^{h,\varepsilon}$ now becomes

$$\mathcal{E}_r(\widetilde{\gamma}_N^{h,\varepsilon}) = \frac{\sqrt{\widetilde{E}\{\widetilde{\gamma}_N^{h,\varepsilon} - E(\widetilde{\gamma}_N^{h,\varepsilon})\}^2}}{E\{\widetilde{\gamma}_N^{h,\varepsilon}\}} = \frac{\sqrt{\eta^{h,\varepsilon} - \gamma_N^2(\varepsilon)}}{\sqrt{N}\gamma_A(\varepsilon)}.$$

where $\eta^{h,\varepsilon} \triangleq E^{h,\varepsilon}\{1_A(X^\varepsilon)M_T^{h,\varepsilon})\}^2$. Note that Jensen's inequality tells us that $\eta^{h,\varepsilon} \ge \gamma_A^2(\varepsilon)$. Thus if the LDP holds for X^ε with good rate function I_x, and if A is an I-continuity set (i.e., $\inf_{f\in A^\circ} I_x(f) = \inf_{f\in \bar{A}} I_x(f) \triangleq I_A$) such that $0 < I_A < \infty$, then we must have

$$\liminf_{\varepsilon\to 0} \varepsilon \log \eta^{h,\varepsilon} \ge 2 \lim_{\varepsilon\to 0} \varepsilon \log \gamma_A(\varepsilon) = -2I_A.$$

If, in particular, the equality $\lim_{\varepsilon\to 0} \varepsilon \log \eta^{h,\varepsilon} = -2I_A$ holds, then we say that the importance sampling estimator $\widetilde{\gamma}^{h,\varepsilon}$ is *asymptotically efficient* or *asymptotically optimal* (cf. Sadowsky-Bucklew [16]). One of the main features of an estimator being asymptotically efficient is that the *critical sample size* $N_\alpha(\varepsilon) \triangleq \inf\{N \in \mathbb{N} : \mathcal{E}_r(\widetilde{\gamma}_N^{h,\varepsilon}) \le \alpha\}$ will grow at most sub-exponentially as $\varepsilon \to 0$, an extremely desirable property in rare event simulation. In other words we have

Theorem 4.1 *An importance sampling estimator $\widetilde{\gamma}_N^{h,\varepsilon}$ is asymptotically optimal if and only if $\lim_{\varepsilon\to 0} N_\alpha(\varepsilon)e^{-\frac{\kappa}{\varepsilon}} \le 0$ for any $\kappa > 0$.*

The remaining question is then how to choose $h \in H_1$ so that the corresponding importance sampling estimator $\widetilde{\gamma}_N^{h,\varepsilon}$ is aymptotically efficient. To this end let us recall the mapping $F_{\tilde{b},\tilde{\sigma}}$ defined by (2.1), and $\tilde{b}$ and $\tilde{\sigma}$ defined by (2.4). Define, for $f \in H_1^x$, $I_x(f) = \frac{1}{2}\int_0^T |\dot{F}_{\tilde{b},\tilde{\sigma}}(f)_s|^2 ds$. Let us now simply denote $F \triangleq F_{\tilde{b},\tilde{\sigma}}$. The following definition of *F-dominating point* is motivated by the notion of "dominating point" from large deviation theory (see, e.g., Ney [14] and Sadowsky-Bucklew[16]), adapted to our setting. For $A \in \mathcal{B}(C[0,T])$, we say that $f^* \in \bar{A} \cap H_1^x$ is an *F-dominating point* of A if

$$\int_0^T \dot{F}(f^*)_s(\dot{F}(f^*)_s - \dot{F}(g)_s)ds \le 0, \qquad \forall g \in \bar{A} \cap H_1^x.$$

We then have

Theorem 4.2 *Suppose that A is an I-continuity set such that $0 < I_A < \infty$. Suppose also that there exists an F-dominating point, f^*, of A. Then the importance sampling estimator $\widetilde{\gamma}^{h^*,\varepsilon}$, where $h^* \triangleq F(f^*)$, is asymptotically efficient.*

References

[1] Barles, G. and Perthame, B., Exit time problems in optimal control and vanishing viscosity solutions of Hamilton-Jacobi equations, SIAM J. Control Optim., Vol. 26, 1988, pp. 1133-1148.

[2] Cvitanić, J. and Ma, J., Hedging options for a large investor and forward-backward SDE's, Ann. Appl. Prob., Vol. 6, No. 2, 1996, pp. 370-398.

[3] Cvitanić, J. and Ma, J., Reflected forward-backward SDEs and obstacle problems with boundary conditions, Preprint, 1997.

[4] Dembo, A. and Zeitouni, O., Large Deviations Techniques and Applications, Jones and Bartlett, Boston, 1993.

[5] Douglas, J., Ma, J. and Protter, P., Numerical methods for forward-backward stochastic differential equations, Ann. Appl. Prob., Vol. 6, No. 3, 1996, pp. 940-968.

[6] Duffie, D., Ma, J. and Yong, J., Black's console rate conjecture, Ann. Appl. Prob., Vol. 5, No. 2, pp. 356-382, 1995.

[7] El Karoui, N, S. Peng, and M.C. Quenez, Backward stochastic differential equation in finance, Mathematical Finance, Vol. 7, No. 1, 1997, pp. 1-71.

[8] Fleming, W. and Soner, M., Controlled Markov Processes and Viscosity Solutions, Springer-Verlag, New York, 1993.

[9] Fournié, E., Lasry, J.M. and Touzi, N., Monte carlo methods for stochastic volatility models, Numerical Methods in Finance, pp. 146-164, Cambridge Univ. Press, Cambridge, 1997.

[10] Fournié, E., Lasry, J.M. and Lions, P.L., Some nonlinear methods for studying far-from-the-money contingent claims. Numerical methods in finance, pp. 115-145, Cambridge Univ. Press, Cambridge, 1997.

[11] Heidelberger, P., Fast simulation of rare events in queueing and reliability models, ACM Trans. on Mod. and Comp. Simulation, Vol. 5, No. 1, 1995, pp. 43-85.

[12] Hu, Y. and Peng, S., Solution of forward-backward stochastic differential equations, Prob. Th. Rel. Fields, Vol. 103, 1995, pp. 273-283.

[13] Ma, J., Protter, P. and Yong, J., Solving forward-backward stochastic differential equations explicitly - a four step scheme, Prob. Th. Rel. Fields., Vol. 98, 1994, pp. 339-359.

[14] Ney, P., Dominating points and the asymptotics of large deviations for random walk on $\mathbf{R}^d$, Ann. Proba., Vol. 11, No. 1, 1983, pp. 158-167.

[15] Pardoux, E. and Tang, S., Forward-backward SDEs and quasilinear parabolic PDEs, Preprint, 1997.

[16] Sadowsky, J.S. and Bucklew, J.A., On large deviations theory and asymptotically efficient Monte Carlo estimation, IEEE Trans. Inf. Theor., Vol. 36, 1990, pp. 579-588.

ON LQG CONTROL OF LINEAR STOCHASTIC SYSTEMS WITH CONTROL DEPENDENT NOISE

John B. Moore

Department of Systems Engineering and
Cooperative Research Center for Robust and Adaptive Systems
Research School of Information Sciences and Engineering
Australian National University
Canberra A.C.T. 0200
AUSTRALIA
John.Moore@anu.edu.au

Xunyu Zhou

Department of Systems Engineering and Engineering Management
The Chinese University of Hong Kong
Shatin N.T.
HONG KONG
xyzhou@se.cuhk.edu.hk

Andrew E.B. Lim

Department of Electrical & Electronic Engineering
University of Melbourne
Parkville Victoria 3052
AUSTRALIA
andrew@ee.mu.OZ.AU

Abstract:

This paper announces initial results of studies on partially observed linear quadratic Gaussian (LQG) models where the stochastic disturbances depend on both states and controls and the measurements may be bilinear in the noise and the states/controls. While the Separation Theorem of standard LQG design does not apply in any strict sense, suboptimal linear state estimate feedback laws are derived based on certain linearizations. The controllers may well be useful for nonlinear stochastic systems where linearized models which include terms bilinear in the noise and states/controls are significantly more accurate than if these terms are set to zero. These controllers are calculated by solving a generalized discrete time Riccati equation. The properties of this equation relating to well posedness of the associated LQG problem are discussed.

1 INTRODUCTION

The classical linear quadratic Gaussian (LQG) control theory for stochastic linear systems assumes that the stochastic disturbances are additive and not control or state dependent [2, 3]. Relaxing this assumption to allow linear state and control dependence in noise leads to a much broader class of stochastic models, which have applications in real-world systems (see [6] for more details). On the other hand, working with this class allows improved approximation by linearization of nonlinear stochastic systems than if the bilinear terms involving the noise are excluded. Recently, LQG theory has been generalized for such a class of linear/bilinear stochastic systems in continuous time [6, 7]. Optimal state feedback control laws are linear and are calculated by solving a so-called *stochastic* matrix Riccati equation which specializes the familiar conventional LQG Riccati equations when the disturbances are independent of the states and controls. The stochastic Riccati equations are by no means as well understood as in the standard case, at least in the continuous time setting. There remains open questions concerning existence and uniqueness of the solutions of these equations. There is also an intriguing property that the control weighting matrix R in a standard quadratic integral cost term need not be positive definite, even in the continuous time case. This property reveals some deep nature of uncertainty. For details see [6, 7].

What is the situation then for the partially observed case? To what extent does the standard LQG methodology [2] with its Separation Theorem apply? Can we achieve useful *linear* state estimate feedback laws?

In this paper the above questions are addressed for discrete time case and some initial thoughts and results are presented. The expectation is that since the models are bilinear in the state and the noise as well as in the control and the noise, some of the virtues of a standard linear Gaussian theory will be lost. Certainly, even if the noise signals are Gaussian, the states and control signals will be generally non-Gaussian. Consequently, optimal (information) state estimators will be infinite dimensional, in general, see for example [4]. Even so, since a conditional linear minimum square error (LMSE) covariance state estimator is known for the models of interest, and is finite dimensional, it makes sense from an implementation point of view to work with such a state estimator and a linear state estimate feedback law, even if such a law is suboptimal.

The conditional LMSE filter has the structure of a Kalman filter, see [2], but with a Kalman gain which is state estimate and control dependent. Likewise, the quadratic state cost, when expressed in terms of state estimates instead of true states, is nonlinear. Appropriate linearizations of the filter equations and cost terms, neglecting higher order terms but allowing terms bilinear in the noise and controls/state estimates in the filter, allows then application of a discrete-time analogy of the recently studied LQG theory in [6]. This leads to an 'optimal' linear state estimate feedback law under assumptions of negligible higher order terms. In practise, this law has some degree of sub-optimality because the neglected higher order terms may be significant to some extent. However, the neglected terms do not include terms bilinear in the innovations (prediction errors) and the state estimates/control, so there is a chance for improved performance over the standard LQG approach which neglects these terms as well as higher order terms.

The paper is organized as follows. In Section 2 an optimal feedback controller is derived for a completely observed discrete time LQG with state and control dependent noise. As in the standard case, solving a discrete time Riccati equation is a key step in calculating the optimal controller. In fact, the associated Riccati equation is a generalization of the standard discrete time Riccati equation. The existence properties of this equation and its relationship to the well posedness of the control problem is discussed in Section 3. Section 4 is concerned with an approximate Kalman filter for the partially observed LQG model. Finally, suboptimal linear state estimate feedback laws are obtained in Section 5 by combining the results in Sections 2 and 4.

2 DISCRETE TIME LQG RESULTS

In this section, we derive discrete time versions of those in [6] which will be useful in a later section.

Consider the discrete time stochastic signal model

$$x_{k+1} = (A_k + w_k^A \Delta A_k)x_k + (B_k + w_k^B \Delta B_k)u_k + w_k, \tag{1}$$

where $x_k \in \mathbf{R}^n$ is the state, $u_k \in \mathbf{R}^m$ is the control, and w_k^A, $w_k^B \in \mathbf{R}^1$ are noise terms, assumed here to be martingale increments on $\mathcal{G}_{k-1}$, where $\mathcal{G}_{k-1}$ is the σ-algebra generated by past noise terms up to w_{k-1}^A, w_{k-1}^B, w_{k-1}. The covariances are assumed to be

$$E[(w_{k-1}^A)^2|\mathcal{G}_{k-1}] = E[(w_{k-1}^B)^2|\mathcal{G}_{k-1}] = 1, \quad E[w_k w_k'|\mathcal{G}_{k-1}] = Q_k,$$

and

$$E[w_{k-1}^A w_{k-1}^B|\mathcal{G}_{k-1}] = \rho_k^{AB}, \quad E[w_{k-1} w_{k-1}^A|\mathcal{G}_{k-1}] = \rho_k^A, \quad E[w_{k-1} w_{k-1}^B|\mathcal{G}_{k-1}] = \rho_k^B.$$

Generalizations of the dependent noise terms $w_k^A \Delta A_k$ and $w_k^B \Delta B_k$ to the case of non-scalar noise is immediate by working with terms $\sum_{i=1}^N w_k^{A^i} \Delta A_k^i$ and $\sum_{i=1}^N w_k^{B^i} \Delta B_k^i$.

The performance index of the problem is given by the standard quadratic sum cost

$$J_T = E\left\{\sum_{k=0}^{T-1}(x_k' Q_k^c x_k + u_k' R_{k+1}^c u_k) + x_T' Q_T^c x_T - x_0' Q_0^c x_0\right\}. \tag{2}$$

In this model, all the A_k, ΔA_k, etc.. are (deterministic) matrices with appropriate dimensions, Q_k^c and Q are non-negative definite matrices, and R_k^c are symmetric matrices (could be indefinite, as in standard discrete time LQG theory). .

Let us solve the above stochastic optimal control problem in two different cases. The results derived below will be applied in Section 4 for partially observed models.

Case I: $\rho_k^A = \rho_k^B = 0$.

Let us consider first the case when $w_k \perp w_k^A, w_k^B$, so that $\rho_k^A = \rho_k^B = 0$. In this case, the optimal control takes the form

$$u_k = K_k^c \, x_k, \tag{3}$$

where

$$\begin{aligned} K_k^c &= -(\Omega_{k+1}^c)^{-1} L_{k+1}^c, \\ L_{k+1}^c &= B_k' S_{k+1} A_k + \rho_k^{AB} \, \Delta B_k' \, S_{k+1} \, \Delta A_k, \\ \Omega_{k+1}^c &= B_k' \, S_{k+1} \, B_k + \Delta B_k' \, S_{k+1} \, \Delta B_k + R_{k+1}^c. \end{aligned}$$

Here, S_k is the solution of a backward matrix Riccati equation

$$\begin{cases} S_k = A_k' \, S_{k+1} \, A_k - L_{k+1}^c{}' \, (\Omega_{k+1}^c)^{-1} \, L_{k+1}^c + (Q_k^c + \Delta A_k' \, S_{k+1} \, \Delta A_k), \\ S_T = Q_T^c. \end{cases} \tag{4}$$

By standard completion of squares arguements, it can be shown that (1)-(2) is well posed if Ω_{k+1}^c is positive definite. In this case, the control law (2)-(4) is the unique optimal control.

In continuous time LQG theory, it is generally required that R is strictly positive definite for the problem to be well posed. Recent results by Chen, Li and Zhou [6] for the continuous time problem show that R can have negative eigenvalues if the diffusion term in the system equations depend on the control. It is interesting to note therefore that in the discrete time problem, the control weighting matrices R_k can have negative eigenvalues and the problem remain well posed, *even if the bilinear terms ΔA_k and ΔB_k are all zero*!

Case II: $\rho_k^A \neq 0$, $\rho_k^B \neq 0$.

In the event that w_k and (w_k^A, w_k^B) are correlated so that ρ_k^A, $\rho_k^B \neq 0$, then the optimal control requires not only the state feedback term as in (3) but also an external input as

$$u_k = K_k^c \, x_k + b_k, \tag{5}$$

where b_k is calculated by linear backward recursions. For details, the reader should consult [5].

3 DISCRETE TIME RICCATI EQUATION

In the continuous time LQG problem, a standard assumption is that $R(t)$ is strictly positive definite. In the paper by Chen, Li and Zhou [6], it is shown that this assumption is not necessary when the diffusion term depends on the control. In this section, we examine the effect of the terms ΔA_k and ΔB_k on the well posedness of the LQG problem (1)-(2).

Recall that the LQG problem (1)-(2) is well posed if and only if $\Omega_k^c \geq 0$ for every k. Note once again that it is possible for the standard LQG problem (ie. $\Delta A_k = 0$ and $\Delta B_k = 0$) to be well posed with either $Q_k < 0$ or $R_k < 0$ (but obviously not both). We show in this section that if $\Delta A_k \neq 0$ or $\Delta B_k \neq 0$, then Q_k and R_k can be made 'more negative' and with the associated problem still remaining well posed. Bounds on the allowable decrease are also derived for certain special cases.

Before doing this however, we need to introduce some notation. Let $\mathcal{K} = \{(S_0, \cdots, S_T) | S_j \in \mathbf{R}^{n \times n}, \text{symmetric}\}$, $\mathcal{Q} = \{(Q_0, \cdots, Q_T) | Q_j \in \mathbf{R}^{n \times n}, \text{symmetric}\}$ and $\mathcal{P} = \{(R_1, \cdots, R_T) | R_j \in \mathbf{R}^{m \times m}, \text{symmetric}\}$. Given a sequence $\bar{R}^c = (\bar{R}_1^c, \cdots, \bar{R}_T^c) \in \mathcal{P}$ of control weights and $\bar{Q}^c = (\bar{Q}_0^c, \cdots, \bar{Q}_T^c) \in \mathcal{Q}$ of state weights, the standard discrete time Riccati equation

$$\left\{ \begin{array}{lcl} S_k & = & A_k' S_{k+1} A_k - A_k' S_{k+1} B_k (\bar{R}_{k+1}^c + B_k' S_{k+1} B_k)^{-1} B_k' S_{k+1} A_k + \bar{Q}_k^c \\ S_T & = & \bar{Q}_T^c \end{array} \right. \tag{6}$$

gives rise to a sequence $(S_0, \cdots, S_T) \in \mathcal{K}$. Hence, we can define a mapping $\psi : \mathcal{P} \times \mathcal{Q} \to \mathcal{K}$ which maps a sequence of control weights $\bar{R}^c = (\bar{R}_1^c, \cdots, \bar{R}_T^c) \in \mathcal{P}$ and state weights $\bar{Q}^c = (\bar{Q}_0^c, \cdots, \bar{Q}_T^c) \in \mathcal{Q}$ to the solution $\psi(\bar{Q}^c, \bar{R}^c) = (\psi_0(\bar{Q}^c, \bar{R}^c), \cdots, \psi_T(\bar{Q}^c, \bar{R}^c)) \in \mathcal{K}$ of (6).

Suppose now that $\bar{Q}^c = Q^c \in \mathcal{Q}$ is given (and fixed) while $\bar{R}^c$ is the variable. In this case, we shall write $\psi(Q^c, \bar{R}^c)$ simply as $\psi(\bar{R}^c)$. It follows that the associated (standard) LQG problem is solvable if and only if

$$\bar{R}_{k+1}^c + B_k' \psi_{k+1}(\bar{R}^c) B_k \geq 0. \tag{7}$$

Our result for the case $\Delta A_k = 0$ is as follows:

Theorem 1 *Let (Q_k^c, R_k^c) be given. If $\Delta A_k = 0$, then the problem (1)-(2) corresponding to (Q_k^c, R_k^c) is well posed if and only if*

$$R_{k+1}^c \geq \bar{R}_{k+1}^c - \Delta B_k' \psi_{k+1}(\bar{R}^c) \Delta B_k \tag{8}$$

for some $\bar{R}_k^c$ such that $(Q_k^c, \bar{R}_k^c)$ satisfies (6)-(7).

Note in particular that if $\Delta B_k \neq 0$, then the control weighting matrices $\bar{R}_k^c$ can be made 'more negative' and the problem (1)-(2) still remains well posed. The bound on this change is given by (8).

In Theorem 1, we assume that $\bar{Q}^c$ is given and fixed. However, it can be shown that $\bar{Q}_k^c$ can be made 'more negative' if $\Delta A_k = 0$ and $\Delta B_k \neq 0$ [5]. The allowable bounds on this change is still an open question.

Consider now the case when $\Delta A_k \neq 0$ but $\Delta B_k = 0$. Let $\bar{R}^c = R^c \in \mathcal{P}$ be fixed. Let $\psi : \mathcal{Q} \to \mathcal{K}$ be a mapping such that $\psi(\bar{Q}^c) = \psi(\bar{Q}^c, R^c)$ is the solution of the standard discrete time Riccati equation

$$\begin{cases} S_k &= A_k' S_{k+1} A_k - A_k' S_{k+1} B_k (R_{k+1}^c + B_k' S_{k+1} B_k)^{-1} B_k' S_{k+1} A_k + \bar{Q}_k^c \\ S_T &= \bar{Q}_T^c. \end{cases} \tag{9}$$

In this case, the associated (standard) LQG problem is solvable if and only if

$$R_{k+1}^c + B_k' \psi_{k+1}(\bar{Q}^c) B_k \geq 0. \tag{10}$$

In much the same way as the case $\Delta A_k = 0$, $\Delta B_k \neq 0$, the following result can be shown.

Theorem 2 *Let* (Q_k^c, R_k^c) *be given. If* $\Delta B_k = 0$, *then the problem (1)-(2) corresponding to* (Q_k^c, R_k^c) *is well posed if and only if*

$$Q_k^c \geq \bar{Q}_k^c - \Delta A_k' \psi_{k+1}(\bar{Q}^c) \Delta A_k \tag{11}$$

for some $\bar{Q}_k^c$ *such that* $(R_k^c, \bar{Q}_k^c)$ *satisfies (9)-(10).*

As in the case of Theorem 1, Theorem 2 shows how much 'more negative' the matrices $\bar{Q}_k^c$ can be made when $\Delta A_k \neq 0$ and $\Delta B_k = 0$. It can also be shown that if $\Delta A_k \neq 0$ and $\Delta B_k = 0$, $\bar{R}_k^c$ can be made 'more negative' [5]. The allowable bounds on this change is still an open question. Similarly, the effect of both $\Delta A_k \neq 0$ and $\Delta B_k \neq 0$ is still unresolved.

4 STATE ESTIMATION

In this section, we first define a partially observed signal model. Next, we apply the known Kalman filter theory to yield a linear minimum variance state estimator, which is then linearized further so that the filter is linear in the states and control, and bilinear in the innovations (prediction errors) and the states/controls.

Consider the following partially observed model:

$$\begin{cases} x_{k+1} &= (A_k + w_k^A \, \Delta A_k)\, x_k + (B_k + w_k^B \, \Delta B_k)\, u_k + w_k, \\ y_k &= (C_k + w_k^c \, \Delta C_k)\, x_k + v_k, \end{cases} \tag{12}$$

where $y_k \in \mathbf{R}^p$. Here w_k^c, v_k are martingale increments, each orthogonal to w_k^A, w_k^B, w_k, and $E[v_k \, v_k'] = R_k$.

Linear conditional minimum variance state estimator: Applying standard filtering results [2] yields the estimator

$$\begin{cases} \hat{x}_{k+1} &= A_k \, \hat{x}_k + B_k \, u_k + K_k(\hat{x}_k, u_k)\, \nu_k, \\ \nu_k &= y_k - C_k \, \hat{x}_k, \end{cases} \tag{13}$$

where the gain $K_k(\hat{x}_k, u_k)$ is given in terms of a coupled matrix Riccati equation as follows

$$K_k(\hat{x}_k, u_k) = L_k \, \Omega_k(\hat{x}_k, u_k)^{-1}, \tag{14}$$

with

$$\begin{aligned} L_k &= A_k\,\Sigma_k\,C_k', \\ \Omega_k(\hat{x}_k, u_k) &= C_k\,\Sigma_k\,C_k' + R_k + \Delta C_k\,(\Sigma_k + \hat{x}_k\,\hat{x}_k')\,\Delta C_k', \end{aligned} \tag{15}$$

and

$$\left\{ \begin{array}{l} \Sigma_{k+1} = A_k\,\Sigma_k\,A_k' - L_k\,\Omega(\hat{x}_k,\,u_k)^{-1}\,L_k' + Q_k + \Delta A_k\,(\Sigma_k + \hat{x}_k\,\hat{x}_k')\,\Delta A_k' \\ \qquad +\Delta B_k\,u_k\,u_k'\,\Delta B_k', \\ \Sigma_0 = E[x_0\,x_0']. \end{array} \right. \tag{16}$$

Here $\hat{x}_k$ is the best linear estimate conditioned on $\mathcal{Y}_{k-1}$, the σ-algebra generated by $y_0, \cdots, y_{k-1}$, where best is in a minimum error variance sense. The associated conditional error covariance is $\Sigma_k = E[(x_k - \hat{x}_k)(x_k - \hat{x}_k)'|\mathcal{Y}_{k-1}]$. Notice that the dependence of the noise on states and controls in our model (12) leads to an error covariance which depends on the past measurements (and controls), and in turn leads to a filter gain $K_k(\,\cdot\,,\cdot\,)$ which is dependent on the past measurements (and controls). Now this dependency of $K_k(\cdot\,,\cdot)$ on $\hat{x}_k$, u_k is by no means affine, but in order to proceed to a control law based on the LQG theory of Section 2, we must linearize $K_k(\,\cdot\,,\cdot\,)$ in $\hat{x}_k$ and u_k.

A filter bilinear in the innovations: Consider a linearization of $K_k(\,\cdot\,,\cdot\,)$, via a Taylor expansion, for simplicity in the $p = 1$ case

$$K_k(\hat{x}_k, u_k) = K_k + K_k^x\,\hat{x}_k + K_k^u\,u_k + o(\|\hat{x}_k\|,\,\|u_k\|).$$

Neglecting the quadratic and higher order terms in $\hat{x}_k$, u_k leads to an approximate filter

$$\begin{aligned} \hat{x}_{k+1} &\approx A_k\,\hat{x}_k + B_k\,u_k + (K_k + K_k^x\,\hat{x}_k + K_k^u\,u_k)\nu_k \\ &= (A_k + K_k^x\nu_k)\hat{x}_k + (B_k + K_k^u\nu_k)u_k + K_k\nu_k. \end{aligned} \tag{17}$$

5 STATE ESTIMATE FEEDBACK

The approach taken in an LQG control design is taken here, namely to consider the state estimator (13) (or in our case the approximation (17)) as a state space signal model with state $\hat{x}_k$, and to re-organize the control performance index J_T of (2) in terms of $\hat{x}_k$, rather than x_k. Noting (13) and the definition for Σ_k, we have a re-organization of J_T as

$$J_T = \sum_{k=0}^{T-1}[\,\hat{x}_k'\,Q_k^c\,\hat{x}_k + u_k'\,R_{k+1}^c\,u_k + \mathrm{tr}(Q_k^c\,\Sigma_k)]. \tag{18}$$

Actually, Σ_k is perhaps best written as $\Sigma_k(\hat{x}_k\,\hat{x}_k', u_k\,u_k')$ since it is dependent on $\hat{x}_k\,\hat{x}_k'$ and $u_k\,u_k'$. Now a Taylor Series expansion leads to

$$\Sigma_k \approx \Sigma_k^0 + \Sigma_k^x\,\hat{x}_k\,\hat{x}_k' + \Sigma_k^u\,u_k\,u_k', \tag{19}$$

being linear in $\hat{x}_k\, \hat{x}'_k$ and $u_k\, u'_k$. Thus (18) under (19) becomes

$$J_T \approx \sum_{k=0}^{T-1} [\hat{x}'_k\, (Q^c_k + \Sigma^x_k)\, \hat{x}_k + u'_k (R^c_{k+1} + \Sigma^u_k)\, u_k + \mathrm{tr}(Q^c_k\, \Sigma^0_k)]. \qquad (20)$$

Now the optimization of (20) under (17) can be tackled using the optimal LQG results of Section 2 with $w^A = w^B = w$. Thus

$$u^{opt}_k \approx K^c_k\, \hat{x}_k + b_k\,, \qquad (21)$$

where $\hat{x}_k$ is derived from the filter (17). Also K^c_k are derived from an approximate specialization of (3)-(4) in which $\Delta A_k = K^x_k$, $\Delta B_k = K^u_k$. The term b_k is derived by solving backward recursions for b_T, b_{T-1}, $\cdots$, b_0 [5].

References

[1] B.D.O. Anderson and J.B. Moore. *Optimal Control: Linear Quadratic Methods.* Prentice Hall, New Jersey, 1989.

[2] B.D.O. Anderson and J.B. Moore. *Optimal Filtering.* Prentice Hall, New Jersey, 1978.

[3] M.H.A. Davis. *Linear Estimation and Stochastic Control* Chapman and Hall, London, 1977.

[4] R.J. Elliott, L Aggoun and J.B. Moore. *Hidden Markov Models: Estimation and Control.* Springer-Verlag, 1995.

[5] J.B. Moore, X.Y. Zhou and A.E.B. Lim. Suboptimal LQG control of linear systems with control dependent noise. (submitted).

[6] S.P. Chen, X.J. Li and X.Y. Zhou. *Stochastic linear quadratic regulators with indefinite control weight costs.* SIAM J. Contr. Optim., to appear.

[7] A.E.B. Lim and X.Y. Zhou. *Optimal stochastic LQR control with integral quadratic constraints and indefinite control weights.* (preprint)

RADIAL SYMMETRY OF CLASSICAL SOLUTIONS FOR BELLMAN EQUATIONS IN ERGODIC CONTROL

Hiroaki Morimoto and Yasuhiro Fujita

Department of Mathematical Sciences, Faculty of Science,
Ehime University, Matsuyama 790, Japan
Department of Mathematics, Faculty of Science,
Toyama University, Toyama 930, Japan

1 INTRODUCTION

We are concerned with the d-dimensional Bellman equation of the form:

$$\lambda = \frac{1}{2}\Delta\phi(x) + F(D\phi(x)) + h(x), \qquad x \in \mathbf{R}^{\mathbf{d}}, \tag{1.1}$$

where

$$F(\xi) = \min\{(\xi, p) : |p| \leq 1\} = -|\xi|, \tag{1.2}$$

and $|\cdot|, (,)$, and D denote the norm, the inner product of vectors, and the gradient respectively. We are given a convex function $h(x)$ with polynomial growth, and the unknown is the pair of a constant λ and a C^2-function $\phi(x)$ on $\mathbf{R}^{\mathbf{d}}$.

The aim of the present paper is to study Bellman equation (1) without Lyapunov-type stability conditions and, in particular, the radial symmetry of $\phi(x)$ in the case that

$$h(x) \text{ is radial, i.e., } h(x) = f(r) \text{ for } r := |x|. \tag{1.3}$$

By the vanishing discount approach, we can show the existence of a unique classical solution of (1), for which the convexity and the polynomial growth property play essential roles. Further, for the radial symmetry, Bellman equation (1) turns out a 1-dimesional simple form which admits an explicit solution. It is shown that the radial symmetry of the Bellman equation is inherited from $h(x)$, and the optimal control is given by a feedback law $-x/|x|$.

2 BELLMAN EQUATIONS OF DISCOUNTED COST CONTROL

We study the existence of a unique solution u_α with polynomial growth to the Bellman equation:

$$\alpha u_\alpha(x) = \frac{1}{2}\Delta u_\alpha(x) + F(Du_\alpha(x)) + h(x), \qquad x \in \mathbf{R^d}, \tag{2.1}$$

where $0 < \alpha < 1/2$. We assume:

$$h : \text{non-negative, convex on } \mathbf{R^d}, \tag{2.2}$$

h satisfies the polynomial growth condition, i.e.,

$$\exists C > 0, m \in \mathbf{N}_+; \quad h(x) \leq C(1 + |x|^m), \quad x \in \mathbf{R^d}, \tag{2.3}$$

$$h \in C^1(\mathbf{R^d}). \tag{2.4}$$

To simplify the notation, we make use of the following quantity:

$$[f]_{\delta,B_r} = \sup_{x \in B_r} |f(x)| + \sup_{x,y \in B_r\ x \neq y} \frac{|f(x) - f(y)|}{|x - y|^\delta},$$

where $f(x)$ is the bounded Hölder continuous function with exponent δ on a ball $B_r = B_r(0)$ of $\mathbf{R^d}$. Let $t_n \in C_c^\infty(\mathbf{R^d})$ be a sequence such that $t_n = 1$ on B_n, 0 outside B_{2n} and $0 \leq t_n \leq 1, |Dt_n| \leq C/n$, and set $h_n = t_n h \in C_c^1(\mathbf{R^d})$. It is clear that $h_n \to h$ and $h_n \leq h$.

Now, let us consider the Bellman equation:

$$\alpha u_n(x) = \frac{1}{2}\Delta u_n(x) + F(Du_n(x)) + h_n(x), \qquad x \in \mathbf{R^d}. \tag{2.5}$$

Theorem 2.1. *Under (5), (6) and (7), equation (8) admits a unique solution* $u_n \in C_0(\mathbf{R^d}) \cap C^2(\mathbf{R^d})$, *which satisfies*

$$\sup_n [u_n]_{\delta,B_r} < \infty, \tag{2.6}$$

$$\sup_n \sum_i [D_i u_n]_{\delta,B_r} < \infty, \tag{2.7}$$

$$\sup_n \sum_{i,j} [D_{ij} u_n]_{\delta,B_r} < \infty, \tag{2.8}$$

for some $0 < \delta < 1$.

Proof. We shall give a brief sketch of the proof. It is well known [2] that, for every $n \in \mathbf{N}_+$, equation (8) has a unique solution u_n of the form:

$$u_n(x) = \inf\{E[\int_0^\infty e^{-\alpha t} h_n(x(t))dt] : |p(t)| \leq 1\}, \tag{2.9}$$

where $x(t)$ is a solution of the stochastic differential equation

$$dx(t) = p(t)dt + dw(t), \qquad x(0) = x \in \mathbf{R^d},$$

defined on some probability space $(\Omega, \mathcal{F}, P; \{\mathcal{F}_t\})$ carrying a d-dimensional standard Brownian motion $w(t)$, and the infimum is taken over the class of all $\mathcal{F}_t$-progressively measurable processes $p(t)$ with $|p(t)| \leq 1$.

Multiplying both sides of (8) by $u_n^\nu \Theta \geq 0$ for $\nu \geq 1$ and integrating over $\mathbf{R^d}$, we have

$$2\alpha \int u_n^{\nu+1}\Theta dx + \int |Du_n|^2 \nu u_n^{\nu-1}\Theta dx \leq 2\int h_n u_n^\nu \Theta dx, \tag{2.10}$$

where $\Theta(x) = e^{-\theta|x|}$ and $0 < \theta < 2$. By (13) and Hölder's inequality,

$$\alpha(\int u_n^{\nu+1}\Theta dx)^{1/(\nu+1)} \leq (\int |h|^{\nu+1}\Theta dx)^{1/(\nu+1)}.$$

Taking $\nu = 1$ in (13), we get

$$\sup_n \int (u_n^2 + |Du_n|^2)\Theta dx < \infty,$$

and thus,

$$\sup_n(|u_n|_{L^2(B_r)} + |Du_n|_{L^2(B_r)}) < \infty \qquad \text{for each } r > 0.$$

By the regularity result [5, Thm 8.8, p.183], there exists $C > 0$ such that

$$|u_n|_{W^{2,2}(B_r)} \leq C(|u_n|_{W^{1,2}(B_{r+1})} + |\Delta u_n|_{L^2(B_{r+1})}).$$

Therefore, we get by (8)

$$\sup_n |u_n|_{W^{2,2}(B_r)} < \infty.$$

By the Sobolev inequality [4, Thm IX.16, p.169], assuming $d > 2$, we have

$$\sup_n |Du_n|_{L^q(B_{r+1})} \leq \sup_n C|Du_n|_{W^{1,2}(B_{r+1})} < \infty, \qquad \frac{1}{q} = \frac{1}{2} - \frac{1}{d}.$$

We know by [1] that

$$|u_n|_{W^{2,q}(B_r)} \leq C(|u_n|_{W^{1,q}(B_{r+1})} + |\Delta u_n|_{L^q(B_{r+1})}).$$

Hence,

$$\sup_n |u_n|_{W^{2,q}(B_r)} < \infty.$$

By a bootstrap argument, we can obtain

$$\sup_n |u_n|_{W^{2,k}(B_r)} < \infty \qquad \text{for all } k > d.$$

Using the Sobolev inequality again, we have

$$\sup_n |u_n|_{L^\infty(B_r)} \leq \sup_n C|u_n|_{W^{1,k}(B_r)} < \infty.$$

Now, we apply the Morrey theorem [4, Thm IX.12, p.166 or p.169]:

$$|u_n(x) - u_n(y)| \leq C|u_n|_{W^{1,k}(B_r)}|x-y|^\delta, \qquad \forall x, y \in B_r, \qquad \delta = 1 - d/k,$$

to obtain (9). Similarly, we have (10). We can easily see by (7) that the derivative $D_i u_n$ satisfies

$$\alpha D_i u_n = \frac{1}{2}\Delta D_i u_n + (DF(Du_n), DD_i u_n) + D_i h_n.$$

Since $DF(\xi)$ is bounded, we have by virtue of [1]

$$\sup_n |Du_n|_{W^{2,k}(B_r)} < \infty.$$

Thus, by the same argument as above, we deduce (11).

Theorem 2.2. *Assume (5), (6), (7). Then there exists a unique solution $u_\alpha \in C^2(\mathbf{R^d})$ of equation (4) such that u_α is convex and satisfies*

$$0 \leq \alpha u_\alpha(x) \leq C(1 + |x|^{m+3}), \qquad x \in \mathbf{R^d}, \tag{2.11}$$

for some constant $C > 0$.

Proof. By Theorem 2.1, it is evident that the sequences $\{u_n\}, \{Du_n\}$ and $\{\Delta u_n\}$ are uniformly bounded and equi-continuous on every B_r. By the Ascoli-Arzelà theorem, we have

$$\begin{aligned} u_n &\to u_\alpha \in C^2(\mathbf{R^d}), \\ Du_n &\to Du_\alpha, \\ \Delta u_n &\to \Delta u_\alpha \qquad \text{uniformly on } B_r, \end{aligned}$$

taking a subsequence if necessary. Passing to the limit in (8), we can obtain (4).

Also, we can show (14) by (12) and

$$u_\alpha(x) = \inf\{E[\int_0^\infty e^{-\alpha t} h(x(t))dt] : |p(t)| \leq 1\}.$$

Thus the convexity of u_α is immediate.

Remark. In case of $d = 1$, the theorem is verified without (7), because h is Lipschitz continuous and (10) implies (11).

3 LIMIT AT INFINITY AND POLYNOMIAL GROWTH

We consider the limit of the solution u_α to Bellman equation (4) as $|x| \to \infty$ and also the polynomial growth property of $u_\alpha - \min u_\alpha$, denoted by v_α. We make the following assumption:

$$\text{There exists } C_0 > 0 \text{ such that } h(x) \geq C_0|x|. \tag{3.1}$$

Our objective in this section is to prove the following result.

Theorem 3.1.
We assume (5), (6), (7), (15). Then we have

$$u_\alpha(x) \quad \to \quad \infty \quad \text{as} \quad |x| \quad \to \quad \infty \quad \textit{uniformly in } \alpha,$$
$$v_\alpha(x) \le C(1+|x|^{m+1}),$$

for some constant $C > 0$.

Proof. By an elementary calculation, we can show that the assertions of Theorem 3.1 hold in case of d = 1. We prove the theorem, comparing (4) with

$$\alpha v_i = \frac{1}{2}\Delta v_i + F_i(Dv_i) + h_i, \quad i = 1, 2, \tag{3.2}$$

where

$$F_1(\xi) = \sum_{j=1}^{d} \min_{|p_j|\le 1} (p_j \xi_j), \qquad h_1(x) = \sum_{j=1}^{d} \eta_1 |x_j|,$$
$$F_2(\xi) = \sum_{j=1}^{d} \min_{|p_j|\le 1/d} (p_j \xi_j), \qquad h_2(x) = \sum_{j=1}^{d} \{\eta_2(|x_j|^m + 1) - \alpha \min u_\alpha\},$$

and each positive constant η_i will be chosen later. We can see that equation (16) admits a solution v_i of the form:

$$v_i(x) = \sum_{j=1}^{d} w_j^{(i)}(x_j),$$

for the solution $w_j^{(i)}(x_j)$ to (16) in the case d = 1. Hence, the following relations are fulfilled:

$$v_1(x) \quad \to \quad \infty \quad \text{as} \quad |x| \quad \to \quad \infty \quad \text{uniformly in } \alpha,$$
$$v_2(x) - \min v_2 \le \sum_j (w_j^{(2)}(x_j) - \min w_j^{(2)}) \le C(1+|x|^{m+1}),$$

for sufficiently large $\eta_2 > 0$. By the same line as (12), we have

$$v_\alpha(x) = \inf\{E[\int_0^\infty e^{-\alpha t}(h(x(t)) - \alpha \min u_\alpha)dt] \ : \ |p(t)| \le 1\},$$
$$v_1(x) = \inf\{E[\int_0^\infty e^{-\alpha t} h_1(x(t))dt] \ : \ |p_j(t)| \le 1\},$$
$$v_2(x) - \min v_2 = \inf\{E[\int_0^\infty e^{-\alpha t}(h_2(x(t)) - \alpha \min v_2)dt] \ : \ |p_j(t)| \le 1/d\}.$$

Since $\{p : |p_j| \le 1/d\} \subset \{p : |p| \le 1\} \subset \{p : |p_j| \le 1\}$, we can obtain

$$v_1 \le u_\alpha, \qquad v_\alpha \le v_2 - \min v_2,$$

for a convenient choice of each η_i such that $h_1 \le h$ and $h - \alpha \min u_\alpha \le h_2 - \alpha \min v_2$. Thus the theorem is established.

4 A-PRIORI ESTIMATES FOR APPROXIMATION

For the approximation problem of (1), we consider here the gradient estimates of v_α satisfying

$$\alpha v_\alpha = \frac{1}{2}\Delta v_\alpha + F(Dv_\alpha) + (h - \alpha \min u_\alpha). \tag{4.1}$$

Theorem 4.1. *Assume (5), (6), (7), (15). Then there exists $0 < \delta < 1$ such that*

$$\sup_{0<\alpha<1/2} [v_\alpha]_{\delta,B_r} < \infty, \tag{4.2}$$

$$\sup_{0<\alpha<1/2} \sum_i [D_i v_\alpha]_{\delta,B_r} < \infty, \tag{4.3}$$

$$\sup_{0<\alpha<1/2} \sum_{i,j} [D_{ij} v_\alpha]_{\delta,B_r} < \infty, \tag{4.4}$$

for every $r > 0$.

Proof. Multiplying both sides by $v_\alpha\Theta$ and integrating over $\mathbf{R}^\mathbf{d}$, we have

$$2\int \alpha v_\alpha^2 \Theta dx - \int (\Delta v_\alpha) v_\alpha \Theta dx - 2\int F(Dv_\alpha) v_\alpha \Theta dx = 2\int (h - \alpha \min u_\alpha) v_\alpha \Theta dx.$$

where Θ is as in the proof of Theorem 2.1. The second term of the left-hand side can be rewritten as

$$\int (Dv_\alpha, D(v_\alpha\Theta))dx \geq \int [|Dv_\alpha|^2 - \theta v_\alpha |Dv_\alpha|]\Theta dx.$$

By the choice of θ with $0 < \theta < 2$, we get

$$\int |Dv_\alpha|^2 \Theta dx \leq 2\int (h - \alpha \min u_\alpha) v_\alpha \Theta dx.$$

Since

$$v_\alpha(\gamma_\alpha) - v_\alpha(x) \geq (Dv_\alpha(x), \gamma_\alpha - x)$$
$$h(\gamma_\alpha) \leq \alpha v_\alpha(\gamma_\alpha) + \alpha \min u_\alpha \leq \alpha u_\alpha(0)$$

for $\gamma_\alpha :=$ arg min v_α, we have

$$0 \leq v_\alpha(x) \leq C|Dv_\alpha(x)|(|x| + 1). \tag{4.5}$$

Then, by the Schwarz inequality

$$\int |Dv_\alpha|^2 \Theta dx \leq 2C\Big(\int (h - \alpha \min u_\alpha)^2 (|x| + 1)^2 \Theta dx\Big)^{1/2} \Big(\int |Dv_\alpha|^2 \Theta dx\Big)^{1/2}.$$

Therefore, we deduce

$$\sup_\alpha \int |Dv_\alpha|^2 \Theta dx < \infty.$$

Now, we can find a constant $C > 0$ independent of α such that

$$|Dv_\alpha|_{L^2(B_r)} \leq C,$$

and, by (21)

$$|v_\alpha|_{L^2(B_r)} \leq C.$$

Again, using the regularity result and (17), we can obtain

$$\sup_\alpha |v_\alpha|_{W^{2,2}(B_r)} < \infty.$$

Finally, by the same bootstrap argument as the proof of Theorem 2.1,

$$\sup_\alpha |v_\alpha|_{W^{2,k}(B_r)} < \infty \qquad \text{for all } k > d.$$

Further, repeating the proof of Theorem 2.1, we can deduce (18), (19), (20).

5 BELLMAN EQUATION OF ERGODIC CONTROL

We shall show the existence of a unique solution $(\lambda, \phi) \in \mathbf{R} \times C^2(\mathbf{R^d})$ to the Bellman equation

$$\lambda = \frac{1}{2}\Delta\phi(x) + F(D\phi(x)) + h(x), \qquad x \in \mathbf{R^d}. \tag{5.1}$$

Theorem 5.1.
We assume (5), (6), (7), (15). Then there exists a subsequence $\alpha \to 0$ such that

$$\begin{aligned} \alpha \min u_\alpha &\to \lambda \in \mathbf{R}_+, \\ v_\alpha(x) &\to \phi(x) \in C^2(\mathbf{R^d}) \quad \textit{uniformly on each } B_r. \end{aligned}$$

The limit (λ, ϕ) satisfies Bellman equation (22), and furthermore

$$\phi : \textit{convex on } \mathbf{R^d}, \tag{5.2}$$

$$\phi(x) \to \infty \quad \text{as} \quad |x| \to \infty, \tag{5.3}$$

$$0 \leq \phi(x) \leq C(1 + |x|^{m+1}), \qquad (C > 0 : \textit{const.}). \tag{5.4}$$

Proof. The proof follows from Theorem 4.1 and the Ascoli-Arzelà theorem.

Remark. We can show the uniqueness of the solution (λ, ϕ) (ϕ up to a constant) of (22) with (23), (24) and (25). For the proof, we refer to [6].

6 RADIAL SYMMETRY

In this section, we study the radial symmetry of $\psi(x)$ of the Bellman equation

$$\lambda = \frac{1}{2}\Delta\psi(x) + F(D\psi(x)) + f(|x|), \quad x \in \mathbf{R^d}, \tag{6.1}$$

raplacing (22) in case of (3). Without loss of generality, by the assumptions on h, we may assume

$$\begin{aligned} &f(r): \text{ non-decreasing on } [0,\infty), \quad f(\infty) = \infty, \\ &|f(r)| \le C(1 + r^m), \quad r \ge 0, \\ &f \in C^1([0,\infty)). \end{aligned} \tag{6.2}$$

We are involved in a reduction to the 1-dimensional equation:

$$\mu = \frac{1}{2}(\varphi''(r) + \frac{d-1}{r}\varphi'(r)) - |\varphi'(r)| + f(r), \quad r > 0. \tag{6.3}$$

Lemma 6.1. *Assume (27). Then there exists a unique solution* $(\mu, \varphi) \in \mathbf{R} \times C^2([0,\infty))$ *(φ up to a constant) of (28) such that*

$$\varphi'(r): \textit{ polynomial growth}, \quad \varphi'(0+) = 0, \quad \varphi'(r) > 0 \textit{ for } r > 0. \tag{6.4}$$

Moreover, the solution (μ, φ) is given by

$$\mu = \int_0^\infty f(s)s^{d-1}e^{-2s}ds \Big/ \int_0^\infty s^{d-1}e^{-2s}ds, \tag{6.5}$$

$$\varphi'(r) = \frac{2e^{2r}}{r^{d-1}} \int_0^r (\mu - f(s))s^{d-1}e^{-2s}ds. \tag{6.6}$$

Proof. Define (μ, φ) by (30) and (31). By an elementary manipulation, we see that $\varphi'(r) > 0$, $\varphi'(0+) = 0$ and

$$\lim_{r\to\infty} \frac{\varphi'(r)}{r^m} = \lim_{r\to\infty} \frac{2(\mu - f(r))r^{d-1}}{(m+d-1)r^{m+d-2} - 2r^{m+d-1}} \le C. \tag{6.7}$$

Thus (29) is verified. Now, multiplying (31) by $r^{d-1}e^{-2r}$ and differentiating both sides, we can get (28).

Conversely, equation (28) turns out a linear differential eqution under (29). Hence we can solve this equation to obtain (30) and (31).

Theorem 6.2. *We assume (3), (27). Then equation (26) admits a solution* $(\lambda, \psi) \in \mathbf{R} \times C^2(\mathbf{R^d})$ *given by*

$$\lambda = \mu, \tag{6.8}$$

$$\psi(x) = \varphi(|x|). \tag{6.9}$$

Further, ψ fulfills (25).

Proof. Define (λ, ψ) by (33) and (34). By a simple calculation, we have

$$\Delta\psi(x) = \varphi''(r) + \frac{d-1}{r}\varphi'(r),$$
$$|D\psi(x)| = |\varphi'(r)|.$$

Hence (λ, ψ) satisfies

$$\lambda = \frac{1}{2}\Delta\psi(x) + F(D\psi(x)) + f(|x|), \quad x \in \mathbf{R}^d \setminus \{0\}. \tag{6.10}$$

By (28) and (29), we see that the limit of $|D\psi(x)|$ exists as $|x| \to 0$, and then $|D\psi(x)|$ and $\Delta\psi(x)$ are bounded on every B_r. By the regularity result [1]

$$|\psi|_{W^{2,k}(B_r)} \leq C(|\psi|_{W^{1,k}(B_{r+1})} + |\Delta\psi|_{L^k(B_{r+1})}) < \infty \quad \text{for} \quad k > d.$$

Moreover, differentiating (35), we get

$$0 = \frac{1}{2}\Delta D_i\psi + (DF(D\psi), DD_i\psi) + f'(|x|)D_i|x|,$$

which implies $\Delta D_i\psi \in L^k(B_r)$. Applying the regularity result again, we have

$$\psi \in W^{3,k}(B_r), \qquad k > d.$$

Therefore, by the imbedding theorem [5], we deduce $\psi \in C^2(B_r)$ and hence (26). We remark by (32) that ψ of (34) fulfills the polynomial growth condition.

7 AN APPLICATION TO ERGODIC CONTROL

We shall study the ergodic control problem to minimize the cost

$$J(p) = \limsup_{T\to\infty} \frac{1}{T} E[\int_0^T h(x(t))dt]$$

over all $p \in \mathcal{P}$ subject to the state equation

$$dx(t) = p(t)dt + dw(t), \qquad x(0) = x,$$

where $\mathcal{P}$ denotes the set of all progressively measurable $\mathcal{F}_t$-adapted processes $p(t)$ such that

$$|p(t)| \leq 1, \qquad \lim_{T\to\infty} \frac{1}{T} E[|x(t)|^{m+1}] = 0 \quad \text{for the response } x(t) \text{ to } p(t).$$

Now, let us consider the stochastic differential equation

$$dx^*(t) = G(D\psi(x^*(t)))dt + dw(t), \qquad x^*(0) = x,$$

where

$$G(z) = \begin{cases} -z/|z| & \text{if } z \in \mathbf{R^d}\backslash\{0\}, \\ 0 & \text{if } z = 0. \end{cases}$$

Lemma 7.1. *For any $n \in \mathbf{N}_+$, there exists $C > 0$ such that*

$$E[|x^*(t)|^{2n}] \leq C(1+t).$$

Proof. The proof follows from the relation:

$$E[|x^*(t)|^{2n}] = |x|^{2n} + E[\int_0^t \{-2n|x^*(s)|^{2n-1} + n(2n+d-2)|x^*(s)|^{2n-2}\}dt].$$

Theorem 7.2. *We make the assumptions of Theorem 6.2. Then the optimal control $p^*(t)$ is given by*

$$p^*(t) = G(D\psi(x^*(t))) = G(x^*(t)),$$

and the value by

$$J(p^*) = \lambda.$$

References

[1] Agmon, S., Douglis, A. and Nirenberg, L. (1959). Estimates near the boundary for solutions of elliptic partial differential equations satisfying general boundary conditions I. *Comm. Pure Appl. Math.* 12 : 623-727.

[2] Bensoussan, A. (1982). *Stochastic Control by Functional Analysis Methods.* North-Holland, Amsterdam.

[3] Bensoussan, A. and Frehse, J. (1992). On Bellman equations of ergodic control in R^n. *J. reine angew. Math.* 429 : 125-160.

[4] Brezis, H. (1992). *Analyse Fonctionnelle;Théorie et applications.* Masson, Paris.

[5] Gilbarg, D. and Trudinger, N.S. (1983) *Elliptic Partial Differential Equations of Second Order.* Springer, Berlin.

[6] Morimoto, H. and Okada, M. Some results on Bellman equation of ergodic control. *SIAM J. Control Optim.* to appear.

OPEN PROBLEMS ON BACKWARD STOCHASTIC DIFFERENTIAL EQUATIONS *

Shige Peng

School of Mathematics and System Science
Shandong University, 250100, Jinan, China

1 BASIC RESULTS ON BACKWARD STOCHASTIC DIFFERENTIAL EQUATIONS

Recent years, many interesting problems in the theory of backward stochastic differential equations (in short, BSDE) have been solved. Others still remain open. In this paper, we will discuss those related to the stochastic control theory.

Let $(\Omega, \mathcal{F}, P)$ be a probability space, let W_t, $t \geq 0$ be a standard Brownnian motion. For the sake of simplification, W_t is of 1-dimension. We are also limited to the interval $[0,T]$, with a finite constant $T > 0$. (see Peng [1991], Duffie, Ma and Yong [1994], Hu [1995], Chen [1997], Peng and Shi [1998] for infinite horizon cases) . All processes are assumed to be $\mathcal{F}_t$-adapted. Here the filtration $\{\mathcal{F}_t\}_{t\geq 0}$ is the natural one generated by the Brownian motion $W_\cdot$. It is the information we have possessed up to time t. We will denote by $M^2(0,T;\mathbf{R}^n)$ the set of all $\mathbf{R}^n$-valued square-integrable $\mathcal{F}_t$-adapted processes.

We are given ξ, an $\mathcal{F}_T$-measurable $\mathbf{R}^m$-valued square-integrable random vector, and a function $f(y,z,t,\omega) : \mathbf{R}^m \times \mathbf{R}^m \times [0,T] \times \Omega \longmapsto \mathbf{R}^m$ such that, for each $(y,z) \in \mathbf{R}^m \times \mathbf{R}^m$, $f(x,y,\cdot) \in M^2(0,T;\mathbf{R}^m)$. The problem of backward stochastic differential equation is to look for a pair of processes $(y_\cdot, z_\cdot) \in M^2(0,T;\mathbf{R}^m)^2$, that solves

$$\begin{aligned} -dy_t = & \ f(y_t, z_t, t)dt - z_s dW_t, \quad t \in [0,T], \\ y_T = & \ \xi. \end{aligned} \tag{1.1}$$

*This work was supported by NSF of China.

Bismut [1973] studied the case when f is linear in (y, z). Pardoux and Peng [1990] obtained a general result: if f is Lipschitz in (y, z), then the solution $(y_t, z_t)_{0 \le t \le T}$ of BSDE (1.1) exists and is unique.

Peng [1991a] observed a mutual relation between the solution $u(x,t)$: $\mathbf{R}^n \times [0,T] \longmapsto \mathbf{R}^m$ of the the system of parabolic PDE

$$\begin{aligned} -\frac{\partial u^k}{\partial t} &= \frac{1}{2}\sum_{i,j=1}^{n}[\sigma\sigma^T]_{ij}\frac{\partial^2 u^k}{\partial x_i \partial x_j} + \sum_{i=1}^{n} b_i \frac{\partial u^k}{\partial x_i} + f(u, \nabla u\sigma, x), \\ u^k(x,T) &= \Phi^k(x) \quad k = 1, \cdots, m. \end{aligned} \tag{1.2}$$

and the solution of $(y_s^{x,t}, z_s^{x,t})$, $s \in [t,T]$ of BSDE (1) in the time interval $[t,T]$, with the following special form of f and ξ: $f(y,z,\cdot) = g(y,z,x^{x,t}(\cdot))$, $\xi = \Phi(x_T^{x,t})$, where $x^{x,t}(s)$, $s \in [t,T]$, is an n-dimensional diffusion process with the initial condition (x,t). It evolves according to the following SDE

$$dx_s = b(x_s, s)ds + \sigma(x_s, s)dW_s, \quad s \in [t,T], \quad x_t = x. \tag{1.3}$$

The mutual relation is: $u(x^{x,t}(s), s) = y^{x,t}(s)$, $\sigma^T \nabla u(x^{x,t}(s), s) = z^{x,t}(s)$. Particularly

$$u(x,t) = y^{x,t}(t), \quad \sigma^T \nabla u(x,t) = z^{x,t}(t). \tag{1.4}$$

This can be formally checked by simply applying Itô's formula to $u(x_t, t)$. For $m = 1$, Peng [1992b] and then Pardoux and Peng [1992] showed that the function u defined by (1.4) is the unique viscosity solution of PDE (1.2) (see Crandall, Ishii and Lions [1992] for the theory of viscosity solution) Pardoux and Peng,[1992] showed that if coefficients are smooth, say C^3, then $u(x,t)$ defined in (1.4) is the $C^{2,1}$ solution of (1.2) (m is allowed to be bigger than 1). An interesting problem is, for $m \ge 2$, to find a suitable notion of the solution of PDE of "viscosity type"(1.2) that characterizes u in (1.4). Barles and Lesigne [1997] studied L^2-space PDE method. El Karoui, Peng and Quenez [1997] introduced a notion of system of quasi linear PDE (1.2). Pardoux and Rao [1997] discussed the viscosity solution for the system of PDE (1.2) with some additional assumptions such as f^i does not depends on z^j for $i \neq j$. But the general problem still remains an open.

We give two interesting special cases of formula (1.4) (for $m = 1$). The first one is $f(y,s) = c(x_s^{x,t})y + g(x_s^{x,t})$.

The function $u(x,t)$ defined by (1.4) is just the well-known Feynman-Kac formula. It is for this reason that we call (1.4) Nonlinear Feynman-Kac Formula. The second case is the well-known Black & Scholes formula of option pricing. In this case $f = ay + bz$. [EPQ,1997] gives many further applications of BSDE in finance.

A useful tool in BSDE is Comparison Theorem introduced in Peng [1992b] and improved in Pardoux, Peng [1992], El Karoui, Peng and Quenez [1997].

Briefly speaking, if f and ξ are dominated respectively by f' and ξ', then the corresponding solution (y, z) is also dominated by (y', z'): $y_t \le y'_t$. Many difficulties concerning the existence theorem in multi-dimensional case ($m \ge 2$) are due to the lack of this important property.

2 STOCHASTIC HAMILTON-JACOBI-BELLMAN EQUATIONS

We consider the following standard control system driven by a control process $\alpha_\cdot \in M^2(0, T; U)$ (called an admissible control), its value must be chosen in a given subset $U \subset \mathbf{R}^k$. The state variable evolves according to

$$\begin{aligned} dX_s^{x,t} &= b(X_s^{x,t}, \alpha_s, s)ds + \sigma(X_s^{x,t}, \alpha_s, s)dW_s, \quad s \in [t, T], \\ X_t^{x,t} &= x. \end{aligned} \tag{2.1}$$

Some assumptions are needed to ensure the existence and uniqueness of the above stochastic differential equation. We are given the following utility function:

$$J_{x,t}(\alpha(\cdot)) = \mathbf{E}^{\mathcal{F}_t}\{\int_t^T f(X_s^{x,t}, \alpha_s)ds + \Phi(X_T^{x,t})\}, \tag{2.2}$$

where $f(x, \alpha, t)$ and $\Phi(x)$ are real-valued functions. Our objective is to maximize the utility function over the set of admissible control. The maximum value is called the value function:

$$u(x, t) = \operatorname{ess\,sup}\{J_{x,t}(\alpha(\cdot)); \alpha(\cdot) \in M^2(0, T; U)\} \tag{2.3}$$

A classic situation is when all coefficients b, σ, f and Φ are deterministic functions of x, α and t. In this case u is a deterministic function of (x, t). It is the unique viscosity solution of the following Hamilton-Jacobi-Bellman (HJB in short) equation:

$$\begin{aligned} -\frac{\partial u}{\partial t} &= \sup\nolimits_{v \in \mathcal{U}} \left\{ \frac{1}{2}\langle D^2 u\sigma(x, \alpha), \sigma(x, \alpha)\rangle + \langle Du, b(x, \alpha, t\rangle + f(x, \alpha, t)] \right\}, \\ u(x, T) &= \Phi(x). \end{aligned} \tag{2.4}$$

But if one of the coefficients are random, e.g., for each x, $\Phi = \Phi(x, \omega)$ is $\mathcal{F}_T$-measurable), the value function $u(x, \cdot)$, for each x, becomes an $\mathcal{F}_t$-adapted process. It is no longer a solution of HJB equation (2.4). Peng [1992a] pointed out that, provided every thing is "smooth enough", the value function $u(x, \cdot)$ together with $v(x, \cdot)$, another function-valued $\mathcal{F}_t$-adapted process, consists the solution of the following backward stochastic PDE (Verification Theorem):

$$\begin{aligned} -du &= \sup\nolimits_{v \in \mathcal{U}}\{\frac{1}{2}\langle D^2 u\sigma(x, \alpha, t), \sigma(x, \alpha, t)\rangle + \langle Du, b(x, \alpha, t\rangle \\ &\quad + \langle Dv, \sigma(x, \alpha)\rangle + f(x, \alpha, t)\}dt - v(x, t)dW, \\ u(x, T) &= \Phi(x). \end{aligned} \tag{2.5}$$

This equation is called stochastic HJB equation. The problem is, in most circumstances, one can not expect that the pair $(u, v)(x, \cdot)$ is 'smooth enough'. Peng [1992a] studied the case where the diffusion term is independent of the control variable: $\sigma = \sigma(x,t)$. An existence and uniqueness theorem has been obtained. A difficult problem is when $\sigma = \sigma(x, \alpha, t)$. Two problems are: (i) To introduce a generalized notion of viscosity solutions of (2.5). (ii) To develop a $C^{2+\alpha}$-theory.

3 STOCHASTIC MATRIX-VALUED RICCATI EQUATION

A typical situation of the above optimization is the so-called linear quadratic case, i.e., $U = \mathbf{R}^k$, b and σ are linear functions, f and Φ are quadratic functions:

$$b_t(x,\alpha) = A_t x + B_t\alpha, \quad \sigma_t(x,\alpha) = C_t x + D_t\alpha \tag{3.1}$$

$$f_t(x,\alpha) = -\frac{1}{2}[\langle R_t x, x\rangle + \langle N_t\alpha, \alpha\rangle], \quad \Phi(x) = -\frac{1}{2}\langle Qx, x\rangle = .. \tag{3.2}$$

where A_t, B_t, C_t D_t, R_t, N_t and Q are bounded and matrix-valued vith suitable dimentions. $R.$, and Q are symmetric and nonnegative. $N.$ is strictly positive. It is easy to check that the value function u is also quadratic in x: $u(x,t) = \frac{1}{2}\langle Y_t x, x\rangle$. Consequently, v is also quadratic: $v(x,t) = \frac{1}{2}\langle Z_t x, x\rangle$. Here Y_t and Z_t are $n \times n$-symmetric matrix-valued processes. Y_t is nonnegative. From (14) one can formally write:

$$-dY_t = F(Y_t, Z_t, t)dt - Z_t dW_t, \quad Y_T = Q, \tag{3.3}$$

where $F(Y, Z, t)$ is an $n \times n$-valued function:

$$\begin{aligned} - \quad & F(Y,Z,t) = A_t^T Y + YA_t + C_t^T Z + ZC_t + C_t^T YC_t + R_t \\ & -(YB_t + ZD_t + D_t YC_t^T)(N_t + D_t^T YD_t)^{-1}(YB_t + ZD_t + D_t YC_t^T)^T .. \end{aligned} \tag{3.4}$$

The problem is to prove the existence and uniqueness of (3.3). When $D = 0$, the problem is solved, see Bismut [1976], Peng [1992a,1993]. The case $n = 1$, $D \neq 0$ is solved ([OT,1997]).

4 GENERAL MAXIMUM PRINCIPLE UNDER RECURSIVE UTILITIES

The above stochastic optimization problem can also be treated by the method of stochastic maximum principle of Pontryagin's type. Let H be the "stochastic Hamiltonian" introduced in Bismut [1973] (see also Bensoussan [1981]) defined by $H = \langle p, b(x,\alpha,t)\rangle + \langle q, \sigma(x,\alpha,t)\rangle + f(x,\alpha,t)$. Based on the Bisnut's 'local maximum principle', Peng [1990] obtained the following 'general stochastic maximum principle':

$$\begin{aligned}\sup_{\alpha\in U} & \left\{H(x_t,\alpha,p_t,q_t-P_t\sigma(x_t,\alpha_t),t)-\frac{1}{2}\langle P(t)\sigma(x_t,\alpha),\sigma(x_t,\alpha)\rangle\right\}\\ \leq & \ H(x_t,\alpha_t,p_t,q_t-P_t\sigma(x_t,\alpha_t),t)-\frac{1}{2}\langle P(t)\sigma(x_t,\alpha),\sigma(x_t,\alpha_t)\rangle\end{aligned} \tag{4.1}$$

where the pair $(p_\cdot,q_\cdot)\in M^2(0,T;\mathbf{R})^2$ is the solution of the following linear BSDE

$$-dp_t=\partial_x H(x_t,\alpha_t,p_t,q_t,t)dt-q_t dW_t,\quad p_T=\Phi(x_T). \tag{4.2}$$

and (P_t,Q_t) solves the following $n\times n$-matrix valued BSDE:

$$\begin{aligned}-dP_t= & \ [b_x^T P+Pb_x+\sigma_x^T Q+Q\sigma_x+\sigma_x^T P\sigma_x\\ & +H_{xx}(x_t,\alpha_t,p_t,q_t,t)]dt-Q_t dW_t,\\ P_T= & \ \Phi_{xx}(x_T).\end{aligned} \tag{4.3}$$

Recently, Duffie, Epstein [1992a,b] and other economists are introduced the notion of recursive utilities in continuous time (see also Duffie, Lions [1992]). El Karoui, Peng and Quenez [1997] give the following general form of recursive utility:

$$J_{x_0,0}(\alpha(\cdot))=y_0 \tag{4.4}$$

where y_0 is solved via the following BSDE

$$-dy_t=f(x_t,\alpha_t,y_t,z_t,t)dt-z_t dW_t,\quad y_T=\Phi(x_T). \tag{4.5}$$

It is easy to check that, when f only depends on (x,α): i.e.,
$f=f(x,\alpha,t)$, then $J(\alpha(\cdot))$ is the classical utility function of type (2.2). Local maximum principle under the recursive utility is given in Peng [P,1993] (see also Xu [1994], Wu and Xu [1997]. The corresponding 'global maximum principle' for the case where f depends nonlinearly on z is open, except for some special case (see Peng and Xu [1998]).

5 EXISTENCE OF BSDE WITH NON LIPSCHITZ COEFFICIENTS

We return to a general BSDE of form (1.1). Since [PP,1990] many efforts have been made to relax the assumption of the uniform Lipschitz condition imposed on f. For 1-dimensional case, i.e., $m=1$, see Lepeltier, San Martin [1996]; Kobylansky [1997] shows that the linear growth condition with respect to (y,z) can be relaxed to the quadratic growth condition provided the prescribed terminal value $y_T=\xi$ is bounded (see also [HLS,1998] for more general situation). This is a very interesting result. But the boundness of ξ excludes a very interesting application, i.e., the problem of risk sensitive control (see

Bensousssan and Nagai[1996], Bielecki and Pliska [1998]). This problem can be naturally formulated as a BSDE with f quadratic in z. In this domain one often interested in the situations where $\xi \in L^2$. Risk sensitive control problems also indicate a totally open topic in BSDE: the problem of explosions.

Another challenge in BSDE is for multi-dimensional case: $m \geq 2$. The problem of the existence with coefficient f only continuous in (y, z) becomes very hard (a Markov case was treated in Hamadame Lepeltier and Peng [1997]). Similar problem appears in situation of the quadratic growth condition. The techniques that successfully applied in 1-dimensional case fails since the lack of the comparison theorem. Multi-dimensional case has many interesting applications. For example: an important open questions in fining the equilibrium strategies in non-zero-sum differential games is formulated as a BSDE where f is only continuous (see [EPQ,1997], [HLP,1997]). The quadratic growth condition relates to harmonic mappings in differential geometry (see Darling [1995]).

6 FORWARD-BACKWARD STOCHASTIC DIFFERENTIAL EQUATIONS

Forward-Backward Stochastic Differential Equations (in short, FBSDE) is derived from many stochastic optimization problems. For fully coupled situation, a local existence and uniqueness result was obtained in Antonelli [1992]

(see also [P,1992c]). The first global existence and uniqueness results, i.e., in an arbitrarily large interval $[0, T]$, was given in Ma and Yong [MY,1995], Ma, Protter and Yong [1994], which is now well known as "four-step-approach". Recently, Pardoux and Tang [1997] also give a new approach.

Here we discuss a different global existence and uniqueness approach given in Hu and Peng [1994] with a continuation method (see also Peng [1991a]). This approach is generalized in Peng and Wu [1996], Yong [1997] (Bridge Method). Let b, σ and f be three $\mathbf{R}^n$-valued function of $\zeta = (x, y, z) \in (\mathbf{R}^n)^3$. We assume that the mapping A defined by $A(\zeta) = (-f, b, \sigma)(\zeta) : (\mathbf{R}^n)^3 \longmapsto (\mathbf{R}^n)^3$ is Lipschitz in ζ and is monotonic:

$$\langle A(\zeta) - A(\zeta'), \zeta - \zeta' \rangle \leq -\beta \|\zeta - \zeta'\|^2$$

We also assume that Φ is Lipschitz in x and is monotonic: $\langle \Phi(x) - \Phi(x'), x - x' \rangle \geq 0$. For each fixed $(x, t) \in \mathbf{R}^n \times [0, T]$, we consider the following problem of FBSDE: to find a triple of processes $(X_s^{x,t}, Y_s^{x,t}, Z_s^{x,t})$, $s \in [t, T]$, that solves

$$\begin{aligned} dX_s^{x,t} &= b(X_s^{x,t}, Y_s^{x,t}, Z_s^{x,t})ds + \sigma(X_s^{x,t}, Y_s^{x,t}, Z_s^{x,t})dW_s, \quad X_t^{x,t} = x. \\ -dY_s^{x,t} &= f(X_s^{x,t}, Y_s^{x,t}, Z_s^{x,t})ds - Z_s^{x,t}dW_s, \quad Y_T^{x,t} = \Phi(X_T^{x,t}). \end{aligned} \tag{6.1}$$

By Hu and Peng [1994], this equation has a unique solution. We then can define the following two (deterministic) function of (x, t),

$$u(x, t) = Y_t^{x,t}, \quad v(x, t) = Z_t^{x,t}. \tag{6.2}$$

If the pair (u, v) is sufficiently smooth, then it is the solution of the following PDE

$$\begin{aligned} -\frac{\partial u}{\partial t} = & \frac{1}{2}[\sigma\sigma^T(x,u,v)]_{ij}\frac{\partial^2 u}{\partial x_i \partial x_j} + b_i(x,u,v)\frac{\partial u}{\partial x_i} + f(x,u,v), \\ v(x,t) = & \sigma^T(x,u,v)\nabla u, \quad (x,t) \in \mathbf{R}^n \times [0,T]. \end{aligned} \tag{6.3}$$

with a Cauchy type of terminal condition: $u(x,T) = \Phi(x)$. An interesting problem is how to find a suitable notion of the solution (u, v) of this PDE that coincides with (6.2). A more general case is when b, σ, f and Φ are random, e.g., for each $x \in \mathbf{R}^n$, $\Phi(x, \omega)$ is $\mathcal{F}_T$ measurable. In this situation the pair $(u(x,\cdot), v(x,\cdot))$ defined in (6.2) is an $(\mathbf{R}^n)^2$-valued $\mathcal{F}_t$-adapted process. The following backward stochastic PDE is then in order:

$$\begin{aligned} -du = & \{\frac{1}{2}[\sigma\sigma^T(x,u,v)]_{ij}\frac{\partial^2 u}{\partial x_i \partial x_j} + b_i(x,u,v)\frac{\partial u}{\partial x_i} \\ & +\sigma^T(x,u,v)\nabla v + f(x,u,v)\}dt - vdW_t, \\ v(x,t) = & \sigma^T(x,u,v)\nabla u, \quad (x,t) \in \mathbf{R}^n \times [0,T]. \end{aligned} \tag{6.4}$$

with the terminal condition: $u(x,T) = \Phi(x)$. The question is to prove the existence and uniqueness of (6.4).

References

Antonelli, F.(1992): 'Backward-Forward Stochastic Differential Equations" *Annals of Applied Probability* **3**. 777-793 (1993).

Bensoussan, A.,(1981) "Lecture on Stochastic Control", LNM No. **972**, *Nonlinear Filtering and Stochastic Control.*

Bensoussan, A., and Nagai, H., (1996): "Min-Max Characterization of Small Noise Limit on Risk–Sensitive Control" *SIAM J. Contr. and Optim.*, **35**(4), 1093–1115.

Barles, G., Lesigne, E., (1997): "SDE, BSDE and PDE", *Backward Stochastic Differential Equation, El Karoui, N., Mazliak, L. (Editors),* , pp47–82, Longman.

Bielecki, T.R., Pliska S.R., (1998) " Risk Sensitive Dynamic Asset Menagement", to appear in *J. Appl. and Optim.*.

Bismut, J.M. (1973): "Conjugate Convex Functions in Optimal Stochastic Control," *J.Math. Anal. Apl.*, **44**, pp.384-404.

Bismut, J.M. (1978): "Contrôle des systèmes linéaires quadratiques : applications de l'intégrale stochastique," *Sémin.Proba. XII.*, Lect. Notes in Mathematics, **649**, pp.180-264, Springer.

Black, F., and M.Scholes (1973): "The Pricing of Options and Corporate Liabilities," *Journal of Political Economy,***3**, pp.637-654.

Chen Z., Epstein, L., (1998): "Ambiguity, Risk and Asset Returns in Continuous Time", Preprints.

Crandall, M.G., Ishii, H., and Lions, P.L., (1992): "User's Guide to Viscosity Solutions of Second Order Partial Differential Equations", *Bull. AMS.*, **27**, 1–67.

Darling, R.W.R., (1995): "Constructing Gamma–Martingales with Prescribed Limits, Using Backward SDE", *Ann. Prob.*, **23**, 1234–1261.

Duffie, D., and L.Epstein (1992a): " Stochastic Differential Utility," *Econometrica*, **60**, (2), 353–394.

Duffie, D., and L.Epstein (1992b): " Asset Pricing with Stochastic Differential Utility" , *Review of Financial Studies*, **5**, 411-436.

Duffie, D., J.Ma, and J.Yong (1994): "Black's Consol Rate Conjecture", *Ann.Applied probability*, **5**, 356–382.

El Karoui, N., Mazliak, L. (Editors), (1997): *Backward Stochastic Differential Equation*, Longman.

El Karoui, N., S.Peng, and M.C.Quenez (1994): "Optimization of Utility Functions" *Working Paper*, Paris VI University.

Fleming, W.F., and Rishel (1975): *Deterministic and Stochastic Control,* Springer-Verlag.

Hamadène, S., and Lepeltier, J.P., (1995): "Zero-Sum Stochastic Differential Games and Backward Equations", *Systems and Control Letters,* **24,** 259-263.

Hamadène, S., and Lepeltier, J.P., (1995): "Backward Equations, Stochastic Control and Stochastic Differential Games", *Stoch. and Stoch. Reports*, **54,** 221-231.

Haussmann,U.G., (1976) "General Necessary Conditions for Optimal Con- trol of Stochastic system", *Math.Programming study* **6**.

Hu, Y. and Peng, S., (1995): "Solution of Forward–Backward Stochastic Differential Equations", *Prob. and Related Fields*, **103**, 273–283.

Kobylansky, M.,(1997): BSDEs and PDEs, The Case with Quadratic Growth, Preprint.

Krylov, N.(1980): *Controlled Diffusion Processes*, New York. Springer Verlag.

Lepeltier, J.P., and San Martin, J., (1996): "Backward Stochastic Differential Equations with Continuous Generators", *Proba. and Stat. Letters*, to appear.

Li, X., and Tang, Sh., (1995) "General Necessary Conditions for Partially Observed Optimal Stochastic Controls", *J. Appl. Prob.*, **32**, 1118–1137.

Ma, J., P.Protter, and J.Yong (1994): "Solving Forward-backward Stochastic Differential Equations Explicitly - a four step Scheme", *Proba. Theory Relat. Fields*, **98**, 339–359 .

Ma, J., and J.Yong (1995): "Solvability of Forward-backward Stochastic Differential Equations and Nodal Set of Hamilton-Jacobi-Bellman Equations" *Annals of Mathematics.*

Pardoux, E. and S.Peng (1990): "Adapted Solution of a Backward Stochastic Differential Equation", *Systems and Control Letters*, **14**, 55–61.

Pardoux, E. and S.Peng (1992): " Backward Stochastic Differential equations and Quasilinear Parabolic Partial Differential Equations" *Lect. Notes in CIS* ,**176**, 200–217, Springer-Verlag.

Peng, S. (1990): "A General Stochastic Maximum Principle for Optimal Control Problems" *SIAM J. Control and Optimization.* **28** (4), 966–979.

Peng, S. (1991a): "Probabilistic Interpretation for Systems of Quasilinear Parabolic PDE and applications", *Stoch. and Stoch. Reports*, **37**, 61–74.

Peng, S. (1991b): "A Nonlinear Feynman–Kac Formula and Applications" *Proceedings of Symposium of System Sciences and Control theory, Chen & Yong ed.* 173–184, World Scientific, Singapore.

Peng, S. (1992a): "Stochastic Hamilton-Jacobi-Bellman Equations", *SIAM J. Control and Optimization* **30** (2), 284–304.

Peng, S. (1992b): "A Generalized Dynamic Programming Principle and Hamilton-Jacobi-Bellman equation", *Stochastics*, **38**, 119–134.

Peng, S. (1993): "Backward Stochastic Differential Equation and It's Application in Optimal Control", *Appl. Math. & Optim.*, **27**,125–144.

Peng, S. (1997b): "Duplicating and Pricing Contingent Claims in Incomplete Markets", to appear.

Tang, S., (1997): "The Maximum Principle for Partially Observed Optimal Control of Stochastic Differential Equations", *SIAM J. Contr. and Optim.*, to appear.

Tang, S., and Li, X., (1994): "Necessary Conditions for Optimal Control of Stochastic Systems with Random Jumps", *SIAM J. Contr. and Optim.* **32**(5), 1447-1475.

Yong, J., (1997): "Finding adapted solutions of forward-backward stochastic differential equations — method of continuation" *Prob. Theory and Rel. Fields*, **107**, 537-572.

COMPARISON THEOREM OF SOLUTIONS TO BSDE WITH JUMPS, AND VISCOSITY SOLUTION TO A GENERALIZED HAMILTON-JACOBI-BELLMAN EQUATION

SITU Rong*

Department of Mathematics
Zhongshan University
Guangzhou 510275, China
mcsstr@zsu.edu.cn

Abstract: Consider the following controlled FSDE and BSDE:

$$dx_s^{t,x,v} = b(s, x_s^{t,x,v}, v_s)ds + \sigma(s, x_s^{t,x,v}, v_s)dw_s + \int_Z c(s, x_{s-}^{t,x,v}, v_s, z)\widetilde{N}_k(ds, dz),$$

$x_t^{t,x,v} = x \in R^d$, $s \in [t, T]$;

$$-dy_s^{t,x,v} = f(s, x_s^{t,x,v}, y_s^{t,x,v}, q_s^{t,x,v}, p_s^{t,x,v}, v_s)ds - q_s^{t,x,v} dw_s - \int_Z p_s^{t,x,v}(z)\widetilde{N}_k(ds, dz),$$

$y_T^{t,x,v} = \Phi\left(x_T^{t,x,v}\right)$, $s \in [t, T]$,

where $v_s \in$ Œ, Œ is the admissible control set of all square integrable $\Im_s$−predictable processes valued in a compact set $U \subset R^m$. Let $u(t,x) = \sup_{v(.)} y_s^{t,x,v}|_{s=t}$. Introduce the Hamiltonian function $H(t, x, r, \beta, A, \eta) : [0, T] \times R^d \times R^1 \times R^d \times R^{d\otimes d} \times C_b^1(R^d) \longmapsto R^1$, as

$$H(t, x, r, \beta, A, \eta) = \sup_{v \in U} \left\{ \begin{array}{c} f(t, x, r, \sigma^*(t, x, v)\beta, \eta(x + c(t, x, v, \cdot)) - \eta(x), v) \\ + \langle \beta, b(t, x, v)\rangle + \frac{1}{2}Tr.(\sigma(t, x, v)\sigma^*(t, x, v)A) \\ + \int_Z [\eta(x + c(t, x, v, z)) - \eta(x) - \eta'(x) \cdot c(t, x, v, z)]\pi(dz) \end{array} \right\}$$

Then by means of our comparison theorem of solutions to the BSDE with jumps under appropriate condition we show that $u(t, x)$ is a viscosity solution of the following H-J-B equation:

$$\partial_t u(t, x) + H(t, x, u(t, x), Du(t, x), D^2u(t, x), u(t, \cdot)) = 0.$$

*This work is supported in part by The Chinese National Natural Science Foundation No. 79790130 and The Foundation of Zhongshan University Front Research.

1 COMPARISON THEOREMS OF SOLUTIONS TO BSDE WITH JUMPS

Backward stochastic differential equations (BSDE) have important applications in the financial market and optimal stochastic control.[2],[6],[10],[11] The comparison theorem of solutions to BSDE is a powerful tool to deal with such kind of problems.[2],[6],[11] In this paper we shall investigate the case with jumps. Consider the following two BSDE: $i = 1, 2$, $t \in [0, T]$

$$y_t^i = Y^i + \int_t^T b^i(s, y_s^i, q_s^i, p_s^i, \omega)ds - \int_t^T q_s^i dw_s - \int_t^T \int_Z p_s^i(z)\tilde{N}_k(ds, dz), \quad (1)$$

where w_t is a $r-$dimensional standard Brownian Motion (BM), $k(\cdot)$ is a Poisson point process, $\tilde{N}_k(ds, dz)$ is the Poisson martingale measure generated by $k(\cdot)$ satisfying

$$\tilde{N}_k(ds, dz) = N_k(ds, dz) - \pi(dz)ds,$$

where $\pi(.)$ is a σ-finite measure on a measurable space $(Z, \Re(Z))$, $N_k(ds, dz)$ is the Poisson counting measure generated by $k(\cdot)$, Y^i is $\Im_T-$measurable, and $\Im_t = \sigma(w_s, k_s, s \leq t)$ (means the completion of the $\sigma-$algebra).

For the precise definition of solution (1) we need the following denotation.

$$L^2_{\Im_t}(R^1) = \left\{ \begin{array}{c} f(t,\omega) : f(t,\omega) \text{ is } \Im_t - \text{adapted}, R^1 - \text{valued such that} \\ E\int_0^T |f(t,\omega)|^2 dt < \infty \end{array} \right\};$$

$L^2_{\Im_t}(R^{1\otimes r})$ is defined similarly.

$$F^2_{\Im_t}(R^1) = \left\{ \begin{array}{c} f(t,z,\omega) : f(t,z,\omega) \text{ is } R^1 - \text{valued}, \Im_t \text{ -predictable} \\ \text{such that } E\int_0^T \int_Z |f(t,z,\omega)|^2 \pi(dz)dt < \infty \end{array} \right\}.$$

Definition 1. (y_t, q_t, p_t) is said to be a solution of (1), say, for $i = 1$, iff $(x_t, q_t, p_t) \in L^2_{\Im_t}(R^1) \times L^2_{\Im_t}(R^{1\otimes r}) \times F^2_{\Im_t}(R^1)$, and it satisfies (1) for $i = 1$.

From definition 1 it is seen that for discussing the solution of (1) we need to assume that b satisfies the following assumption

(A) $b : [0, T] \times R^1 \times R^{1\otimes r} \times L^2_{\pi(.)}(R^1) \times \Omega \to R^1$ is jointly measurable, $\Im_t-$adapted, where

$$L^2_{\pi(.)}(R^1) = \left\{ \begin{array}{c} f(z) : f(z) \text{ is } R^1 - \text{valued, and} \\ \|f\|^2 = \int_Z |f(z)|^2 \pi(dz) < \infty \end{array} \right\}.$$

We have the following comparison theorem for solutions to (1).

Theorem 1. Assume that

1° $b^1(t, y, q, p, \omega) \geq b^2(t, y, q, p, \omega)$

2° $b^1(t, y, q, p, \omega) = \beta(t, y, q, \omega) + \int_Z C_t(z, \omega)p(z)\pi(dz)$,

$|\beta(t, y_1, q_1, \omega) - \beta(t, y_2, q_2, \omega)| \leq k_0 (|y_1 - y_2| + |q_1 - q_2|)$,

where $k_0 \geq 0$ is a constant, and $C_t(z, \omega)$ satisfies the condition

(B) $C_t(z) \geq -1$, $C_t(z) \in F^2_{\Im_t}(R^1)$,

3° $Y^i \in \Im_T, E|Y^i|^2 < \infty, i = 1, 2$, and $Y^1 \geq Y^2$.

Then $P - a.s.$

$$y_t^1 \geq y_t^2, \text{for all } t \in [0, T].$$

Theorem 1 can be shown by using the following lemma, which is given in [8].

Lemma 1. Under assumption in Theorem 1 and additional assumption that $\int_Z \left|1 - \sqrt{1 + C_t(z)}\right|^2 \pi(dz)dt \leq k_0$;

the conclusion of Theorem 1 is true.

Proof.

Take $Z_m \uparrow Z$ such that $\pi(Z_m) < \infty$. Denote

$\widetilde{N}_k^{(m)}(dt, dz) = \widetilde{N}_k(dt, dz \cap Z_m)$, etc.

Then for each $m = 1, 2, \cdots$ there exists a unique solution (x_t^m, q_t^m, p_t^m) satisfying the following BSDE [9],[10]

$$y_t^m = E(Y|\Im_T^{w,\widetilde{N}_k^{(m)}}) + \int_t^T (a_s(\omega)y_s^m + b_s(\omega)q_s^{m*} + \int_Z C_s(z)p_s^m(z)\pi^{(m)}(dz)$$
$$+f_0(s,\omega))ds - \int_t^T q_s^m dw_s - \int_t^T \int_Z p_s^m \widetilde{N}_k^{(m)}(ds, dz)$$
$$= E(Y|\Im_T^{w,\widetilde{N}_k^{(m)}}) + \int_t^T (a_s(\omega)y_s^m + b_s(\omega)q_s^{m*} + \int_{Z_m} C_s(z)p_s^m(z)\pi(dz)$$
$$+f_0(s,\omega))ds - \int_t^T q_s^m dw_s - \int_t^T \int_{Z_m} p_s^m \widetilde{N}_k(ds, dz),$$

where $Y = Y_1 - Y_2$, q_s^{m*} means the transpose of q_s^m,

$$a_t(\omega) = I_{(y_t^1 \neq y_t^2)}(\widetilde{\beta}(t, y_t^1, q_t^1, \omega) - \widetilde{\beta}(t, y_t^2, q_t^1, \omega))(y_t^1 - y_t^2)^{-1},$$
$$b_{it}(\omega) = I_{(q_{it}^1 \neq q_{it}^2)}(\widetilde{\beta}(t, y_t^2, \widetilde{q}_{i-1,t}^2, \omega) - \widetilde{\beta}(t, y_t^2, \widetilde{q}_{i,t}^2, \omega))(q_{it}^1 - q_{it}^2)^{-1}, i = 1, ..., r,$$
$$f_0(t,\omega) = b^1(t, y_t^2, q_t^2, p_t^2, \omega) - b^2(t, y_t^2, q_t^2, p_t^2, \omega) \geq 0,$$

where $\widetilde{q}_{i,t}^2 = (q_{1t}^2, ..., q_{it}^2, q_{(i+1)t}^1, ..., q_{rt}^1)$, $q_t^2 = (q_{1t}^2, ..., q_{rt}^2)$, etc.

Since $(\widehat{y}_t, \widehat{q}_t, \widehat{p}_t) = (y_t^1 - y_t^2, q_t^1 - q_t^2, p_t^1 - p_t^2)$ satisfies a similar BSDE as follows:

$\widehat{y}_t = Y + \int_t^T (a_s(\omega)\widehat{y}_s + b_s(\omega)\widehat{q}_s + \int_Z C_s(z)\widehat{p}_s\pi(dz) + f_0(s,\omega))ds - \int_t^T \widehat{q}_s dw_s - \int_t^T \int_Z \widehat{p}_s \widetilde{N}_k(ds, dz)$.

Hence by Ito's formula to $|y_t^m - \widehat{y}_t|^2$ and Gronwall's inequality one has that

$$E(|y_t^m - \widehat{y}_t|^2 + \tfrac{1}{2}\int_t^T (|q_s^m - q_s|^2 + \int_Z |p_s^m I_{Z_m}(z) - p_s|^2 \pi(dz))ds$$
$$\leq \widetilde{k}_0 E\left|E(Y|\Im_T^{w,\widetilde{N}_k^{(m)}}) - Y\right|^2.$$

Since

$$\Im_T^{w,\widetilde{N}_k^{(m)}} \subset \Im_T^{w,\widetilde{N}_k^{(m+1)}},$$

and

$$\Im_T^{w,\widetilde{N}_k} = \vee_m \Im_T^{w,\widetilde{N}_k^{(m)}}.$$

Hence by Levy's theorem (Ikeda[4], Theorem 1.6.6, pp31) as $m \to \infty$,

$$E(Y|\Im_T^{w,\widetilde{N}_k^{(m)}}) \to Y, \ a.s.,$$

and $\left\{ E(|Y|^2 \Big| \Im_T^{w,\widetilde{N}_k^{(m)}}) \right\}_{m=1}^{\infty}$ is uniformly integrable. Hence as $m \to \infty$

$$E\left|E(Y|\Im_T^{w,\widetilde{N}_k^{(m)}}) - Y\right|^2 \to 0.$$

Therefore as $m \to \infty$

$$E|y_t^m - \widehat{y}_t|^2 \to 0.$$

However, by Lemma 1 it is true that as $t \in [0, T]$

$$y_t^m \geq 0, \forall m.$$

Hence $P - a.s.$

$\widehat{y}_t \geq 0$, for all $t \in [0, T]$. □

Sometimes we need to consider the following BSDE with jumps, for example, in the application to the financial market[1], which is a little bit different from (1): $i = 1, 2$,

$$y_t^i = Y^i + \int_t^T b^i(s, y_s^i, q_s^i, p_s^i, \omega)ds - \int_t^T q_s^i dw_s - \int_t^T p_s^i d\tilde{N}_k(s),\ 0 \leq t \leq T, \quad (1)'$$

where $b^i : [0, T] \times R^1 \times R^{1\otimes r} \times R^1 \times \Omega \to R^1$ is jointly measurable, $\Im_t$–adapted, $\tilde{N}_k(t)$ is a centralized Poisson process, such that

$$d\tilde{N}_k(t) = \tilde{N}_k(dt, Z),$$

and we assume that $\pi(Z) = 1$, i.e.

$$\tilde{N}_k(t) = N_k(t) - t,$$

$N_k(t)$ is a Poisson process with $EN_k(t) = t$. In this case we can even remove the condition (B) in 2° of Theorem 1 as follows:

Theorem 1'. Assume that

1° $b^1(t, y, q, p, \omega) \geq b^2(t, y, q, p, \omega)$,

2° $\left|b^1(t, y_1, q_1, p_1, \omega) - b^1(t, y_2, q_2, p_2, \omega)\right| \leq k_0 \left(|y_1 - y_2| + |q_1 - q_2| + |p_1 - p_2|\right)$.

3° $Y^i \in \Im_T, E\left|Y^i\right|^2 < \infty, i = 1, 2$, and $Y^1 \geq Y^2$.

Then $P - a.s.$

$$y_t^1 \geq y_t^2, \text{ for all } t \in [0, T].$$

To show Theorem 1' we need the following preparation. Consider the following simplier RSDE:

$$y_t = Y + \int_t^T (a_s(\omega)y_s + b_s(\omega)q_s^T + c_s(\omega)p_s + f_0(\omega))ds - \int_t^T q_s dw_s - \int_t^T p_s d\tilde{N}_k(s),$$

$0 \leq t \leq T$,

where we assume that a_t, b_t are $\Im_t$–adapted, c_t is $\Im_t$–predictable, and they are all bounded, i.e. there exists a constant $k_0 \geq 0$ such that

$$|a_t| + |b_t| + |c_t| \leq k_0.$$

Proposition 1. If

$$f_0(\omega), Y \geq 0,$$

then $P - a.s.$

$$y_t \geq 0, \text{ for all } t \in [0, T].$$

To show Proposition 1 let us use the Girsanov measure transformation and to make an additional assumption that

$$(B_1) : c_t > -1.$$

The general case can be reduced to this simple case by using finite steps of induction. Let

$$z_t = \exp\left(\int_0^t b_s \cdot dw_s - \tfrac{1}{2}\int_0^t |b_s|^2 ds\right) \cdot \prod_{0<s\leq t}(1 + c_s \Delta N_s) e^{\int_0^t c_s d(\tilde{N}_S - N_S)}.$$

Then under addition condition (B_1) the following formula defines a new probability measure:

$$d\tilde{P} = z_T dP.$$

By Ito's formula it is not difficult to show that

$$y_t = E_{\tilde{P}}\left[Y e^{\int_t^T a_s ds} + \int_t^T e^{\int_t^u a_s ds} f_0(u) du \,|\Im_t\right].$$

Hence the conclusion of Proposition 1 is derived. For the general case $c_t > -1 - k_0$ we can write

$$c_t = \sum_{i=1}^{n_0} c_i(t)$$

such that all $c_i(t) > -1$. Then through finite steps of transformation we can arrive at the conclusion. Now Theorem 1' is not difficult to derive from Proposition 1 as the proof of Theorem 1.

2 OPTIMAL CONTROL AND BELLMAN PRINCIPLE

Now we will apply the comparison theorem of solutions to BSDE (Theorem 1) to discuss the optimal control problem for the BSDE systems with jumps, which is corresponding to some optimal recursive utility function problem in the financial market.[2],[11] Consider the following controlled Forward SDE (FSDE):

$$dx_s^{t,x,v} = b(s, x_s^{t,x,v}, v_s)ds + \sigma(s, x_s^{t,x,v}, v_s)dw_s + \int_Z c(s, x_{s-}^{t,x,v}, v_s, z)\tilde{N}_k(ds, dz),\ x_t^{t,x,v} = x \in R^d,\ s \in [t,T]; \quad (2)$$

where $b(s,x,v)$ is non-random, etc., and $v(\cdot) \in Œ$, Œ is the admissible control set of all square integrable $\Im_s$−predictable processes valued in a compact set $U \subset R^m$; i.e.

$$Œ = \left\{ \begin{array}{c} v = v(t,\omega):\ \text{it is } \Im_t - \text{predictable, } v(t,\omega) \in U \text{ such that} \\ E\int_0^T |v(t,\omega)|^2 dt < \infty \end{array} \right\}.$$

It is well known that FSDE (2) has a unique solution $x_s^{t,x,v} \in R^d, s \in [t,T]$, provided that the following conditions are satisfied:[7]

$$\text{(H1)} \begin{cases} |b(t,x,v)|^2 + |\sigma(t,x,v)|^2 + \int_Z |c(t,x,v,z)|^2 \pi(dz) \\ \quad \le k_0(1 + |x|^2 + |v|^{2\alpha}), \\ |b(t,x,v) - b(t,x',v')|^2 + |\ \sigma(t,x,v) - \sigma(t,x',v')|^2 \\ \quad + \int_Z |c(t,x,v,z) - c(t,x',v',z)|^2 \pi(dz) \\ \quad \le k_0(|x-x'|^2 + |v-v'|^{2\alpha}), \alpha \in (0,1], \end{cases}$$

where $b \in R^d, \sigma \in R^{d\otimes r}, c \in R^d$. Now consider the following BSDE:

$$-dy_s^{t,x,v} = f(s, x_s^{t,x,v}, y_s^{t,x,v}, q_s^{t,x,v}, p_s^{t,x,v}, v_s)ds - q_s^{t,x,v}dw_s - \int_Z p_s^{t,x,v}(z)\tilde{N}_k(ds,dz),\ y_T^{t,x,v} = \Phi\left(x_T^{t,x,v}\right),\ s \in [t,T]. \quad (3)$$

It is also known that (3) has a unique solution $(y_s^{t,x,v}, q_s^{t,x,v}, p_s^{t,x,v})$, provided that the following conditions are satisfied:[9],[10]

$$\text{(H2)} \begin{cases} |f(t,x,0,0,0,v)|^2 + |\Phi(x)|^2 \le k_0(1+|x|^2+|v|^2), \\ |f(t,x,y,q,p,v) - f(t,x',y',q',p',v')|^2 + |\ \Phi(x) - \Phi(x')|^2 \\ \le k_0(|x-x'|^{2\alpha} + |y-y'|^2 + |q-q'|^2 + \|p-p'\|^2 + |v-v'|^{2\alpha}), \\ \alpha \in (0,1], \end{cases}$$

where $f, \Phi \in R^1$. Let

$$J(t,x,v(\cdot)) = y_s^{t,x,v}\ |_{s=t}\ ,$$
$$u(t,x) = \sup_{v(.)\in Œ} J(t,x,v(\cdot)). \quad (4)$$

In the financial market (2), (3) and (4) can be interpreted as the price processes of stocks, the wealth process of an investor, and the optimal recursive utility function, respectively. Our object here is to discuss (4). Let us denote the solution of (3) by

$$y_s^{t,x,v} = G_{s,T}^{t,x,v}(\Phi\left(x_T^{t,x,v}\right)),$$

and introduce the following assumption:

(H3) $f(t,x,y,q,p,v)$ satisfies the condition 2° in Theorem 1 for each $v \in U, x \in R^d$.

We have

Theorem 2. (Bellman Principle). Under assumption (H1) - (H3) $\forall 0 \leq \delta < T-t$

$u(t,x) = \sup_{v(\cdot)\in Œ} G^{t,x,v}_{t,t+\delta}(u\left(t+\delta, x^{t,x,v}_{t+\delta}\right))$.

Theorem 2 tells us that to find the optimal value functional on $[t,T]$, we can divide it into two steps: First, find out the optimal value functional $u\left(t+\delta, x^{t,x,v}_{t+\delta}\right)$on $[t+\delta, T]$. Second, set $u\left(t+\delta, x^{t,x,v}_{t+\delta}\right)$ as the terminal value then find out the optimal value functional on $[t, t+\delta]$. Theorem 2 can be proved by using our Comparison theorem of solutions to BSDE with jumps (Theorem 1), the following lemmas and similar approaches as that in [6].

Lemma 2. Under assumption (H1) $\forall x, x' \in R^d, \forall v(\cdot), v'(\cdot) \in Œ$,

$E^{\Im_t} \sup_{t\leq s\leq T} |x_s|^2 \leq k_T < \infty$,

$E^{\Im_t} \sup_{t\leq s\leq T} \left|x^{t,x,v}_s - x^{t,x',v'}_s\right|^2 \leq k_T(|x-x'|^2 + E^{\Im_t} \int_t^T |v(s)-v'(s)|^{2\alpha} ds)$.

Lemma 3. Under assumption (H1)-(H2) $\forall x, x' \in R^d, \forall v(\cdot), v'(\cdot) \in Œ$,

$\left|y^{t,x,v}_t\right|^2 + E^{\Im_t} \int_t^T (|q^{t,x,v}_s|^2 + \|p^{t,x,v}_s\|^2) ds \leq k_T(1+|x|^2)$,

$\left|y^{t,x,v}_t - y^{t,x',v'}_t\right|^2 + E^{\Im_t} \int_t^T (\left|q^{t,x,v}_s - q^{t,x',v'}_s\right|^2 + \left\|p^{t,x,v}_s - p^{t,x',v'}_s\right\|^2) ds$

$\leq k_T(|x-x'|^{2\alpha} + (E^{\Im_t} \int_t^T |v_s - v'_s|^{2\alpha} ds)^\alpha + E^{\Im_t} \int_t^T |v_s - v'_s|^{2\alpha} ds)$.

3 VISCOSITY SOLUTION OF HJB EQUATION

Now introduce the Hamiltonian function $H(t,x,r,\beta,A,\eta) : [0,T] \times R^d \times R^1 \times R^d \times R^{d\otimes d} \times C^1_b(R^d) \longmapsto R^1$, as

$$H(t,x,r,\beta,A,\eta) = \sup_{v\in U} \left\{ \begin{array}{c} f(t,x,r,\sigma^*(t,x,v)\beta, \eta(x+c(t,x,v,\cdot)) - \eta(x), v) \\ + \langle \beta, b(t,x,v)\rangle + \frac{1}{2} Tr.(\sigma(t,x,v)\sigma^*(t,x,v)A) \\ + \int_Z [\eta(x+c(t,x,v,z)) - \eta(x) - \eta'(x)\cdot c(t,x,v,z)]\pi(dz) \end{array} \right\} \quad (5)$$

By means of our comparison theorem of solutions to the BSDE with jumps under appropriate condition one can show that $u(t,x)$ is a viscosity solution of the following H-J-B equation:

$$\partial_t u(t,x) + H(t,x,u(t,x), Du(t,x), D^2u(t,x), u(t,\cdot)) = 0. \quad (6)$$

Let us introduce the definition of viscosity solution as follows:

Definition 2. A continuous function $u(t,x)$ is called a viscosity sub-solution (super-solution) of (6), iff $\forall \varphi \in C^{1,3}_b([0,T] \times R^d)$ (the totality of all real functions, which has a bounded continuous first derivative with respect to t, and up to the third derivatives with respect to x) at all minimum point (maximum point) (t,x) of function $\varphi - u$ one has

$\partial_t\varphi(t,x) + H(t,x,u(t,x), D\varphi(t,x), D^2\varphi(t,x), \varphi(t,\cdot)) \geq 0 \ (\leq 0)$.

If $u(t,x)$ is both a viscosity super-solution and sub-solution, then it is called a viscosity solution.

Since now actually H is concerned with the jump coefficient c, and we need some $\alpha-$Hölder continuous of the function in supremum in (5) for x. Hence we put one more assumption

(H4) one of the following conditions is satisfied:

(i) $|b(s,x,v)| + |\sigma(s,x,v)| + \|c(s,x,\cdot,v)\| + \|c_x(s,x,z,v)\| \leq k_0, \int_Z |c| \, \pi(dz) \leq k_0$;

(ii) $\|c_x(s,x,z,v)\| \leq k_0(1+|x|), \int_Z |c|\,\pi(dz) \leq k_0(1+|x|)$,
$E(\int_0^T \int_Z |c(s,x_{s-}(\omega),z,v)|^2 N_k(ds,dz))^p \leq k_0(1+E\int_0^T |x_s(\omega)|^{2p}\,ds), p = 1,\cdots,4$;
for any R^d−valued $\Im_t$−adapted cadlag process $x_s(\omega)$ holds, where c_x means the gradient of c with respect to x, which is assumed to be existed in both cases.

We have

Theorem 3. Under assumption (H1)-(H4) $u(t,x)$ defined by (4) is a viscosity solution of (6).

To show Theorem 3 we need several lemmas. For a given φ let us denote

$$F(s,x,y,q,p,v) = \partial_s\varphi(s,x) + \pounds(s,x,v)\varphi(s,x)$$
$$+f(s,x,y+\varphi(s,x),q+D\varphi(s,x)\sigma(s,x,v),p+\varphi(s,x+c(s,x,\cdot,v))-\varphi(s,x),v),$$

where

$$\pounds(s,x,v) = \langle D\varphi, b(t,x,v)\rangle + \tfrac{1}{2}tr[\sigma(t,x,v)\sigma^*(t,x,v)D^2\varphi]$$
$$+\int_Z(\varphi(s,x+c(s,x,z,v)) - \varphi(s,x) - \langle D\varphi, c(t,x,z,v)\rangle)\pi(dz).$$

For the property of F we have the following

Lemma 4. $F(s,x,y,q,p,v)$ still satisfies the Lipschitzian condition for (y,q,p), i.e. the second condition in (H2) for (y,q,p) holds. Moreover, in case (i) of (H4)

$$|F(s,x_1,y,q,p,v) - F(s,x_2,y,q,p,v)|$$
$$\leq k_0(|x_1-x_2|^\alpha + |x_1-x_2|), \tag{7}$$

in case (ii) of (H4)

$$|F(s,x_1,y,q,p,v) - F(s,x_2,y,q,p,v)|$$
$$\leq k_0(1+\max(|x_1|,|x_2|))(|x_1-x_2|^\alpha + |x_1-x_2|), \tag{8}$$

where $\alpha \in (0,1]$ comes from assumption (H2). Furthermore, in case (ii) of (H4) holds, one has that

$$E\sup_{t\leq T}|x_t|^{2p} \leq k_0, \forall p = 1,\cdots,4. \tag{9}$$

By using Lemma 4, Comparison theorem (Theorem 1), and Theorem 2 one can show theorem 3 similar to [6].

References

[1] Bardhan, I. and Chao, X. (1993). Pricing options on securities with discontinuous returns. *Stoch. Proc. Appl. 48,* 123-137.

[2] El Karoui, N. Peng, S., Quenez, M.C. (1997). Backward stochastic differential equations in finance. *Math. Finance 1,* 1-71.

[3] Grandall, M.G. and Lions, P.L. (1983). Viscosity solutions of Hamilton-Jacobi-Bellman equations. *Trans. Amer. Math. Soc. 227,* 1-42.

[4] Ikeda, N. and Watanabe, S. (1989). *Stochastic Differential Equations and Diffusion Processes.* North Holland Pub. Co.

[5] Ishii, H. (1989). On uniqueness and existence of viscosity solutions of Fully non-linear second order PDE's. *Comm. Pure Appl. Math. 42,* 15-45.

[6] Peng, S. (1992). A generalized dynamic programming principle and Hamilton-Jacobi-Bellman equation. *Stochastic & Stochastic Reports 38,* 119-134.

[7] Situ Rong. (1991). Theory and applications of stochastic differential equations. *Contemporary Math. 118 AMS,* 261-280.

[8] Situ Rong. (1996). Comparison theorem of solutions to backward stochastic differential equations and its applications. *Proceedings of CSIAM The 3rd Conference on System and Control Mathematics.* Dailian, 46-50.

[9] Situ Rong. (1997). On solutions of backward stochastic differential equations with jumps and applications. *Stoch. Proc. Applic.* 66, 209-236.

[10] Tang, S. and Li, X. (1994). Necessary conditions for optimal control of stochastic systems with random jumps. *SIAM J. Control and Optim.* 32:5, 1447-1475.

[11] Yan, J.A. Peng, S. et al. (1997). *Topics on Stochastic Analysis.* Scien. Press. Beijing.

MULTIVARIATE CONSTRAINED PORTFOLIO RULES: DERIVATION OF MONGE-AMPÈRE EQUATIONS

Srdjan Stojanovic
Department of Mathematical Sciences
University of Cincinnati
Cincinnati, OH 45221-0025
U.S.A.
srdjan@math.uc.edu
http://math.uc.edu/~srdjan/

1 STATEMENT OF THE PROBLEM

Consider following classical problem in mathematical finance (see [16, 8, 4, 17, 3]).

$N+1$ assets are traded continuously. The $N+1$st of these assets, called *bank account*, has the value $B(t)$ at time t, which evolves according to the equation

$$dB(t) = r(t)B(t)dt, \quad B(0) = b \in [0, \infty). \tag{1.1}$$

N assets, called *stocks*, are subject to the systematic risk, and have the price-per-share $P_{(i)}(t)$ at time t, which evolve according to the equations

$$dP_{(i)}(t) = b_{(i)}(t)P_{(i)}(t)dt + \sum_{j=1}^{N} \sigma_{(ij)}(t)P_{(i)}(t)dW_j(t)$$
$$P_{(i)}(0) = p_{(i)} \in [0, \infty). \tag{1.2}$$

for $i = 1, ..., N$. In (1.1,1.2) $r(t) \geq 0$ (the *interest rate*), $b_{(i)}(t)$ (the *appreciation rates*), $\sigma_{(ij)}(t)$ (the *volatility coefficients*), are given deterministic functions of time $t \in [0, T]$, and $W_j(t)$ are independent 1-dimensional Brownian motions.

We shall assume, throughout this paper, that matrix

$$\sigma(t) \quad \text{is invertible, for every } t \in [0, T). \tag{1.3}$$

Further, we consider an economic agent who can decide, at each instant $t \in [0,T)$, knowing the amount of his current wealth $X(t) \in [0,\infty)$, and knowing functions $r, b_{(i)}$ and $\sigma_{(ij)}$, the following:

how much money $\pi_{(i)}(t, X(t)) \in (-\infty, +\infty)$ to invest in the ith stock. Vector function

$$\pi = (\pi_{(1)}, ..., \pi_{(N)})^T : (t,x) \mapsto R^N \tag{1.4}$$

is called a *portfolio rule*.

Then (see [8]), the agent's wealth $X(t) = X^{\pi}(t)$ evolves according to the equation

$$\begin{aligned} dX(t) = \{r(t)X(t) + \pi^T(t, X(t))[b(t) - r(t)1_N]\}\, dt \\ +\pi^T(t, X(t))\sigma(t)dW(t) \\ X(0) = x \in [0,\infty). \end{aligned} \tag{1.5}$$

In (1.5) π^T denotes the transpose of π, while

$$1_N = (1, ..., 1)^T \in R^N. \tag{1.6}$$

It is assumed that if the wealth of the agent becomes zero at some time, the evolution stops there.

For a given portfolio rule $\pi(\cdot,\cdot)$, we define the expected *payoff*

$$E_{(0,x)}\left\{e^{-\int_0^T \beta(\tau)d\tau}\varphi(X^{\pi}(T))\right\}. \tag{1.7}$$

In (1.7), $E_{(t,x)}$ denotes the expectation given the condition that at time t the wealth of the agent was equal to x. Also, $\beta(t) \geq 0$ is the *discount factor* (or the *inflation rate*), while φ is the *utility function* for the terminal wealth $X^{\pi}(T)$.

So, the payoff is the expectation of the present value of the utility of the final amount of the wealth of the agent.

Different stochastic control problems can be considered for the dynamics (1.5) and payoff (1.7), depending on the set of admissible strategies $\mathcal{AS}$. We shall consider following problems.

Problem 0. *Find the optimal (feedback) strategy*

$$\tilde{\pi}_0(\cdot,\cdot) \in \mathcal{AS}_0 = \{\pi(\cdot,\cdot)\,|\pi_{(i)}(t,x) \in (-\infty,+\infty)\ \ \forall(t,x)\ \ 1 \leq i \leq N\} \tag{1.8}$$

such that the expected payoff is maximized on $\mathcal{AS}_0$.

This is a classical problem (see [8]).

Let function $a(t,x)$ be given.

Problem 1. *Find the optimal (feedback) strategy*

$$\tilde{\pi}_1(\cdot,\cdot) \in \mathcal{AS}_1 = \left\{\pi(\cdot,\cdot)\,\middle|\,\sum_{i=1}^{N}\pi_{(i)}(t,x)\mu_{(i)}(t,x) = a(t,x)\ \ \forall(t,x)\right\} \tag{1.9}$$

such that the expected payoff is maximized on $\mathcal{AS}_1$.

Let functions $a_1(t,x) \le a_2(t,x)$ be given.

Problem 2. *Find the optimal (feedback) strategy*

$$\tilde{\pi}_2(\cdot,\cdot) \in \mathcal{AS}_2$$
$$= \left\{ \pi(\cdot,\cdot) \;\middle|\; a_1(t,x) \le \sum_{i=1}^{N} \pi_{(i)}(t,x)\mu_{(i)}(t,x) \le a_2(t,x) \quad \forall (t,x) \right\} \tag{1.10}$$

such that the expected payoff is maximized on $\mathcal{AS}_2$.

We notice that constrained portfolio selection has been studied (see [19, 2, 10], with applicable results mainly in the case when portfolio is one-dimensional.

2 MONGE-AMPÈRE-TYPE EQUATION OF THE PORTFOLIO SELECTION PROBLEM UNDER THE CONSTRAINT $\sum_{I=1}^{N} \pi_{(I)}(T,X)\mu_{(I)}(T,X) = A(T,X)$

As it is usual in stochastic control, one considers the *value function*

$$V_i(t,x) = \sup_{\pi \in \mathcal{AS}_i} E_{(t,x)} \left\{ e^{-\int_t^T \beta(\tau)d\tau} \varphi\left(X^{\pi}(T)\right) \right\} \tag{1.11}$$

$i = 0,1,2$. Then, by the stochastic calculus,[1] and dynamic programming arguments, the value function $V_0(t,x)$ is the solution[2] of the following Dirichlet problem for the Hamilton-Jacobi-Bellman partial differential equation:

$$\begin{cases} V_{0t}(t,x) + \sup\limits_{\pi \in R^N} \Big[\frac{1}{2} \|\sigma(t)^T\pi\|^2 V_{0xx}(t,x) \\ +(r(t)x + \pi^T(b(t) - r(t)1_N))V_{0x}(t,x)\Big] - \beta(t)V_0(t,x) = 0 \end{cases} \tag{1.12}$$

$$t \in [0,T),\; x \in (0,\infty)$$

$$V_0(T,x) = \varphi(x), \quad x \in (0,\infty) \tag{1.13}$$

$$V_0(t,0) = 0, \quad t \in [0,T), \tag{1.14}$$

while, V_1 and V_2 are solutions of

$$\begin{cases} V_{1t}(t,x) \\ \quad + \sup\limits_{\{\pi \in R^N : \sum_{i=1}^{N} \pi_{(i)}\mu_{(i)}(t,x) = a(t,x)\}} \Big[\frac{1}{2} \|\sigma(t)^T\pi\|^2 V_{1xx}(t,x) \\ + \left(r(t)x + \pi^T(b(t) - r(t)1_N)\right) V_{1x}(t,x)\Big] - \beta(t)V_1(t,x) = 0 \end{cases} \tag{1.15}$$

and

$$\begin{cases} V_{2t}(t,x) \\ + \sup_{\left\{\pi \in R^N : a_1(t,x) \leq \sum_{i=1}^N \pi_{(i)}\mu_{(i)}(t,x) \leq a_2(t,x)\right\}} \left[\frac{1}{2}\|\sigma(t)^T\pi\|^2 V_{1xx}(t,x)\right. \\ \left. + \left(r(t)x + \pi^T(b(t) - r(t)1_N)\right) V_{1x}(t,x)\right] - \beta(t)V_1(t,x) = 0 \end{cases} \tag{1.16}$$

respectively, with the same boundary value as V_0. We do not address here the issue of the "boundary condition" at infinity, i.e., the behavior of $V(t,x)$ as $x \to \infty$. This is very important issue also in numerical calculations, when one has to truncate the problem to the finite domain, and then to impose a boundary condition.

We shall need the following notation. If $\mathbf{u}$ and $\mathbf{v}$ are vectors in R^N, then

$$\text{orth}_{\mathbf{v}}\mathbf{u} = \mathbf{u} - \text{proj}_{\mathbf{v}}\mathbf{u}. \tag{1.17}$$

where $\text{proj}_{\mathbf{v}}\mathbf{u} = \frac{\mathbf{u}^T\mathbf{v}}{\|\mathbf{v}\|^2}\mathbf{v}$.

Theorem 1. *Let V be a classical solution of (1.15,1.13,1.14). Then, V solves the following* Monge-Ampère-type equation

$$V_t(t,x)V_{xx}(t,x) - \frac{1}{2}\left\|\text{orth}_{[\sigma(t)^{-1}\mu(t,x)]}\left[\sigma(t)^{-1}(b(t) - r(t)1_N)\right]\right\|^2 V_x(t,x)^2$$
$$+\frac{1}{2}\frac{a(t,x)^2}{\|\sigma(t)^{-1}\mu(t,x)\|^2}V_{xx}(t,x)^2$$
$$+\left[a(t,x)\frac{(\sigma(t)^{-1}\mu(t,x))^T\sigma(t)^{-1}(b(t) - r(t)1_N)}{\|\sigma(t)^{-1}\mu(t,x)\|^2} + r(t)x\right]V_x(t,x)V_{xx}(t,x)$$
$$-\beta(t)V(t,x)V_{xx}(t,x) = 0 \tag{1.18}$$

Moreover, the optimal portfolio rule *is given by*

$$\tilde{\pi}(t,x) = -(\sigma(t)^{-1})^T\left[\frac{V_x(t,x)}{V_{xx}(t,x)}\text{orth}_{[\sigma(t)^{-1}\mu(t,x)]}\left[\sigma(t)^{-1}(b(t) - r(t)1_N)\right]\right.$$
$$\left. -a(t,x)\frac{\sigma(t)^{-1}\mu(t,x)}{\|\sigma(t)^{-1}\mu(t,x)\|^2}\right]. \tag{1.19}$$

Remark 1. *So, the problem of constrained portfolio selection is completely solved provided one has an efficient, stable,*[3] *and reliable numerical method for finding* proper *solution of (1.18,1.13,1.14). Proper in a sense that it coincides with the solution of the Hamilton-Jacobi-Bellman equation, i.e., coincides with the value function, since one should notice that uniqueness, in general, does* not *hold either for (1.18,1.13,1.14), nor for (1.25,1.13,1.14), and (1.31,1.13,1.14). So, the crucial property of any numerical algorithm attempting to solve the* *above equation is the ability of a selection of the proper solution. Such an algorithm was developed by the author (see [18]).*

Remark 2. *The equation (1.18), and therefore also equations (1.12), (1.15), and (1.16), become* degenerate, *when* $\varphi_{xx} \not< 0$. *An application would be the probability maximization of reaching certain level (say* $x = 1$*) of wealth (see [9]), in which case* $\varphi(x) = 0, x < 1$, *and* $\varphi(x) = 1, x \geq 1$. *Problems like that can be solved numerically nevertheless. The terminal condition (1.13) is* not *taken in a continuous way if* $\varphi_{xx} > 0$ *at some* x *(in the sense of distributions), rather concave envelope of* φ *is taken.*

Remark 3. *The superiority of the equation (1.18) over (1.15), in the view of the author, is manifold. For example:*

1) while equations (1.12), (1.15) , and (1.16) are vastly different, that is barely observable from their formulations, as opposed to equations (1.25), (1.18), and (1.31). The difficulty is, therefore, hidden, which does not, as a rule, simplify the matters;

2) the richness of the "calculus structure" is fully implemented in formulations (1.25), (1.18), and (1.31), as opposed to the Hamilton-Jacobi-Bellman formulation, which contributes to the accuracy and efficency of the numerical algorithm, to say the least.

Consider now the special case when $N = 2$. We use the following notation:

$$\sigma(t) = \sigma = \begin{pmatrix} \sigma_{11} & \sigma_{12} \\ \sigma_{21} & \sigma_{22} \end{pmatrix}, \quad b(t) = b = \begin{pmatrix} b_1 \\ b_2 \end{pmatrix}, \quad 1_2 = \begin{pmatrix} 1 \\ 1 \end{pmatrix}, \tag{1.20}$$

$$a(t,x) = a, \quad r(t) = r, \quad \beta(t) = \beta, \quad V(t,x) = V, \quad \tilde{\pi}(t,x) = \begin{pmatrix} \tilde{\pi}_1 \\ \tilde{\pi}_2 \end{pmatrix}. \tag{1.21}$$

Corollary 1. *Let* $N = 2$, $\mu = 1_2$. *Let* V *be a classical solution of (1.15,1.13,1.14). Then,* V *solves the following* Monge-Ampère-type equation

$$V_t V_{xx} - \frac{1}{2} \frac{(b_1 - b_2)^2}{(\sigma_{11} - \sigma_{21})^2 + (\sigma_{22} - \sigma_{12})^2} (V_x)^2$$

$$+ \frac{1}{2} \frac{a^2(\sigma_{11}\sigma_{22} - \sigma_{21}\sigma_{12})^2}{(\sigma_{11} - \sigma_{21})^2 + (\sigma_{22} - \sigma_{12})^2} (V_{xx})^2$$

$$+ \frac{b_1(\sigma_{21}(\sigma_{21} - \sigma_{11}) + \sigma_{22}(\sigma_{22} - \sigma_{12})) + b_2(\sigma_{12}(\sigma_{12} - \sigma_{22}) + \sigma_{11}(\sigma_{11} - \sigma_{21}))}{(\sigma_{11} - \sigma_{21})^2 + (\sigma_{22} - \sigma_{12})^2}$$

$$\cdot a V_x V_{xx}$$

$$+ r(x - a) V_x V_{xx} - \beta V V_{xx} = 0 \tag{1.22}$$

Moreover, the optimal portfolio rule *is equal to*

$$\tilde{\pi}_1 = \frac{1}{(\sigma_{11} - \sigma_{21})^2 + (\sigma_{22} - \sigma_{12})^2}$$

$$\cdot\left((b_2-b_1)\frac{V_x}{V_{xx}}+a((\sigma_{21}(\sigma_{21}-\sigma_{11})+\sigma_{22}(\sigma_{22}-\sigma_{12}))\right) \tag{1.23}$$

$$\tilde{\pi}_2=\frac{1}{(\sigma_{11}-\sigma_{21})^2+(\sigma_{22}-\sigma_{12})^2}$$
$$\cdot\left((b_1-b_2)\frac{V_x}{V_{xx}}+a((\sigma_{12}(\sigma_{12}-\sigma_{22})+\sigma_{11}(\sigma_{11}-\sigma_{21}))\right) \tag{1.24}$$

3 CONSTRAINT OF THE FORM $A_1(T,X)\leq\sum_{I=1}^{N}\pi_{(I)}(T,X)\mu_{(I)}(T,X)\leq A_2(T,X)$

Proposition 1. *Let V_0 be a classical solution of (1.12,1.13,1.14). Then, V_0 solves the following* Monge-Ampère-type equation

$$\begin{aligned}&V_{0t}(t,x)V_{0xx}(t,x)-\frac{1}{2}\|\sigma(t)^{-1}(b(t)-r(t)1_N)\|^2V_{0x}(t,x)^2\\&+r(t)xV_0(t,x)_xV_{0xx}(t,x)-\beta(t)V_0(t,x)V_{0xx}(t,x)=0\end{aligned} \tag{1.25}$$

Moreover, the optimal portfolio rule is given by

$$\tilde{\pi}_0=-(\sigma(t)^{-1})^T\sigma(t)^{-1}(b(t)-r(t)1_N)\frac{V_{0x}(t,x)}{V_{0xx}(t,x)}. \tag{1.26}$$

It is not difficult now to solve the optimal multivariate constrained portfolio selection problem, when the constraint has the form

$$a_1(t,x)\leq\sum_{i=1}^{N}\pi_{(i)}(t,x)\mu_{(i)}(t,x)\leq a_2(t,x) \tag{1.27}$$

for some given functions

$$a_1(t,x)\leq a_2(t,x) \tag{1.28}$$

and $\mu(t,x)=(\mu_{(1)}(t,x),...,\mu_{(N)}(t,x))^T$ as before.

For each $t\in[0,T)$ define (nonlinear) differential operator $A=A(t)$

$$\begin{aligned}A&:u\mapsto Au\\ \text{domain}(A)&=\{u\in C^2|u_{xx}<0\}\end{aligned} \tag{1.29}$$

by the following elaborate formula

$$Au = \begin{cases} -\frac{1}{2}\|\sigma^{-1}(b - r1_N)\|^2 (u_x)^2 & \text{if} \quad \begin{array}{l} a_1 \leq -\mu^T {\sigma^{-1}}^T \sigma^{-1} \\ \cdot(b - r1_N)\frac{u_x}{u_{xx}} \leq a_2 \end{array} \\ \begin{array}{l} -\frac{1}{2}\|\mathrm{orth}_{[\sigma^{-1}\mu]}[\sigma^{-1}(b - r1_N)]\|^2 \\ \cdot(u_x)^2 + \frac{1}{2}\frac{(a_1)^2}{\|\sigma^{-1}\mu\|^2}(u_{xx})^2 \\ + \frac{a_1(\sigma^{-1}\mu)^T\sigma^{-1}(b-r1_N)}{\|\sigma^{-1}\mu\|^2}u_x u_{xx} \end{array} & \text{if} \quad \begin{array}{l} a_1 > -\mu^T {\sigma^{-1}}^T \sigma^{-1} \\ \cdot(b - r1_N)\frac{u_x}{u_{xx}} \end{array} \\ \begin{array}{l} -\frac{1}{2}\|\mathrm{orth}_{[\sigma^{-1}\mu]}[\sigma^{-1}(b - r1_N)]\|^2 \\ \cdot(u_x)^2 + \frac{1}{2}\frac{(a_2)^2}{\|\sigma^{-1}\mu\|^2}(u_{xx})^2 \\ + \frac{a_2(\sigma^{-1}\mu)^T\sigma^{-1}(b-r1_N)}{\|\sigma^{-1}\mu\|^2}u_x u_{xx} \end{array} & \text{if} \quad \begin{array}{l} a_2 < -\mu^T {\sigma^{-1}}^T \sigma^{-1} \\ \cdot(b - r1_N)\frac{u_x}{u_{xx}} \end{array} \end{cases} \tag{1.30}$$

Theorem 2. *Let V_2 be a classical solution of (1.16,1.13,1.14). Then, V_2 solves the following* Monge-Ampère-type equation

$$V_{2t}(t,x)V_{2xx}(t,x) + A(t)V_2(t,x) + r(t)xV_{2x}(t,x)V_{2xx}(t,x) \\ -\beta(t)V_2(t,x)V_{2xx}(t,x) = 0 \tag{1.31}$$

Moreover, the optimal portfolio rule is given by

$$\tilde{\pi}_2 = \begin{cases} -{\sigma^{-1}}^T\sigma^{-1}(b - r1_N)\frac{V_{2x}}{V_{2xx}} & \text{if} \quad \begin{array}{l} a_1 \leq -\mu^T {\sigma^{-1}}^T \sigma^{-1} \\ \cdot(b - r1_N)\frac{V_{2x}}{V_{2xx}} \leq a_2 \end{array} \\ -{\sigma^{-1}}^T \left[\begin{array}{l} \mathrm{orth}_{[\sigma^{-1}\mu]}[\sigma^{-1}(b - r1_N)] \\ \cdot\frac{V_{2x}}{V_{2xx}} - \frac{a_1}{\|\sigma^{-1}\mu\|^2}\sigma^{-1}\mu \end{array}\right] & \text{if} \quad \begin{array}{l} a_1 > -\mu^T {\sigma^{-1}}^T \sigma^{-1} \\ \cdot(b - r1_N)\frac{V_{2x}}{V_{2xx}} \end{array} \\ -{\sigma^{-1}}^T \left[\begin{array}{l} \mathrm{orth}_{[\sigma^{-1}\mu]}[\sigma^{-1}(b - r1_N)] \\ \cdot\frac{V_{2x}}{V_{2xx}} - \frac{a_2}{\|\sigma^{-1}\mu\|^2}\sigma^{-1}\mu \end{array}\right] & \text{if} \quad \begin{array}{l} a_2 < -\mu^T {\sigma^{-1}}^T \sigma^{-1} \\ \cdot(b - r1_N)\frac{V_{2x}}{V_{2xx}} \end{array} \end{cases}$$

Notes

1. cf. [6, 11]
2. cf. [8, 5]
3. Stability is very important since the optimal portfolio rule depends on $\frac{V_x}{V_{xx}}$.

References

[1] L. A. Caffarelli and X. Cabre, *Fully Nonlinear Elliptic Equations*, AMS, Providence (1995).

[2] J. Cvitanić and I. Karatzas, Hedging contingent claims with constrained portfolios, *Ann. Appl. Probab.*, **3** 652-681 (1993).

[3] M. H. A. Davis and A. R. Norman, Portfolio selection with transaction costs, *Math. Oper. Res.*, **15** 676-713 (1990).

[4] W. H. Fleming and R. W. Rishel, *Deterministic and Stochastic Control*, Springer-Verlag, New York (1975).

[5] W. H. Fleming and H. M. Soner, *Controlled Markov Processes and Viscosity Solutions*, Springer-Verlag, New York (1993).

[6] A. Friedman, *Stochastic Differential Equations and Applications, Vol. 1 & 2*, Academic Press, New York (1976).

[7] D. Gilbarg and N. S. Trudinger, *Elliptic Partial Differential Equations of Second Order*, Springer-Verlag, Berlin (1983).

[8] I. Karatzas, *Lectures on the Mathematics of Finance*, CRM Monograph Series Volume 8, AMS, Providence (1991).

[9] I. Karatzas, Adaptive control of a diffusion to a goal and parabolic Monge-Ampère-type equation, *Preprint* (1997).

[10] I. Karatzas and S. G. Kou, Pricing contingent claims with constrained portfolios, *Ann. Appl. Probab.*, **6** 321-394 (1996).

[11] I. Karatzas and S. E. Shreve, *Brownian Motion and Stochastic Calculus*, Springer-Verlag, New York (1996).

[12] N. V. Krylov, *Controlled Diffusion Processes*, Springer-Verlag, New York (1980).

[13] N. V. Krylov, *Nonlinear Elliptic and Parabolic Equations of the Second Order*, D. Reidel Publ. Comp., Dordrecht (1987).

[14] H. J. Kushner and P. G. Dupuis, *Numerical Methods for Stochastic Control Problems in Continuous Time*, Springer-Verlag, New York (1992).

[15] P. L. Lions, Sur les équations de Monge-Ampère, *Manuscripta Math.*, **41** 1-43 (1983).

[16] R. C. Merton, Optimum consumption and portfolio rules in a continuous-time model, *J. Econometric Theory*, **3** 373-413 (1971).

[17] R. C. Merton, *Continuous Time Finance*, Blackwell, Cambridge MA (1992).

[18] S. Stojanovic, Optimal portfolio diversification under no-short-selling and budget constraints and market index tracking, *Mathematica* package, http://math.uc.edu/~srdjan/description/ann.htm

[19] T. Zariphopoulou, Consumption-investment models with constraints, *SIAM J. Control & Optimiz.*, **32** 59-85 (1994).

LIMITATIONS AND CAPABILITIES OF FEEDBACK FOR CONTROLLING UNCERTAIN SYSTEMS*

L.-L. Xie and L. Guo

Institute of Systems Science
Chinese Academy of Sciences
Beijing, 100080, P.R.China

Lguo@iss03.iss.ac.cn

Abstract: Feedback is used primarily for reducing the effects of uncertainties on the performance of control systems, and the understanding of its limitations and capabilities is fundamental. This paper will present some preliminary results in this direction, showing that for a large class of discrete-time uncertain nonlinear stochastic systems, the growth rate of nonlinearities (GRN) is critical: if the GRN is faster than linear, then feedback stabilization is impossible in general even for systems with linear uncertain parameters; if, however, the GRN is linear, then an asymptotically optimal feedback can be constructed for a class of nonparametric uncertain systems.

1 INTRODUCTION

The main objective of using feedback in a control system design is to reduce the effects of the system uncertainties on the control performance. The uncertainties of a system usually stem from two sources: structure uncertainties and external disturbances (noises). In general, the later is easier to cope with than the former. A fundamental question in the area of systems and control is: What is the limitations and capabilities of feedback for controlling uncertain systems? This is a conundrum which only a few control areas could shed some light on. Robust control and adaptive control are two such areas where structure uncertainties of a system are the main concern in the controller design. Robust control usually requires that the true system lies in a small ball centered by a known nominal model (cf.[12]), whereas adaptive control does not need such a prerequest (cf.[1] [7]). Although much progress has been made in these

*Please address all correspondence to Professor Lei Guo. This work was supported by the National Natural Science Foundation of China and the National Key Project of China.

two areas over the past two decades, the understanding of the fundamental question concerning about capabilities and limitations of feedback is far from being complete.

For linear finite dimensional systems with uncertain parameters, a well-developed theory of adaptive control exists today, both for stochastic systems (cf.[1] [4] [6]) and for deterministic systems with small unmodelled dynamics (cf.[7]). This theory can easily be generalized to nonlinear systems with linear unknown parameters and with linearly growing nonlinearities (cf.[9]). However, the generalization of the existing theory to systems with nonlinearities having nonlinear growth rates is possible only for continuous-time system (cf.[8]), not for discrete-time systems (cf.[5]). This is a fundamental difference between adaptive feedback control of discrete-time and continuous-time systems.

All the above mentioned results are concerned with parametric models with linear uncertain parameters. This is of course only a special situation. The more challenging problem is to control uncertain nonparametric systems. We now give it a little more detailed account. Let $f(\cdot)$ be an unknown nonparametric function describing the nonlinear dynamics of a control system. Various approximation techniques exist in the literature (e.g. Volterra series, fuzzy and neural nets, wavelets, etc.), which basically state that for x in a compact set, $f(\cdot)$ can be uniformly approximated by parametric functions of the form

$$g(\theta, x) \triangleq \sum_{i=1}^{N} a_i \sigma(b_i^T x),$$

where $\sigma(\cdot)$ is a known "basis" function, a_i's and b_i's are unknown parameters or weights denoted by θ, and N is an integer reflecting the complexity/accuracy of approximation.

Thus, one may conceive that the above explicit parametric model $g(\theta, x)$ can be used in adaptive control instead of using the original nonparametric model $f(\cdot)$. This natural idea has indeed attracts considerable attention from researches (e.g. [11]). This approach, though attractive and effective in some applications, has several fundamental limitations/difficulties. First, in order to ensure that x (which usually represents the system state or output signals) lies in a compact set for reliable approximation, stability of the system must be established first, and the parametric model provides little (if any) help in this regard; Second, searching for the optimal parameters a_i's and b_i's usually involves in global nonlinear optimization, of which a general efficient scheme is still lacking by now, and moreover, the on-line combination of the estimation and control (adaptive control) will further complicate the problem; Third, no matter how large the approximation complexity N is, there always exists an approximation error in the model, which will inevitably prevent the parametric-model-based control to be optimal in general. Hence, it is of advantages to consider the nonparametric model $f(\cdot)$ directly, and it is natural to use the nonparametric estimation methods which are well-developed in the mathematical statistics literature (cf. [2] and the references therein). To the best of the authors' knowledge, the first concrete theoretical result on nonparametric adaptive control seems to have been obtained by Oulidi (see [2]), who proved that the diminishingly excited certainty equivalence nonparametric adaptive control is asymptotically optimal for systems with bounded noises.

The understanding of the limitations and capabilities of feedback for controlling uncertain systems is inextricably linked to the puzzling question: how much of the uncertainties can be predicted and reduced based on the available information? Clearly, it is more convenient to answer this "predictability" question in the stochastic framework, since the concept of "conditional expectation" in probability theory is a natural, suitable and powerful tool.

In this paper, we shall present some preliminary yet concrete results on the limitations and capabilities of feedback in the presence of structure uncertainties of the systems to be controlled. We shall first show that even in the case of linear uncertain parameters, feedback control may still not be able to stabilize the system if the nonlinearities have a nonlinear growth rate in general. Then we will show that if the linear growth rate constraint is imposed, then a feedback controller can be constructed to control a large class of nonparametric uncertain systems in an asymptotically optimal way, without resorting to any external excitations.

2 LIMITATIONS OF FEEDBACK FOR STABILIZING UNCERTAIN NONLINEAR SYSTEMS

Consider the following typical discrete-time polynomial nonlinear regression model

$$y_{t+1} = \theta_1 y_t^{b_1} + \theta_2 y_t^{b_2} + \cdots + \theta_n y_t^{b_n} + u_t + w_{t+1}, \quad t \geq 0, \tag{1.1}$$

where y_t and u_t are the system output and input signals respectively, θ_i $(1 \leq i \leq n)$ are unknown parameters and w_t is the noise signal.

Assume that

(A1) b_i $(1 \leq i \leq n)$ are nonnegative real numbers making (1.1) meaningful and satisfying

$$b_1 > b_2 > \cdots > b_n > 0; \tag{1.2}$$

(A2) $\{w_t\}$ is a Gaussian white noise sequence with distribution $N(0,1)$;

(A3) The unknown parameter vector $\theta \triangleq [\theta_1, \cdots, \theta_n]^\tau$ is independent of $\{w_t\}$ and has a Gaussian distribution $N(\bar{\theta}, I_n)$.

Our objective is to study the global stabilizability of (1.1) under the above conditions. First, we give a precise definition of stabilizability.

Definition 2.1 Let $\sigma\{y_i, 0 \leq i \leq t\}$ be the σ-field generated by the observations y_i, $0 \leq i \leq t$. The system (1.1) is said to be a.s. globally stabilizable, if there exists a feedback control

$$u_t \in \mathcal{F}_t^y \triangleq \sigma\{y_i, 0 \leq i \leq t\}, \quad t = 0, 1, \cdots \tag{1.3}$$

such that for any initial condition $y_0 \in R^1$, $\limsup_{T\to\infty} \frac{1}{T}\sum_{t=1}^{T} y_t^2 < \infty$, a.s.

Remark 2.1 We remark that the global stabilization of (1.1) is a trivial task in either the case where θ is known or the case where the noise is free (i.e. $w_t \equiv 0$). To be precise, if θ were known, we can put $u_t \equiv -\left(\theta_1 y_t^{b_1} + \theta_2 y_t^{b_2} + \cdots + \theta_n y_t^{b_n}\right)$, which gives $y_{t+1} \equiv w_{t+1}$, and the system is stabilized since

$$\limsup_{T\to\infty} \frac{1}{T}\sum_{t=1}^{T} y_t^2 = \limsup_{T\to\infty} \frac{1}{T}\sum_{t=1}^{T} w_t^2 < \infty.$$

In the case where θ is unknown but the noise is free ($w_t \equiv 0$), we can obtain the true value of the parameter θ by solving n linear independent equations. For example, if in the first $(n+1)$ steps, we choose $\{u_t, 0 \le t \le n\}$ to be independently identically distributed random variables with probability density function $p(x)$, then it is not difficult to prove the nonsingularity of the following matrix:

$$A \triangleq \begin{bmatrix} y_1^{b_1} & y_1^{b_2} & \cdots & y_1^{b_n} \\ y_2^{b_1} & y_2^{b_2} & \cdots & y_2^{b_n} \\ \vdots & \vdots & \ddots & \vdots \\ y_n^{b_1} & y_n^{b_2} & \cdots & y_n^{b_n} \end{bmatrix} \tag{1.4}$$

Hence the true value of the parameter θ can easily be obtained by solving the following linear equation derived by rearranging (1.1):

$$A \cdot \theta = [y_2 - u_1, y_3 - u_2, \cdots, y_{n+1} - u_n]^{\tau}.$$

Then again we can take the control as $u_t = -\left(\theta_1 y_t^{b_1} + \theta_2 y_t^{b_2} + \cdots + \theta_n y_t^{b_n}\right)$ for $t > n$, which globally stabilizes the noise-free system. For more general parametric-strict-feedback models with no noises, a similar approach can also be applied to design a globally stabilizing adaptive controller regardless of the growth rate of the nonlinearities.

Unfortunately, the main drawback of the above approach is that the resulting adaptive controller is not robust with respect to noises. In fact, the presence of noises will change the stabilizability of discrete-time nonlinear systems dramatically if the growth rate of the nonlinearities is faster than linear, as will be shown by the following theorem together with its corollaries.

Theorem 2.1 ([10]) Under Assumptions (A1)–(A3), the system (1.1) is **not** a.s. globally stabilizable by feedback whenever the following inequality

$$P(z) < 0, \quad z \in (1, b_1) \tag{1.5}$$

has a solution, where $P(z)$ is a polynomial defined by

$$P(z) = z^{n+1} - b_1 z^n + (b_1 - b_2) z^{n-1} + \cdots + (b_{n-1} - b_n) z + b_n. \tag{1.6}$$

To understand the implications of Theorem 2.1, we now give some detailed discussions on the inequality (1.5).

Corollary 2.1 ([10]) Let b_i $(1 \le i \le n)$ satisfy

$$b_1 > 1 \quad \text{and} \quad 0 < b_i - b_{i+1} \le \frac{\sqrt{b_1}}{2}(\sqrt{b_1} - 1)^2, \quad 1 \le i \le n-1,$$

then (1.5) has a solution whenever $n \ge 2\log\left(\frac{\sqrt{b_1}+1}{\sqrt{b_1}-1}\right) \Big/ \log b_1$. Consequently, whenever $b_1 > 1$ and the number of unknown parameters n is suitably large, there always exist $0 < b_n < b_{n-1} < \cdots < b_1$ such that (1.1) is not a.s. globally stabilizable.

Remark 2.2 By Corollary 2.1 we know that the usual linear growth condition imposed on the nonlinear function $f(\cdot)$ of the general control model

$$y_{t+1} = \theta^\tau f(y_t) + u_t + w_{t+1}, \quad \theta \in R^n \tag{1.7}$$

cannot be essentially relaxed in general for global adaptive stabilization, unless additional conditions on the number n and the structure of $f(\cdot)$ are imposed.

Remark 2.3 Let us consider the following continuous-time counter-part model

$$dy_t = [\theta^\tau f(y_t) + u_t]dt + dw_t, \quad t \ge 0, \tag{1.8}$$

where $\theta \in R^p$ is an unknown parameter vector, and $f(x) : R^1 \to R^p$ is a continuous function satisfying the local Lipschitz condition, and $\{w_t\}$ is a standard Brownian motion. Assume that $\|f(x)\| \le L_1 + L_2|x|^k$ for some integer $k > 0$ and for some constants $L_1, L_2 > 0$. Then it can be shown that the following feedback control of nonlinear damping type:

$$u_t = -cy_t - y_t^{2k+1}, \quad c > 0$$

can stabilize the systems regardless of the growth rate of the nonlinearities (measured by k).

The above two remarks demonstrate the fundamental difficulties between feedback stabilizability of discrete- and continuous-time uncertain systems.

Corollary 2.2 Let $b_1 > 2$, then for $n > 1 + 2\log\left(\frac{2}{b_1 - 2}\right) \Big/ \log\left(\frac{b_1}{2}\right)$, (1.5) has a solution for any $\{b_i\}$ satisfying $1 \le b_n < b_{n-1} < \cdots < b_2 < b_1$. On the other hand, if $b_1 \le 2$, then for any n, there always exist $1 \le b_n < b_{n-1} < \cdots < b_2 < b_1$ such that (1.5) has no solution.

Corollary 2.3 For any $n \ge 1$ and any $b_1 > b_2 > \cdots > b_n > 0$,
(i) A necessary condition for (1.5) to have a solution is $\sum_{i=1}^n b_i > 4$;
(ii) A sufficient condition for (1.5) to have a solution is either $b_1 > 4$, or

$$\sum_{i=1}^n b_i > (n+1)(1 + \frac{1}{n})^n.$$

The above three corollaries give us a picture concerning about situations where the nonlinear model (1.1) is not a.s. globally stabilizable by feedback.

3 OPTIMAL FEEDBACK CONTROL OF NONPARAMETRIC UNCERTAIN SYSTEMS

In this section, we are going to show that if the nonlinearities have a certain linear growth rate, then optimal feedback control can be designed even for the following uncertain nonparametric model:

$$y_{t+1} = f(y_t) + u_t + \varepsilon_{t+1}, \tag{1.9}$$

where y_t, u_t and ε_t are the d-dimensional system output signals, input signals and white noises, and $f(\cdot)$ is an unknown nonlinear function.

Our objective is to design a feedback control u_t based on the observations $\{y_i, i \leq t\}$ at each step t, such that the system output $\{y_t\}$ tracks a known reference signal $\{y_t^*\}$ in an optimal way. If $f(\cdot)$ were known, it is obvious that such a controller would take the following form:

$$u_t = -f(y_t) + y_{t+1}^*.$$

Since in the present case, $f(\cdot)$ is unknown, we adopt the nonparametric estimation approach as used in [2], but without resorting to the external excitations used there.

Let $K(\cdot)$ be a nonnegative kernel function satisfying the following conditions:

$$K(0) > 0, \quad \int K(s)ds = 1, \quad \int K^2(s)ds < \infty, \quad \int \|s\|K(s)ds < \infty.$$

Here in our estimation process, let $K(\cdot)$ have a compact support, i.e.,

$$K(s) = 0, \qquad \text{for } \|s\| > A.$$

Let $\delta(\cdot,\cdot)$ be a function shifted from $K(\cdot)$:

$$\delta_j(x, y) \triangleq K(j^a(x - y)), \quad \forall j > 0, \quad \delta_0 = 0, \tag{1.10}$$

where $a \in (0, \frac{1}{2d})$, d is the dimension of the system signals.

The nonparametric estimate of $f(y), y \in \mathbb{R}^d$ at time t is defined by

$$\widehat{f}_t(y) = \begin{cases} N_t^{-1}(y) \sum_{j=1}^{t} \delta_{j-1}(y_{j-1}, y)(y_j - u_{j-1}), & \text{if } N_t(y) > 0; \\ 0 & \text{otherwise,} \end{cases} \tag{1.11}$$

where,

$$N_t(y) \triangleq \sum_{j=1}^{t} \delta_{j-1}(y_{j-1}, y) \tag{1.12}$$

To define the adaptive feedback control, we need to introduce a sequence of truncation bounds denoted by $\{h_t\}$, which is positive, monotonically diverges to infinity, and satisfies

$$h_t = o(\sqrt{\log t}), \text{ as } t \to \infty. \tag{1.13}$$

Now, by the (truncated) certainty equivalence principle, the nonparametric adaptive control can be defined as

$$u_t = -\widehat{f}_t(y_t) I_{(|\widehat{f}_t(y_t)| \le h_t)} + y^*_{t+1} \tag{1.14}$$

where $I_{(\cdot)}$ is the indicator function.

With this control, the closed-loop system equation is

$$y_{t+1} = f(y_t) - \widehat{f}_t(y_t) I_{(|\widehat{f}_t(y_t)| \le h_t)} + y^*_{t+1} + \varepsilon_{t+1}, \tag{1.15}$$

which is obviously a nonlinear dynamical system.

In order to analyze the properties of (1.15), we introduce the following assumptions on the system (1.9):

(A4) The nonlinear function $f(\cdot)$ is Lipschitz continuous, and there exist two constants $\alpha \in (0,1)$ and $\beta \in (0, \infty)$ such that

$$||f(x)|| \le \alpha ||x|| + \beta, \quad \forall x \in \mathbb{R}^d.$$

(A5) $\{\varepsilon_t\}$ is a Gaussian white noise sequence with mean zero and variance $\sigma^2 > 0$.

(A6) The reference signal $\{y^*_t\}$ is bounded.

Theorem 3.1 Consider the control system (1.9) where the nonlinear function $f(\cdot)$ is completely unknown. Let the assumptions (A4)-(A6) be fulfilled. Then the adaptive feedback tracking control defined by (1.14) is asymptotically optimal in the sense that

$$\lim_{T \to \infty} \frac{1}{T} \sum_{t=1}^{T} (y_t - y^*_t)^2 = \sigma^2, \quad a.s..$$

The proof may be found in [3].

4 CONCLUDING REMARKS

In this paper, we have presented several preliminary results concerning limitations and capabilities of feedback for controlling uncertain nonlinear systems. Of course, this is just a standing point towards a more comprehensive theory. Many interesting problems still remain open, among which we only mention the following two: (i) If in Theorem 2.1 the inequality (1.5) has no solution, can we construct a stabilizing feedback? (ii) Is it possible to remove the stringent condition $\alpha \in (0,1)$ in Condition (A4) of Theorem 3.1?

References

[1] H. F. Chen and L. Guo, *Identification and Stochastic Adaptive Control*, Boston, MA: Birkhäuser, 1991.

[2] M. Duflo, *Random Iterative Models*, Berlin: Springer-Verlag, 1997.

[3] L. Guo and L. L. Xie, An optimal nonparametric adaptive control without external excitation, *Preprints of the 14th IFAC World Congress*, Beijing, July 5-9, 1999. (To appear).

[4] L. Guo, Self-convergence of weighted least-squares with applications to stochastic adaptive control, *IEEE Trans. Automat. Contr.*, Vol.41, No.1, pp.79-89, 1996.

[5] L. Guo, On critical stability of discrete-time adaptive nonlinear control, *IEEE Trans. Automat. Contr.*, Vol.42, No.11, pp.1488-1499, 1997.

[6] L. Guo and H. F. Chen, The Astrom-Wittenmark self-tuning regulator revisited and ELS-based adaptive trackers, *IEEE Trans. Automat. Contr.*, Vol.36, No.7, pp.802-812, 1991.

[7] P. A. Ioannou and J. Sun, *Robust Adaptive Control*, NJ: Prentice-Hall, 1996.

[8] M. Krstic, I. Kanellakopoulos and P. V. Kokotovic, *Nonlinear and Adaptive Control Design*, New York: John Wiley & Sons, 1995.

[9] C. Wei and L. Guo, Adaptive control of a class of nonlinear stochastic systems, *Proc. 1996 Chinese Control Conference,* pp.508-512, Qingdao, September 15-20, 1996.

[10] L.L.Xie and L.Guo, Fundamental limitations of discrete-time adaptive nonlinear control, *IEEE Trans. Automat. Contr.*, to appear, 1998.

[11] K.S. Narendra and K.Parthasarathy, Identification and control of dynamical systems using neural networks, *IEEE Trans. Neural Networks,* vol.1, pp.4-27, 1990.

[12] K.Zhou, J.C.Doyle and K.Glover, *Robust and Optimal Control*, NJ: Prentice Hall, 1996.

TIME-SCALE SEPARATION AND STATE AGGREGATION IN SINGULARLY PERTURBED SWITCHING DIFFUSIONS*

G. Yin and M. Kniazeva

Department of Mathematics
Wayne State University
Detroit, MI 48202
gyin, maria@math.wayne.edu

Abstract: This work presents asymptotic results for singularly perturbed switching diffusions consisting of diffusion components and a pure jump component. The states of the pure jump component are divisible into a number of groups having recurrent states. An aggregated process is obtained by collecting all the states in each recurrent group as one state. We show the aggregated process converges weakly to a switching diffusion with generator being an average with respect to the quasi-stationary distribution of the jump process.

1 INTRODUCTION

Recent applications in manufacturing and queueing networks have posed many challenging problems involving singularly perturbed Markovian systems (see [1, 14, 16, 15, 17, 21, 22] and the references therein). For example, in manufacturing systems, one often models the machine capacity by a Markov chain with finite state space to describe the situation that the machines are subject to breakdowns and repairs. Meanwhile, the demand of the underlying product may be considered as a diffusion process to capture the uncertainty of the demand behavior. A direct treatment of large dimensional systems is frequently infeasible from a computational point of view. A viable alternative calls for time-scale separation modeling and aggregation approach so as to break the system into small pieces with manageable size, which naturally leads to the formulation of singularly perturbed system.

In [9], we studied properties of probability distribution of fast varying Markov chain by matched asymptotic expansion. Chains with fast and slow motions

*This research was supported in part by the National Science Foundation under grant DMS-9529738, and in part by Wayne State University.

were consider in [10]; further results were contained in [21, 23, 24] among others. We carried out further investigation to include singularly perturbed diffusion processes and singularly perturbed switching diffusions [6, 7, 8, 19, 20]. This paper continues our effort by treating singularly perturbed switching-diffusion processes with the jump process decomposed into several groups with fast and slow motions. The rest of the paper is arranged as follows. Section 2 gives the formulation. Section 3 presents a number of asymptotic results. Section 4 concludes the paper with brief remarks.

2 FORMULATION

Suppose $T > 0$. We work with finite horizon $[0, T]$. Consider a nonstationary Markov process $Y^\varepsilon(t) = (X(t), \gamma^\varepsilon(t))$, where $X(t)$ is a 1-dimensional diffusion process and $\gamma^\varepsilon(t)$ is a pure jump process.

The state space of the process $Y^\varepsilon(\cdot)$ is $\mathcal{X} = [0,1] \times \mathcal{M} = [0,1] \times \{1, \ldots, m\}$ (i.e., we consider the case of periodic diffusions with state space $[0,1]$ and the jump process has state space $\mathcal{M}$). If $\gamma^\varepsilon(t) = i$, the evolution of $X(t)$ is represented by the differential operator $\mathcal{D}_i$ with $\mathcal{D}_i \widetilde{f} = (1/2)a_i(x,t)(\partial^2/\partial x^2)\widetilde{f} + b_i(x,t)(\partial/\partial x)\widetilde{f}$ for appropriate smooth functions $\widetilde{f}(\cdot)$, $a_i(\cdot)$ and $b_i(\cdot)$. With $X(t) = x$ fixed, the pure jump component $\gamma^\varepsilon(t)$ satisfies $P(\gamma^\varepsilon(t+\delta) = j | \gamma^\varepsilon(t) = i, X(t) = x) = q^\varepsilon_{ij}(x,t)\delta + o(\delta)$, where $o(\delta) \to 0$ as $\delta \to 0$, for each $i, j \in \mathcal{M}$, $q^\varepsilon_{ij}(x,t) \geq 0$ when $j \neq i$, and $q^\varepsilon_{ii}(x,t) = -\sum_{j \neq i} q^\varepsilon_{ij}(x,t)$. Assume that all coefficients $a_i(\cdot)$, $b_i(\cdot)$, and $q^\varepsilon_{ij}(\cdot)$ are sufficiently smooth functions of (x,t). In view of [5], for $i = 1, \ldots, m$, and smooth real-valued function $f(\cdot,\cdot,i)$, the generator L^ε of this switching-diffusion process takes the form

$$\begin{aligned} L^\varepsilon f(x,t,i) = \frac{\partial f(x,t,i)}{\partial t} &+ b_i(x,t)\frac{\partial f(x,t,i)}{\partial x} + \frac{1}{2}a_i(x,t)\frac{\partial^2 f(x,t,i)}{\partial x^2} \\ &+ \sum_{j=1}^m q^\varepsilon_{ij}(x,t) f(x,t,j). \end{aligned} \tag{1.1}$$

The probability density of the process, $p^\varepsilon(x,t) = (p^\varepsilon_1(x,t), \ldots, p^\varepsilon_m(x,t))$ is the solution of the forward equation

$$\frac{\partial p^\varepsilon_i}{\partial t} = \mathcal{D}^*_i p^\varepsilon_i + \sum_{j=1}^m p^\varepsilon_j q^\varepsilon_{ij},\ p^\varepsilon_i(x,0) = g_i(x), \quad i = 1, \ldots, m,$$

where $\mathcal{D}^*_i$ is the adjoint of $\mathcal{D}_i$, $\mathcal{D}^*_i \cdot = \mathcal{D}^*_i(x,t) \cdot = (1/2)(\partial^2/\partial x^2)(a_i(x,t)\ \cdot) - (\partial/\partial x)(b_i(x,t)\ \cdot)$, and $g(x) = (g_1(x), \ldots, g_m(x))$ is the initial distribution for $Y^\varepsilon(t)$.

Assume that $Q^\varepsilon(x,t) = (q^\varepsilon_{ij}(x,t))$ has the form

$$Q^\varepsilon(x,t) = \frac{1}{\varepsilon}\widetilde{Q}(x,t) + \widehat{Q}(x,t), \text{ where } \widetilde{Q}(x,t) = \operatorname{diag}(\widetilde{Q}^1(x,t), \ldots, \widetilde{Q}^l(x,t)) \tag{1.2}$$

such that for each $t \in [0,T]$, and each $k = 1,\dots,l$, $\widetilde{Q}^k(x,t)$ is a generator with dimension $m_k \times m_k$, The state space $\mathcal{M}$ of the jump process admits a decomposition

$$\begin{aligned}\mathcal{M} &= \mathcal{M}_1 \cup \mathcal{M}_2 \cup \cdots \cup \mathcal{M}_l \\ &= \{\varpi_{11},\dots,\varpi_{1m_1}\} \cup \{\varpi_{21},\dots,\varpi_{2m_2}\} \cup \{\varpi_{l1},\dots,\varpi_{lm_l}\}.\end{aligned}$$

The basic assumptions we are using are as follows.

(C1) For each fixed $x \in [0,1]$, $\gamma^\varepsilon(\cdot)$ is a Markov chain with state space $\mathcal{M}$.

(C2) For each $i = 1,\dots,m$,

- $a_i(\cdot) \in C^{2,1}$, (i.e., the second partial derivatives of $a_i(\cdot)$ with respect to x and the partial derivatives with respect to t are continuous), $b_i(\cdot) \in C^{1,1}$;
- $(\partial/\partial t)a_i(x,\cdot)$ and $(\partial/\partial t)b_i(x,\cdot)$ are Lipschitz on $[0,T]$;
- $a_i(x,t) > 0$. for all $(x,t) \in [0,1] \times [0,T]$.

(C3) The function $\widetilde{Q}(\cdot,\cdot) \in C^{1,1}$. For each $x \in [0,1]$, $(\partial/\partial t)\widetilde{Q}(x,\cdot)$ and $\widehat{Q}(x,\cdot)$ are Lipschitz continuous on $[0,T]$.

(C4) For each $x \in [0,1]$, each $t \in [0,T]$, and each $i = 1,\dots,l$, $\widetilde{Q}^i(x,t)$ is weakly irreducible, i.e.,

$$\nu^i(x,t)\widetilde{Q}^i(x,t) = 0,\ \nu^i(x,t)\mathbb{1}_{m_i} = \sum_{j=1}^{m_i} \nu^i_j(x,t) = 1 \tag{1.3}$$

has a unique solution that is nonnegative (i.e., $\nu^i_j(x,t) \geq 0$ for each $j = 1,\dots,m_i$) and is termed quasi-stationary distribution. In the above $\mathbb{1}_{m_i}$ is an m_i-dimensional column vector that has all components being equal to 1.

For basic properties of Markov chains, see [2] via a piecewise-deterministic approach. In what follows, we obtain certain weak convergence results for the model posed. For discussion of weak convergence, see [4, 11]. Related studies about stochastic differential systems are in [18].

3 ASYMPTOTIC PROPERTIES

First, we obtain a result on the asymptotic expansion of the system of interest. The rest of the section mainly concerns about occupation measures. We derive results on a fixed-x process. The idea behind is that when x is fixed, the corresponding process is a Markov chain. Define $\widetilde{\mathbb{1}} = \mathrm{diag}(\mathbb{1}_{m_1},\dots,\mathbb{1}_{m_l})$, and

$$\overline{Q}(x,t) = \mathrm{diag}(\nu^1(t),\dots,\nu^l(t))\widehat{Q}(x,t)\widetilde{\mathbb{1}}. \tag{1.4}$$

3.1 Asymptotic Expansion

The following lemma can be obtained by using the techniques of [20] (Note: $\widetilde{Q}(x,t)\widetilde{\mathbb{1}} = 0$). We omit the proof here.

Lemma 3.1. *Assume* (C1)–(C4). *For* $k = 1, \ldots, l$, *denote by* $\nu^k(x,t)$ *the quasi-stationary distribution of* $\widetilde{Q}^k(t)$. *Then the following assertions hold:*

(i) *For some* $\kappa_0 > 0$,

$$p^\varepsilon(x,t) = v(x,t)\text{diag}(\nu^1(x,t), \ldots, \nu^l(x,t)) + O(\varepsilon + \exp(-\kappa_0 t/\varepsilon)),$$

where $v(x,t) = (v_1(x,t), \ldots, v_l(x,t))$ *satisfies*

$$\begin{aligned} \frac{\partial v(x,t)}{\partial t} &= \overline{\mathcal{D}}^* v(x,t) + v(x,t)\overline{Q}(x,t) \\ v(x,0) &= (g^1(x)\mathbb{1}_{m_1}, \ldots, g^l(x)\mathbb{1}_{m_l}) \end{aligned} \tag{1.5}$$

where $\overline{\mathcal{D}}^*$ *is the "average" of the forward operators w.r.t. the quasi-stationary distribution* (see [19, 20]) *with* $\overline{\mathcal{D}}^* v(x,t) = (\overline{\mathcal{D}}_1^* \; v_1(x,t), \ldots, \overline{\mathcal{D}}_l^* v_l(x,t))$ *and* $g^i(x) \in \mathbb{R}^{1\times m_i}$ *denotes the ith partitioned vector of* $g(x)$.

(ii) *For fixed* $x \in [0,1]$, *denote the transition probability by* $P^\varepsilon(x,t,t_0)$. *Then for some* $\kappa_0 > 0$, $P^\varepsilon(x,t,t_0) = \overline{P}(x,t,t_0) + O(\varepsilon + \exp(-\kappa_0(t-t_0)/\varepsilon))$, *where (using the notation of partitioned matrices)*

$$\overline{P}(x,t,t_0) = \widetilde{\mathbb{1}}V(x,t,t_0)\text{diag}(\nu^1(x,t), \ldots, \nu^l(x,t)), \tag{1.6}$$

and $V(x,t,t_0) = (v_{ij}(x,t,t_0)) \in \mathbb{R}^{l\times l}$ *satisfying the differential equation:*

$$\frac{\partial V(x,t,t_0)}{\partial t} = V(x,t,t_0)\overline{Q}(x,t),\; V(x,t_0,t_0) = I,$$

where $\overline{Q}(x,t)$ *is defined in* (1.4).

3.2 Fixed-x Processes

To proceed, define the "fixed-x" processes $\widetilde{Y}^\varepsilon(t)$ and $\overline{Y}_i^\varepsilon(t)$ as follows. Denote

$$\widetilde{Y}^\varepsilon(t) = Y^\varepsilon(t)\Big|_{X(t)=x} = (x, \gamma^\varepsilon(t)), \quad \text{and } \overline{Y}_i^\varepsilon(t) = Y^\varepsilon(t)\Big|_{X(t)=x,\gamma^\varepsilon(t)\in\mathcal{M}_i}.$$

For fixed x, define $\overline{\gamma}^\varepsilon(t) = i$, if $\gamma^\varepsilon(t) \in \mathcal{M}_i$, for $i = 1, \ldots, l$. Finally define the aggregated process $\overline{Y}^\varepsilon(t)$ by $\overline{Y}^\varepsilon(t) = (X(t), \overline{\gamma}^\varepsilon(t))$. For any $t \in [0,T]$, $x \in [0,1]$, $i = 1, \ldots, l$ and $j = 1, \ldots, m_i$, define

$$\eta_{ij}^\varepsilon(x,t) = E\left(\int_0^t \left(I_{\{\widetilde{Y}^\varepsilon(s)=(x,\varpi_{ij})\}} - \nu_j^i(x,s)I_{\{\overline{Y}_i^\varepsilon(s)\}}\right) ds\right)^2. \tag{1.7}$$

Proposition 3.2. *Suppose that* (C1)-(C4) *hold. Then for* $i = 1, \ldots, l$, $j = 1, \ldots, m_i$, $\sup_{x\in[0,1],\, 0\le t\le T} \eta_{ij}^\varepsilon(x,t) = O(\varepsilon)$.

Sketch of Proof. Recall that for $i = 1, \ldots, l$, $\overline{\gamma}^\varepsilon(t) = i$ iff $\gamma^\varepsilon(t) \in \mathcal{M}_i$. Differentiating $\eta^\varepsilon_{ij}(\cdot)$ with respect to t leads to

$$\begin{aligned}\frac{\partial \eta^\varepsilon_{ij}(x,t)}{\partial t} = 2\Bigg[&\int_0^t P(\widetilde{Y}^\varepsilon(t) = (x, \varpi_{ij}), \widetilde{Y}^\varepsilon(s) = (x, \varpi_{ij}))ds \\ &- \int_0^t \nu^i_j(x,t) P(\overline{Y}^\varepsilon_i(t), \widetilde{Y}^\varepsilon(s) = (x, \varpi_{ij}))ds \\ &- \int_0^t \nu^i_j(x,s) P(\widetilde{Y}^\varepsilon(t) = (x, \varpi_{ij}), \overline{Y}^\varepsilon_i(s))ds \\ &+ \int_0^t \nu^i_j(x,s)\nu^i_j(x,t) P(\overline{Y}^\varepsilon_i(t), \overline{Y}^\varepsilon_i(s))ds\Bigg].\end{aligned} \tag{1.8}$$

We can show that

$$\begin{aligned}&\int_0^t P(\widetilde{Y}^\varepsilon(t) = (x, \varpi_{ij}), \widetilde{Y}^\varepsilon(s) = (x, \varpi_{ij}))ds - \int_0^t \nu^i_j(x,t) P(\overline{Y}^\varepsilon_i(t), \widetilde{Y}^\varepsilon(s) = (x, \varpi_{ij}))ds \\ &\quad = \int_0^t O(\varepsilon + \exp(-\kappa_0(t-s)/\varepsilon))ds = O(\varepsilon), \text{ and}\end{aligned}$$

$$\begin{aligned}&-\int_0^t \nu^i_j(x,s) P(\widetilde{Y}^\varepsilon(t) = (x, \varpi_{ij}), \overline{Y}^\varepsilon_i(s))ds + \int_0^t \nu^i_j(x,s)\nu^i_j(x,t) P(\overline{Y}^\varepsilon_i(t), \overline{Y}^\varepsilon_i(s))ds \\ &\quad = \int_0^t O(\varepsilon + \exp(-\kappa_0(t-s)/\varepsilon))ds = O(\varepsilon).\end{aligned}$$

Therefore,

$$\frac{\partial \eta^\varepsilon_{ij}(x,t)}{\partial t} = 2\int_0^t O(\varepsilon + \exp(-\kappa_0(t-s)/\varepsilon))ds = O(\varepsilon). \tag{1.9}$$

This implies that $\eta^\varepsilon_{ij}(x,t) = O(\varepsilon)$ and the bound is uniform in x and t.

3.3 Weak Convergence of $\overline{Y}^\varepsilon(\cdot)$

We aim at deriving a weak convergence result for $\overline{Y}^\varepsilon(\cdot)$. First, we obtain its tightness.

Proposition 3.3. *Assume* (C1)–(C4). *Then $\overline{Y}^\varepsilon(\cdot)$ is tight in $D^2[0,T]$.*

Proof. We use the Kurtz' criteria. Denote by $\mathcal{F}^\varepsilon_t$ the σ-algebra generated by $\{\overline{Y}^\varepsilon(u), u \le t\}$ and E^ε_t the conditional expectation on $\mathcal{F}^\varepsilon_t$. Then

$$\begin{aligned}E^\varepsilon_t|\overline{Y}^\varepsilon(t+s) - \overline{Y}^\varepsilon(t)|^2 &\le K E^\varepsilon_t[X(t+s) - X(t)]^2 + 2E^\varepsilon_t[\overline{\gamma}^\varepsilon(t+s) - \overline{\gamma}^\varepsilon(t)]^2 \\ &\le K\sum_{k=1}^m E^\varepsilon_t E\left([X(t+s) - X(t)]^2 | \mathcal{F}^\varepsilon_t,\ \gamma^\varepsilon(t) = k\right) \\ &\quad + 2E^\varepsilon_t \int_0^1 E\left([\overline{\gamma}^\varepsilon(t+s) - \overline{\gamma}^\varepsilon(t)]^2 | \mathcal{F}^\varepsilon_t,\ X(t) = x\right)dx.\end{aligned} \tag{1.10}$$

Note that for each $k = 1, \dots, m$, when $\gamma^\varepsilon(t) = k$,

$$X(t) = X(0) + \int_0^t b_k(X(u), u)du + \int_0^t \sqrt{a_k(X(u), u)}dw, \tag{1.11}$$

where $w(\cdot)$ is a standard Brownian motion. By virtue of the boundedness of $a_k(\cdot)$ and in view of a well-known result in stochastic calculus (see [13, p. 125]), $E_t^\varepsilon \left(\int_t^{t+s} \sqrt{a_k(x,u)}dw\right)^2 \le Ks$. Owing to the defining equation (1.11), and using the boundedness of $b_k(\cdot)$, $E\left([X(t+s) - X(t)]^2 | \mathcal{F}_t^\varepsilon,\ \gamma^\varepsilon(t) = k\right) \le K(s^2 + s)$.

As for the last term in (1.10),

$$\begin{aligned}&\int_0^1 E\left([\overline{\gamma}^\varepsilon(t+s) - \overline{\gamma}^\varepsilon(t)]^2 | \mathcal{F}_t^\varepsilon,\ X(t) = x\right) dx \\ &\quad \le K \sum_{i=1}^{l+1} \sum_{j=1}^{m_i} \int_0^1 E\left([\overline{\gamma}^\varepsilon(t+s) - \overline{\gamma}^\varepsilon(t)]^2 | \gamma^\varepsilon(t) = \varpi_{ij},\ X(t) = x\right) dx.\end{aligned} \tag{1.12}$$

Note that

$$\begin{aligned}&E\left([\overline{\gamma}^\varepsilon(t+s) - \overline{\gamma}^\varepsilon(t)]^2 | \gamma^\varepsilon(t) = \varpi_{ij},\ X(t) = x\right) \\ &\quad = \sum_{k=1}^{l} (k-i)^2 P(\overline{\gamma}^\varepsilon(t+s) = k | \gamma^\varepsilon(t) = \varpi_{ij},\ X(t) = x) \\ &\quad \le l^2 \sum_{k \ne i} v_{ik}(t+s, t, x) + O(\varepsilon + \exp(-\kappa_0 s/\varepsilon)).\end{aligned} \tag{1.13}$$

The definition of $V(t, s, x)$ then infers that for each $x \in [0, 1]$, $\lim_{s\to 0} v_{ik}(t + s, t, x) = 0$ for $i \ne k$. Putting the above arguments together, what we have shown is: $\lim_{s\to 0} \lim_{\varepsilon\to 0} E E_t^\varepsilon |\overline{Y}^\varepsilon(t+s) - \overline{Y}^\varepsilon(t)|^2 = 0$. Then the tightness criterion (see [11, Theorem 3, p. 47]) infers that the process $\{\overline{Y}^\varepsilon(\cdot)\}$ is tight in $D^2[0, T]$.

Define an "averaged" generator by

$$\overline{L}\tilde{g}(x, t, i) = \frac{\partial \tilde{g}(x, t, i)}{\partial t} + \overline{\mathcal{D}}_i \tilde{g}(x, t, i) + \sum_{j=1}^{l} \overline{q}_{ij}(x, t) \tilde{g}(x, t, j) \tag{1.14}$$

for any bounded and measurable function $\tilde{g}(\cdot)$ satisfying $\tilde{g}(\cdot, \cdot, i) \in C^{2,1}$ and for $i = 1, \dots, l$, where $\overline{\mathcal{D}}_i$ denotes the adjoint of $\overline{\mathcal{D}}_i^*$. By using the characteristic function $E\exp(\iota(\tau_1 X(t) + \tau_2 \overline{\gamma}(t)))$, where ι is the pure imaginary number with $\iota^2 = -1$, we can show that the martingale problem with operator $\overline{L}$ has a unique solution for each initial condition.

Proposition 3.4. *Under the* (C1)–(C4), $\overline{Y}^\varepsilon(\cdot)$ *converges weakly to* $\overline{Y}(\cdot)$, *a Markov process with* $\overline{L}$ *given by* (1.14).

The main idea of the proof involves the use of martingale problem formulation (see [11]), i.e., we show that $\overline{Y}(t) = (X(t), \overline{\gamma}(t))$ is a solution of the martingale problem with operator $\overline{L}$. To establish the desired result, we show that for any bounded and measurable function $f(\cdot)$ with $f(\cdot,\cdot,\gamma) \in C^{2,1}$ for each $\gamma \in \mathcal{M}$, $f(X(t),t,\overline{\gamma}(t)) - \int_0^t \overline{L}\ f(X(s),s,\overline{\gamma}(s))ds$ is a martingale. To do so, define $\overline{f}(x,t,\gamma) = \sum_{k=1}^{l} f(x,t,k) I_{\{\gamma \in \mathcal{M}_k\}}$, work with the function $\overline{f}(\cdot)$ instead, and show that $\overline{f}(X(t),t,\gamma(t)) - \int_0^t \overline{L}\ \overline{f}(X(s),s,\gamma(s))ds$ is a martingale. To accomplish this, we work with the pre-limit problem, and define a piecewise constant process $\widetilde{X}^\varepsilon(\cdot)$ by $\widetilde{X}^\varepsilon(t) = X(k\delta_\varepsilon)$ for $t \in [k\delta_\varepsilon, (k+1)\delta_\varepsilon)$. We then show that the limit of $\overline{f}(X(t+s), t+s, \gamma^\varepsilon(t+s)) - \overline{f}(X(t),t,\gamma^\varepsilon(t)) - \int_t^{t+s} \overline{L}\,\overline{f}(X(u),u,\gamma^\varepsilon(u))du$ is the same as that of the process with $X(t)$ replaced by $\widetilde{X}(t)$. This enables us to characterize the limit. More details can be found in [19].

References

[1] V. V. Anisimov, Switching processes: Averaging principle, diffusion approximation and applications, *Acta Appl. Math.* **40** (1995), 95-141.

[2] M. H. A. Davis, *Markov Models and Optimization*, Chapman & Hall, London, 1993.

[3] S. D. Eidelman, *Parabolic Systems*, North-Holland, New York, 1969.

[4] S. N. Ethier and T. G. Kurtz, *Markov Processes: Characterization and Convergence*, J. Wiley, New York, 1986.

[5] I. I. Gihman and A. V. Skorohod, *Theory of Stochastic Processes, III*, Springer-Verlag, Berlin, 1979.

[6] A. M. Il'in, R. Z. Khasminskii and G. Yin, Singularly perturbed switching diffusions: rapid switchings and fast diffusions, preprint, 1997.

[7] R. Z. Khasminskii and G. Yin, Asymptotic series for singularly perturbed Kolmogorov-Fokker-Planck equations, *SIAM J. Appl. Math.* **56** (1996), 1766-1793.

[8] R. Z. Khasminskii and G. Yin, On transition densities of singularly perturbed diffusions with fast and slow components, *SIAM J. Appl. Math.* **56** (1996), 1794-1819.

[9] R. Z. Khasminskii, G. Yin and Q. Zhang, Asymptotic expansions of singularly perturbed systems involving rapidly fluctuating Markov chains, *SIAM J. Appl. Math.* **56** (1996), 277-293.

[10] R. Z. Khasminskii, G. Yin and Q. Zhang, Constructing asymptotic series for probability distribution of Markov chains with weak and strong interactions, *Quart. Appl. Math.* **LV** (1997), 177-200.

[11] H. J. Kushner, *Approximation and Weak Convergence Methods for Random Processes, with Applications to Stochastic Systems Theory*, MIT Press, Cambridge, MA, 1984.

[12] O. A. Ladyzhenskaia, V. A. Solonnikov and N. N. Ural'tseva, *Linear and quasi-linear equations of parabolic type*, Trans. Math. Monographs, V. 23, Amer. Math. Soc., Providence, RI, 1968.

[13] R. S. Liptser and A. N. Shiryayev, *Statistics of Random Processes I*, Springer-Verlag, New York, 1997.

[14] Z. G. Pan and T. Başar, H^∞-control of Markovian jump linear systems and solutions to associated piecewise-deterministic differential games, in *New Trends in Dynamic Games and Applications*, G. J. Olsder (Ed.), 61-94, Birkhäuser, Boston, 1995.

[15] R. G. Phillips and P. V. Kokotovic, A singular perturbation approach to modelling and control of Markov chains, *IEEE Trans. Automat. Control* **26** (1981), 1087-1094.

[16] J. R. Rohlicek and A. S. Willsky, The reduction of perturbed Markov generators: An algorithm exposing the role of transient states, *J. ACM* **35** (1988), 675-696.

[17] S. P. Sethi and Q. Zhang, *Hierarchical Decision Making in Stochastic Manufacturing Systems*, Birkhäuser, Boston, 1994.

[18] A. V. Skorohod, *Asymptotic Methods of the Theory of Stochastic Differential Equations*, Trans. Math. Monographs, V. 78, Amer. Math. Soc., Providence, 1989.

[19] G. Yin, Weak convergence of an aggregated process arising from singularly perturbed switching diffusions, preprint, 1998.

[20] G. Yin and M. Kniazeva, Singularly perturbed multidimensional switching diffusions with fast and slow switchings, preprint, 1997.

[21] G. Yin and Q. Zhang, *Continuous-time Markov Chains and Applications: A Singular Perturbations Approach,* Springer-Verlag, New York, 1998.

[22] G. Yin and Q. Zhang (Eds.), *Mathematics of Stochastic Manufacturing Systems*, Proc. 1996 AMS-SIAM Summer Seminar in Applied Mathematics, Lectures in Applied Mathematics, Amer. Math. Soc., Providence, RI, 1997.

[23] Q. Zhang and G. Yin, A central limit theorem for singularly perturbed nonstationary finite state Markov chains, *Ann. Appl. Probab.* **6** (1996), 650-670.

[24] Q. Zhang and G. Yin, Structural properties of Markov chains with weak and strong interactions, *Stochastic Process Appl.* **70** (1997), 181-197.

STOCHASTIC CONTROLS AND FORWARD-BACKWARD SDES *

Jiongmin Yong

Laboratory of Mathematics for Nonlinear Sciences
and Department of Mathematics
Fudan University, Shanghai 200433, China

jyong@ms.fudan.edu.cn

1 INTRODUCTION

Optimal control theory has more than 40 years of history. For deterministic finite dimensional case, the works of Pontryagin *et al* (maximum principle [14]), Bellman (dynamic programming [1]) and Kalman (linear quadratic regulator problem [3]) have been regarded as three main milestones in the field. For inifite dimensional (deterministic) theory, we refer the readers to the book by Li and Yong [4].

For stochastic optimal control theory, we would like to mention the works by Kushner, Haussmann, Bismut, Bensoussan, Fleming, Wonham, and so on. For an extensive discussion and summary, see the forthcoming book by Yong and Zhou [18]. The purpose of this paper is to give a brief survey on the works done by the people from Fudan Group related to stochastic controls and forward-backward stochastic differential equations.

2 MAXIMUM PRINCIPLE

We consider the following controlled stochastic system:

$$dx(t) = b(x(t), u(t))dt + \sigma(x(t), u(t))dW(t), \qquad x(0) = x_0. \tag{2.1}$$

The cost functional is given by:

$$J(u(\cdot)) = E\Big\{ \int_0^T f(x(t), u(t))dt + h(x(T)) \Big\}. \tag{2.2}$$

*This work is supported in part by the NSFC, under grant 79790130, the National Distinguished Youth Science Foundation of China under grant 19725106, the Chinese State Education Commission Science Foundation under grant 97024607, and the Li Foundation.

Control $u(t) \in U$ almost surely, with U being some metric space. The optimal control problem is to minimize cost (2.2) subject to (2.1). We note that the diffusion coefficient σ depends on the control variable u and the control domain U is not necessarily convex. The main difficulty in deriving the necessary conditions for the optimal controls is that the Itô integral only guarantees the estimate: $\int_t^{t+\varepsilon} \sigma dW(s) \sim \sqrt{\varepsilon}$, which is not enough if one use the same argument for deterministic problem to treat the stochastic problem (2.1)–(2.2). Peng proved the following result in 1990 [12].

Theorem 2.1. *Let $(\bar{x}, \bar{u})$ be optimal. Then there exists an adapted solutions (p, q), (P, Q) of:*

$$dp = -H_x(\bar{x}, \bar{u}, p, q)dt + qdW(t), \qquad p(T) = -h_x(\bar{x}(T)), \tag{2.3}$$

$$\begin{aligned} dP = -\big[B^T P + PB + \Sigma^T P\Sigma + \Sigma^T Q + Q^T\Sigma \\ +H_{xx}(\bar{x}, \bar{u}, p, q)\big]dt + QdW(t), \qquad P(T) = -h_{xx}(\bar{x}(T)), \end{aligned} \tag{2.4}$$

such that

$$\Delta H(t) - \frac{1}{2}\Delta\sigma(t)^T P(t)\Delta\sigma(t) \geq 0, \quad \text{a.e. , a.s. ,}$$

where

$$H(x, u, p, q) = -f(x, u) + p^T b(x, u) + q^T \sigma(x, u), \tag{2.5}$$

$$B = b_x(\bar{x}(t), \bar{u}(t)), \tag{2.6}$$

$$\Delta H(t) = H(\bar{x}, \bar{u}(t), p, q) - H(\bar{x}, u, p, q), \tag{2.7}$$

$$\Delta\sigma(t) = \sigma(\bar{x}(t), \bar{u}(t)) - \sigma(\bar{x}(t), u). \tag{2.8}$$

We call (2.3) and (2.4) the first and second adjoint equations, respectively, and (2.5) the maximum condition.

3 RELATIONS BETWEEN MP AND DP

We would like to look at the relation between maximum principle (MP, for short) and dynamic programming (DP, for short). Consider the stochastic optimal control problem on $[s, T]$ with $s \in [0, T)$. We introduce the value function:

$$V(s, y) = \inf_{u(\cdot)} J(s, y; u(\cdot)).$$

The corresponding HJB equation takes the following form:

$$-V_t + \sup_{u \in U} G(x, u, -V_x, -V_{xx}) = 0, \qquad V\big|_{t=T} = h, \tag{3.1}$$

where

$$G(x, u, p, P) = \frac{1}{2}\text{tr}\left[\sigma\sigma^T(x, u)P\right] + p^T b(x, u) - f(x, u). \tag{3.2}$$

Classical result asserts that if $V(t,x)$ is smooth enough and $(\bar{x},\bar{u})$ is optimal, then

$$\begin{aligned} V_t(t,\bar{x}(t)) &= G(\bar{x},\bar{u},-V(t,\bar{x}),-V_{xx}(t,\bar{x}), \\ &= \max_{u\in U} G(\bar{x},u,-V_x(t,\bar{x}),-V_{xx}(t,\bar{x}), \\ V_x(t,\bar{x}(t)) &= -p(t), \\ V_{xx}(t,\bar{x}(t))\sigma(t,\bar{x}(t)) &= -q(t). \end{aligned} \tag{3.3}$$

Since the value function is not necessarily smooth, the above result is jsut formal. Zhou proved the following in 1991 [19]

Theorem 3.1. *Let $(\bar{x},\bar{u})$ be optimal, $V(t,x)$ be the value function, (p,q) and (P,Q) be adapted solutions of adjoint equations. Then*

$$\begin{aligned} D^{1,2,-}_{t+,x}V(t,\bar{x}(t)) &\subseteq (-\infty,\mathcal{H}(t)]\times\{-p(t)\}\times(-\infty,-P(t)], \\ D^{1,2,+}_{t+,x}V(t,\bar{x}(t)) &\supseteq [\mathcal{H}(t),\infty)\times\{-p(t)\}\times[-P(t),\infty), \end{aligned} \tag{3.4}$$

where

$$\begin{aligned} \mathcal{H}(t) = \frac{1}{2}\sigma(\bar{x},\bar{u})^T P\sigma(\bar{x},\bar{u}) + p^T b(\bar{x},\bar{u}) - f(\bar{x},\bar{u}) \\ +\sigma(\bar{x},\bar{u})^T\left[q - P\sigma(\bar{x},\bar{u})\right], \end{aligned} \tag{3.5}$$

and $D^{1,2,\pm}_{t+,x}V$ are the generalized second order right differentials.

4 VERIFICATION THEOREM

Verification theorem gives sufficient conditions for an admissible pair to be optimal. The classical result is as follows: Let v be a solution of HJB equation (3.1). Then

$$v(s,y) \le V(s,y), \tag{4.1}$$

and $(\bar{x},\bar{u})$ is optimal if and only if

$$v_t(t,\bar{x}) = G(\bar{x},\bar{u},-v_x(t,\bar{x}),-v_{xx}(t,\bar{x})).$$

Again, since the value function might be non-smooth, the above result is merely formal. Zhou, Yong and Li proved the following in 1997 [20].

Theorem 4.1. *Let v be a viscosity solution of HJB equation (3.1). Then (4.1) holds and $(\bar{x},\bar{u})$ is optimal if there exists a triple $(\bar{q},\bar{p},\bar{P})\in D^{1,2,+}_{t+,x}v(t,\bar{x}(t))$ $(t\in[0,T])$ such that*

$$E\int_0^T \bar{q}(t)dt \le E\int_0^T G(\bar{x},\bar{u},-\bar{p},-\bar{P})dt. \tag{4.2}$$

We point out that the above condition is only sufficient.

5 FBSDES

Let $(\Omega, \mathcal{F}, P, \{\mathcal{F}_t\}_{t\geq 0})$ be a probability space and $W(\cdot)$ be a d-dimensional standard Brownian motion with $\mathcal{F}_t = \sigma(W(s), s \leq t)$. Consider

$$\begin{aligned} dX(t) &= b(t, X(t), Y(t), Z(t))dt + \sigma(t, X(t), Y(t), Z(t))dW(t), \\ dY(t) &= h(t, X(t), Y(t))dt + \hat{\sigma}(t, X(t), Y(t), Z(t))dW(t), \\ X(0) &= x, \quad Y(T) = G(X(T)). \end{aligned} \tag{5.1}$$

Definition 5.1. An adapted solution of (5.1) is a triple of $\{\mathcal{F}_t\}_{t\geq 0}$-adapted processes (X, Y, Z) satisfying the above in the usual sense of Itô integral.

The key issues here are the following: The Itô integral is *forward*, the second equation is to be solved *backwardly*, and the solution is required to be *forwardly* adapted. Two major motivations of studying the above equations are: stochastic maximum principle, and pricing of European type contingent claims.

Note that FBSDEs are not always solvable. Here is an simple example.

$$\begin{aligned} dX(t) &= Y(t)dt + \sigma(X(t), Y(t), Z(t))dW(t), \\ dY(t) &= -X(t) + Z(t)dW(t), \\ X(0) &= x, \qquad Y(T) = X(T), \end{aligned}$$

where σ is any suitable function. For $T = \frac{\pi}{4} + 2k\pi$, $k \geq 0$ and $x \neq 0$, the above does not have adapted solutions. In fact if $(X(\cdot), Y(\cdot), Z(\cdot))$ were an adapted solution, then $x(t) = EX(t)$, $y(t) = EY(t)$ would satisfy

$$\begin{aligned} \dot{x}(t) &= y(t), \qquad \dot{y}(t) = -x(t), \\ x(0) &= x, \qquad y(T) = x(T), \end{aligned}$$

which is impossible.

We now introduce the *Method of Optimal Control* to approach the solvability of (5.1). To this end, we consider

$$\begin{aligned} dX &= b(X, Y, Z)dt + \sigma(X, Y, Z)dW(t), \\ dY(t) &= h(X, Y, Z)dt + ZdW(t), \\ X(0) &= x, \qquad Y(0) = y, \end{aligned} \tag{5.2}$$

We regard (X, Y) as the state and $Z \in \mathcal{Z}[0,T] \triangleq L^2_{\mathcal{F}}(0, T; \mathbb{R}^{m\times d})$ as the control. Introduce cost functional:

$$J(x, y; Z(\cdot)) \triangleq Ef(X(T), Y(T)), \tag{5.3}$$

with f being Lipschitz and

$$\begin{aligned} f(x,y) &\geq 0, \qquad \forall (x,y) \in \mathbb{R}^n \times \mathbb{R}^m, \\ f(x,y) &= 0, \quad \text{if and only if} \quad y = g(x). \end{aligned} \tag{5.4}$$

Problem (OC). For $(x,y) \in \mathbb{R}^n \times \mathbb{R}^m$, find a $\bar{Z}(\cdot) \in \mathcal{Z}[0,T]$, such that

$$V^0(x,y) \triangleq \inf_{Z(\cdot)\in\mathcal{Z}[0,T]} J(x,y;Z(\cdot)) = J(x,y;\overline{Z}(\cdot)). \tag{5.5}$$

Proposition 5.2. *For $x \in \mathbb{R}^n$, (5.1) admits an adapted solution ifand only if $\exists y \in \mathbb{R}^m$, s.t. $V^0(x,y) = 0$ and there exists an (optimal) control $Z(\cdot) \in \mathcal{Z}[0,T]$, such that*

$$J(x,y;Z(\cdot)) = V^0(x,y)(=0).$$

Thus, we need to look at the following two sub-problems:
(i) Find (x,y) such that $V^0(x,y) = 0$;
(ii) Existence of an optimal control for (x,y) such that $V^0(x,y) = 0$.

Definition 5.3. For $x \in \mathbb{R}^n$, (5.1) is said to be *approximately solvable* if $\forall \varepsilon > 0$, $\exists$ $(X_\varepsilon, Y_\varepsilon, Z_\varepsilon)$, such that

$$\begin{aligned} dX_\varepsilon &= b(X_\varepsilon,Y_\varepsilon,Z_\varepsilon)dt + \sigma(X_\varepsilon,Y_\varepsilon,Z_\varepsilon)dW(t), \\ dY_\varepsilon &= h(X_\varepsilon,Y_\varepsilon,Z_\varepsilon)dt + Z_\varepsilon dW(t), \\ X_\varepsilon(0) &= x, \end{aligned} \tag{5.6}$$

(5.11) $$E\big|Y_\varepsilon(T) - g(X_\varepsilon(T))\big| < \varepsilon.$$

We call $(X_\varepsilon, Y_\varepsilon, Z_\varepsilon)$ an *approximate adapted solution* of (5.1).

Theorem 5.4. *Under some mild conditions, (5.1) is approximately solvable if and only if there exists a $y \in \mathbb{R}^m$, such that $V^0(x,y) = 0$.*

Let us consider a special case.

$$\begin{aligned} dX &= b(X,Y)dt + \sigma(X,Y)dW(t), \\ dY &= h(X,Y)dt + ZdW(t), \\ X(0) &= x, \qquad Y(T) = g(X(T)). \end{aligned} \tag{5.7}$$

All the coefficients could also be time dependent. The point is that functions b, σ and h are independent of Z.

Theorem 5.5. *Let g be bounded in $C^{2+\alpha}(\mathbb{R}^n)$, b, σ, h be C^2 with bounded first and second derivatives and*

$$|b(x,0)| + |\sigma(x,0)| + |h(x,0)| \le L, \quad \forall x \in \mathbb{R}^n. \tag{5.8}$$

Then (5.7) is approximately solvable.

The difficulties that need to be overcome are: (i) σ might be degenerate, and (ii) The control domain ($\mathbb{R}^{m\times d}$) is not compact.

6 EUROPEAN TYPE CONTINGENT CLAIMS

We now look at problem of European type contingent claims, which is very closely related to the BSDEs and/or FBSDEs. Consider the following:

$$dX(t) = b(t, X(t))dt + \sigma(t, X(t))dW(t), \quad X(0) = x_0, \tag{6.1}$$

$$\begin{aligned} dY(t) &= \big[r(t, X(t))Y(t) + \langle\, h(t, X(t)), Z(t)\,\rangle\big]dt \\ &\quad + \langle\, Z(t), \sigma(t, X(t))dW(t)\,\rangle, \qquad Y(0) = y_0. \end{aligned} \tag{6.2}$$

In the above, $X(\cdot)$ is the log-price process, and $Y(\cdot)$ is the wealth process. We do not assume the non-degeneracy of σ. Thus, the market could be *incomplete*. We introduce the following assumption:

(H) There exist $L > 0$, $\lambda \geq 0$, such that

$$\begin{aligned} &|b(t,x) - b(t,\widehat{x})| + |\sigma(t,x) - \sigma(t,\widehat{x})| \leq L|x - \widehat{x}|, \\ &|b(t,x)| + |\sigma(t,x)| + |r(t,x)| + |h(t,x)| + e^{-\lambda|x|}|g(x)| \leq L. \end{aligned} \tag{6.3}$$

For European call option, $g(x) = (e^x - q)^+$ for some $q > 0$. Thus, one can take $\lambda = 1$. Also, we note that assuming $g(x)$ to be bounded will exclude many interesting cases.

Now, we introduce the set of all portfolios:

$$\begin{aligned} \mathcal{Z}[0,T] \stackrel{\Delta}{=} \{ Z : [0,T] \times \Omega \to \mathbb{R}^n \mid\ & Z(\cdot) \text{ is } \{\mathcal{F}_t\}_{t\geq 0}\text{-adapted} \\ & |\langle\, h(\cdot, X(\cdot)), Z(\cdot)\,\rangle|, |\sigma(\cdot, X(\cdot))^T Z(\cdot)| \in L^2_{\mathcal{F}}(0,T;\mathbb{R})\,\}. \end{aligned}$$

We will identify $\xi \in L^2_{\mathcal{F}_T}(\Omega;\mathbb{R})$ with the European contingent claim whose payoff at $t = T$ is ξ.

Definition 6.1. Let $X(\cdot)$ be the strong solution of (6.1). We say that ξ is *hedgeable*, if $Y(T) \geq \xi$, a.s. , and ξ is *replicatable* if $Y(T) = \xi$, a.s. , where $Y(\cdot)$ is the solution of (6.2) for some $(y_0, Z(\cdot)) \in \mathbb{R} \times \mathcal{Z}[0,T]$.

Clearly, replicatability implies hedgeability. We can prove the following facts

(i) Any bounded ξ is always hedgeable, regardless of the completeness of the market;

(ii) When market is complete, any $\xi \in L^2_{\mathcal{F}_T}(\Omega;\mathbb{R})$ is replicatable;

When market is incomplete,

(iii) Some bounded ξ could be not replicatable;

(iv) Some unbounded ξ could be not hedgable.

Based on the above, we now concentrate on contingent claim of form $g(X(T))$ with $g : \mathbb{R}^n \to \mathbb{R}$. Then replicatability of $g(X(T))$ is equivalent to the solvability of the following BSDE:

$$\begin{aligned} dY(t) &= \big[r(t, X(t))Y(t) + \langle\, h(t, X(t)), Z(t)\,\rangle\big]dt \\ &\quad + \langle\, Z(t), \sigma(t, X(t))dW(t)\,\rangle, \qquad Y(T) = g(X(T)). \end{aligned} \tag{6.4}$$

It is a linear BSDE with random coefficients with the following main features:

(i) $\sigma(t, X(t))$ could be degenerate;

(ii) $Y(T)$ depends only on $X(T)$, with $X(\cdot)$ being the solution of an FSDE whose diffusion is **compatible** with that of (6.7).

The following method is an extension of the so-called Four Step Scheme due to Ma-Protter-Yong [5] in studying the solvability of FBSDEs.

Suppose (Y, Z) is an adapted solution of (6.7). Let

$$Y(t) = u(t, X(t)), \qquad t \in [0, T], \text{ a.s. }, \tag{6.5}$$

where $u(\cdot\,,\cdot)$ is undetermined. Applying Itô's formula,

$$\begin{aligned} dY(t) &= d[u(t, X(t))] \\ &= \Big\{u_t(t, X(t)) + \frac{1}{2}\text{tr}\left[\sigma(t, X(t))\sigma(t, X(t))^T u_{xx}(t, X(t))\right] \\ &+ \langle\, b(t, X(t)), u_x(t, X(t))\,\rangle\Big\}dt + \langle\, \sigma(t, X(t))^T u_x(t, X(t)), dW(t)\,\rangle \end{aligned} \tag{6.6}$$

Comparing (6.7) with (6.9), we should have

$$\sigma(t, X)^T u_x(t, X) = \sigma(t, X)^T Z, \tag{6.7}$$

$$\begin{aligned} u_t(t, X) + \frac{1}{2}\text{tr}\left[\sigma(t, X)\sigma(t, X)^T u_{xx}(t, X)\right] + \langle\, b(t, X), u_x(t, X)\,\rangle \\ = r(t, X)u(t, X) + \langle\, h(t, X), Z\,\rangle, \end{aligned} \tag{6.8}$$

$$u(T, X(T)) = g(X(T)). \tag{6.9}$$

Thus, it suffices to solve:

$$u_t + \frac{1}{2}\text{tr}\left[\sigma\sigma^T u_{xx}\right] + \langle\, b - h, u_x\,\rangle - ru = 0, \qquad u(T, x) = g(x). \tag{6.10}$$

If this problem admits a classical solution $u(t, x)$, then

$$Y(t) = u(t, X(t)), \qquad Z(t) = u_x(t, X(t)), \tag{6.11}$$

gives a price and a portfolio replicating the contingent claim $g(X(T))$.

Under some smooth conditions, (6.10) admits a classical solution.

Open Questions:

(1) Large Investor

(2) Random Coefficients

References

[1] R. Bellman, *Dynamic Programming,* Princeton Univ. Press, N.J., 1957.

[2] N. El Karoui, S. Peng, and M. C. Quenez, *Backward stochastic differential equations in finance, Math. Finance,* 7 (1997), 1–71.

[3] R. E. Kalman, *Contributions to the theory of optimal control, Bol. Soc. Math. Mexicana,* 5 (1960), 102–119.

[4] X. Li and J. Yong, *Optimal Control Theory for Infinite Dimensional Systems*, Birkhäuser, Boston, 1995.

[5] J. Ma, P. Protter, and J. Yong, *Solving forward-backward stochastic differential equations explicitly — a four step scheme, Prob. Th. & Rel. Fields*, 98 (1994), 339–359.

[6] J. Ma and J. Yong, *Solvability of forward-backward SDEs and the nodal set of HJB equations, Chin. Ann. Math. Ser. B*, 16 (1995), 279–298.

[7] J. Ma and J. Yong, *Adapted solution of a degenerate backward SPDE, with applications, Stoch. Proc. & Appl.* 70 (1997), 59–84.

[8] J. Ma and J. Yong, *Approximate solvability of forward-backward differential equations,* preprint.

[9] J. Ma and J. Yong, *Forward-Backward Stochastic Differential Equations and Their Applications,* in preparation.

[10] E. Pardoux and S. Peng, *Adapted solution of backward stochastic equation, Syst. & Control Lett.*, 14 (1990), 55–61.

[11] E. Pardoux and S. Tang, *The study of forward-backward stochastic differential equation and its application in quasilinear PDEs,* preprint.

[12] S. Peng, *A general stochastic maximum principle for optimal control problems, SIAM J. Control & Optim.*, 28 (1990), 966-979.

[13] S. Peng and Z. Wu, *Fully coupled forward-backward stochastic differential equations with arbitrarily large time duration,* preprint.

[14] L. S. Pontryagin, V. G. Boltyanskii, R. V. Gamkrelidze and E. F. Mischenko, *Mathematical Theory of Optimal Processes,* Wiley, New York, 1962.

[15] J. Yong, *Finding adapted solutions of forward-backward stochastic differential equations — method of continuation, Prob. Th. & Rel Fields,* 107 (1997), 537–572.

[16] J. Yong, *Linear forward-backward stochastic differential equations, Appl. Math. Optim.*, to appear.

[17] J. Yong, *European type contingent claims in an incomplete market with constrained wealth and portfolio,* preprint.

[18] J. Yong and X. Y. Zhou, *Stochastic Control: Hamiltonian Systems and HJB Equations,* Springer-Verlag, New York, to appear.

[19] X. Y. Zhou, *A unified treatment of maximum principle and dynamic programming in stochastic controls, Stoch. & Stoch. Rep.*, 36 (1991), 137–161.

[20] X. Y. Zhou, J. Yong, and X. Li, *Stochastic verification theorems within the framework of viscosity solutions, SIAM J. Control & Optim.*, 35 (1997), 243–253.

ASYMPTOTICALLY OPTIMAL CONTROLS OF HYBRID LQG PROBLEMS: SUMMARY OF RESULTS

Q. Zhang*

Department of Mathematics
University of Georgia
Athens, GA 30602
qingz@math.uga.edu

G. Yin†

Department of Mathematics
Wayne State University
Detroit, MI 48202
gyin@math.wayne.edu

Abstract: This paper is concerned with hybrid control of a class of linear quadratic Gaussian systems modulated by a finite-state Markov chain. It develops approximation schemes for systems involving singularly perturbed Markov chains with weak and strong interactions. Computation results indicate that our approximation schemes perform quite well.

1 INTRODUCTION

This work develops asymptotically optimal controls of a class of hybrid LQG (linear quadratic Gaussian) systems, in which the systems are modulated by a Markov chain (or Markov jump process). The main feature of the hybrid systems is imbedded in the Markovian jump process. Many such systems arise in various applications in speech recognition, telecommunications, and manufacturing, for treating various system noise and uncertainties. Roughly, a hybrid system shows both "continuous" and "discrete" characteristics. In our formula-

*The research of this author was supported in part by the Office of Naval Research Grant N00014-96-1-0263.

†The research of this author was supported in part by the National Science Foundation under grant DMS-9529738.

tion, random environment continuously (in time) disturbs the system (through the system coefficient matrices), whereas the resulting effect takes place at a set of discrete values (discrete events). It is thus natural to introduce a Markov chain model. Recent research effort on LQG problems has been devoted to the study of hybrid linear quadratic Gaussian (LQG) systems for over a decade; see [1], [2] and the references cited therein for a literature review.

To account for uncertainty, unlike the traditional LQG problem in which the system matrices A and B are fixed, we allow these matrices to depend on a Markov jump process with finite-state space in this work. Thus corresponding to different states of the Markov chain, the system itself displays different configuration. The situation is more involved as compared to the traditional setting. It is typical in many applications that the state space of the Markov chain is very large. In such a case, it is difficult to obtain solutions to the associated Riccati equations. To overcome the difficulty, we use singular perturbation techniques in the modeling, control design, and optimization. The resulting systems naturally display certain two-time-scale behavior, a fast time scale and a slowly varying one. To put this in a manageable framework, we introduce a small parameter $\varepsilon > 0$, and model the underlying system as one involving singularly perturbed Markov chains.

This paper concentrates on asymptotic and near optimality of hybrid systems involving singularly perturbed Markov chains. To treat the systems, we use an averaging approach to analyze the system in which the underlying Markov chain involves weak and strong interactions. The idea is to aggregate the states according to their jump rates and replace the actual system with its average. Using the optimal control of limit (average) problem as a guide, we then construct controls for the actual systems leading to feasible approximation schemes. We show that these approximation schemes give us nearly optimal controls. By focusing on approximate optimality, we succeed to reduce the complexity of the underlying systems drastically. The reduction of dimensionality is the major advantage of the averaging approach. To demonstrate how the average schemes work, we provide a numerical example of a one-dimensional system. This paper only deals with Markov chains with recurrent states. Similar results can be obtained for models with transient states and absorbing states. Proofs of results and technical complements are omitted and referred to [1].

2 PROBLEM FORMULATION

Let $\varepsilon > 0$ be a small parameter. Consider the Markov $\alpha^\varepsilon(t)$ generated by Q^ε, which consists of two parts, a rapidly changing part and a slowly varying one, i.e.,

$$Q^\varepsilon = \frac{1}{\varepsilon}\widetilde{Q} + \widehat{Q}.$$

Consider the linear system

$$\begin{aligned} &dx(t) = [A(\alpha^\varepsilon(t))x(t) + B(\alpha^\varepsilon(t))u(t)]dt + \sigma dw(t), \\ &x(s) = x, \text{ for } s \le t \le T, \end{aligned} \tag{1}$$

where $x(t) \in \mathbb{R}^{n_1}$ is the state, $u(t) \in \mathbb{R}^{n_2}$ is the control, $A(i) \in \mathbb{R}^{n_1 \times n_1}$ and $B(i) \in \mathbb{R}^{n_1 \times n_2}$ are well defined and have finite values for $i \in \mathcal{M}$, and $w(\cdot)$ is a standard Brownian motion. The objective is to find the optimal control $u(\cdot)$ so that the expected quadratic cost function

$$J^\varepsilon(s,x,\alpha,u(\cdot)) = E\left\{\int_s^T [x'(t)M(\alpha^\varepsilon(t))x(t)+u'(t)N(\alpha^\varepsilon(t))u(t)]dt + x'(T)Dx(T)\right\}$$

is minimized, where $M(i)$, $N(i)$ are well defined and have finite values for $i \in \mathcal{M}$, $M(i)$ are symmetric nonnegative definite matrices for $i \in \mathcal{M}$, D and $N(i)$ are symmetric and positive definite for $i \in \mathcal{M}$, E is the expectation given $\alpha(s) = \alpha$ and $x(s) = x$, and the processes $\alpha^\varepsilon(\cdot)$ and $w(\cdot)$ are independent.

3 HJB AND RICCATI EQUATIONS

We consider the optimal LQG control problem by using the dynamic programming approach, and derive the associated Hamilton-Jacobi-Bellman (HJB) and Riccati equations. Let $v^\varepsilon(s,i,x)$ be the value function, i.e., $v^\varepsilon(s,i,x) = \inf_{u(\cdot)} J^\varepsilon(s,i,x,u(\cdot))$. Then v^ε satisfies the following HJB equation:

$$\begin{cases} 0 = \dfrac{\partial v^\varepsilon(s,i,x)}{\partial s} + \min\limits_u \left\{ (A(i)x + B(i)u)' \dfrac{\partial v^\varepsilon(s,i,x)}{\partial x} + x'M(i)x + u'N(i)u \right. \\ \qquad \left. + \dfrac{1}{2}\mathrm{tr}\left(\sigma\sigma' \dfrac{\partial^2 v^\varepsilon(s,i,x)}{\partial x^2}\right) + Q^\varepsilon v^\varepsilon(s,\cdot,x)(i) \right\}, \quad 0 \le s \le T \\ v^\varepsilon(T,i,x) = x'Dx, \; i \in \mathcal{M}, \end{cases}$$

where

$$Q^\varepsilon v^\varepsilon(s,\cdot,x)(i) = \sum_{j \neq i} q_{ij}^\varepsilon (v^\varepsilon(s,j,x) - v^\varepsilon(s,i,x)).$$

Let

$$v^\varepsilon(s,i,x) = x'K^\varepsilon(s,i)x + q^\varepsilon(s,i), \tag{2}$$

for some $m \times m$ matrix K^ε and a scalar function q^ε. Substituting (2) into the Riccati equation and comparing the coefficients of x leads to the following Riccati equation for $K^\varepsilon(s,i)$,

$$\begin{cases} \dot{K}^\varepsilon(s,i) = -K^\varepsilon(s,i)A(i) - A'(i)K^\varepsilon(s,i) - M(i) \\ \qquad\qquad + K^\varepsilon(s,i)B(i)N^{-1}(i)B'(i)K^\varepsilon(s,i) - Q^\varepsilon K^\varepsilon(s,\cdot)(i) \\ K^\varepsilon(T,i) = D, \end{cases} \tag{3}$$

where $Q^\varepsilon K^\varepsilon(s,\cdot)(i) = \sum_{j\neq i} q_{ij}^\varepsilon (K^\varepsilon(s,j) - K^\varepsilon(s,i))$, and the equation for q^ε,

$$\begin{cases} \dot{q}^\varepsilon(s,i) = -\mathrm{tr}(\sigma\sigma' K^\varepsilon(s,i)) - Q^\varepsilon q^\varepsilon(s,\cdot)(i), \\ q^\varepsilon(T,i) = 0. \end{cases} \tag{4}$$

The optimal control $u^{\varepsilon,*}(s,i,x)$ has the form:

$$u^{\varepsilon,*}(s,i,x) = -N^{-1}(i)B'(i)K^\varepsilon(s,i)x. \tag{5}$$

To find the optimal control, one has to solve the Riccati equations. However, in many problems in telecommunication and manufacturing, such solutions are very difficult to obtain due to the large dimensionality. In this case, one has to resort to approximation schemes.

4 APPROXIMATION SCHEMES

Consider the generator of the Markov chain given by

$$Q^{\varepsilon} = \frac{1}{\varepsilon}\widetilde{Q} + \widehat{Q} = \frac{1}{\varepsilon}\begin{pmatrix} \widetilde{Q}^1 & & & \\ & \widetilde{Q}^2 & & \\ & & \ddots & \\ & & & \widetilde{Q}^l \end{pmatrix} + \widehat{Q}, \tag{6}$$

where $\widetilde{Q}^k \in \mathbb{R}^{m_k \times m_k}$ are weakly irreducible, for $k = 1, 2, \ldots, l$, and $\sum_{k=1}^{l} m_k = m$. Let $\mathcal{M}_k = \{s_{k1}, \ldots, s_{km_k}\}$, for $k = 1, \ldots, l$, denote the states corresponding to $\widetilde{Q}^k$ and let $\mathcal{M}$ denote the state space of the underlying chains. Then

$$\mathcal{M} = \mathcal{M}_1 \cup \cdots \cup \mathcal{M}_l = \Big\{ s_{11}, \ldots, s_{1m_1}, \ldots, s_{l1}, \ldots, s_{lm_l} \Big\}.$$

Note that $\widetilde{Q}$ governs the rapidly changing part and $\widehat{Q}$ describes the slowly varying components. The slow and fast components are coupled through weak and strong interactions in the sense that the underlying Markov chain fluctuates rapidly within a single group $\mathcal{M}_j$ and jumps less frequently among groups $\mathcal{M}_k$ and $\mathcal{M}_j$ for $k \neq j$. The states in $\mathcal{M}_k$, $k = 1, \ldots, l$, are not isolated or independent of each other. More precisely, if we consider the states in $\mathcal{M}_k$ as a single "state," then these "states" are coupled through the matrix $\widehat{Q}$, and transitions from $\mathcal{M}_k$ to $\mathcal{M}_j$, $k \neq j$ are possible. By aggregating the states s_{kj} in $\mathcal{M}_k$ as one state k, we obtain an aggregated process $\{\overline{\alpha}^{\varepsilon}(\cdot)\}$ defined by $\overline{\alpha}^{\varepsilon}(t) = k$ when $\alpha^{\varepsilon}(t) \in \mathcal{M}_k$. The process $\overline{\alpha}^{\varepsilon}(\cdot)$ is not necessarily Markovian. However, using certain probabilistic argument, we have shown in [2] that $\overline{\alpha}^{\varepsilon}(\cdot)$ converges weakly to a Markov chain $\overline{\alpha}(\cdot)$ generated by

$$\overline{Q} = \mathrm{diag}(\nu^1, \ldots, \nu^l)\widehat{Q}\mathrm{diag}(\mathbb{1}_{m_1}, \ldots, \mathbb{1}_{m_l}),$$

where $\mathbb{1}_n = (1, \ldots, 1)' \in \mathbb{R}^n$. Based on such a result, we obtain the following theorem.

Theorem 4.1. *For $k = 1, 2, \ldots, l$ and $j = 1, 2, \ldots, m_k$,*

$$K^{\varepsilon}(s, s_{kj}) \to \overline{K}(s, k) \text{ and } q^{\varepsilon}(s, s_{kj}) \to \overline{q}(s, k),$$

uniformly on $[0,T]$ *as* $\varepsilon \to 0$, *where* $\overline{K}(s,k)$ *and* $\overline{q}(s,k)$ *are the unique solutions to the following differential equations: For* $k = 1, 2, \ldots, l$,

$$\begin{cases} \dot{\overline{K}}(s,k) = -\overline{K}(s,k)\left(\sum_{j=1}^{m_k} \nu_j^k A(s_{kj})\right) - \left(\sum_{j=1}^{m_k} \nu_j^k A'(s_{kj})\right)\overline{K}(s,k) \\ \quad -\sum_{j=1}^{m_k} \nu_j^k M^{-1}(s_{kj}) + \overline{K}(s,k)\left(\sum_{j=1}^{m_k} \nu_j^k B(s_{kj}) N^{-1}(s_{kj}) B'(s_{kj})\right)\overline{K}(s,k) \\ \qquad\qquad -\overline{Q}\,\overline{K}(s,\cdot)(k), \\ \overline{K}(T,k) = D, \end{cases}$$

and

$$\begin{cases} \dot{\overline{q}}(s,k) = -\mathrm{tr}\left(\sigma\sigma'\overline{K}(s,k)\right) - \overline{Q}\overline{q}(s,\cdot)(k), \\ \overline{q}(T,k) = 0, \end{cases}$$

respectively.

The convergence of $K^\varepsilon(s,i)$ and $q^\varepsilon(s,i)$ leads to that of

$$v^\varepsilon(s,i,x) = x'K^\varepsilon(s,i)x + q^\varepsilon(s,i),$$

where $K^\varepsilon(s,i)$ and $q^\varepsilon(s,i)$ denote the solutions to the differential equations (3) and (4), respectively. In fact,

$$v^\varepsilon(s,s_{kj},x) \to v(s,k,x), \text{ for } j = 1,\ldots,m_k, \text{ as } \varepsilon \to 0,$$

where

$$v(s,k,x) = x'\overline{K}(s,k)x + \overline{q}(s,k),$$

corresponds to the value function of a limit problem. Let $\mathcal{U}$ denote the control set for the limit problem:

$$\mathcal{U} = \left\{U = (U^1,\ldots,U^l) : U^k = (u^{k1},\ldots,u^{km_k}), u^{kj} \in \mathbb{R}^{n_2}\right\}.$$

Define

$$f(s,k,x,U) = \sum_{j=1}^{m_k} \nu_j^k \left(A(s_{kj})x + B(s_{kj})u^{kj}\right)$$

$$\overline{M}(k) = \sum_{j=1}^{m_k} \nu_j^k M(s_{kj}), \text{ and } \widetilde{N}(k,U) = \sum_{j=1}^{m_k} \nu_j^k \left(u^{kj,\prime} N(s_{kj}) u^{kj}\right).$$

Then it can be shown that $v(s,k,x)$ satisfies the following HJB equations:

$$\begin{cases} 0 = \dfrac{\partial v(s,k,x)}{\partial s} + \min_{U\in\mathcal{U}}\left\{ f(s,k,x,U)\dfrac{\partial v(s,k,x)}{\partial x} + x'\overline{M}(k)x + \widetilde{N}(k,U) \right. \\ \qquad \left. +\dfrac{1}{2}\mathrm{tr}\left(\sigma\sigma'\dfrac{\partial^2 v(s,k,x)}{\partial x^2}\right) + \overline{Q}v(s,\cdot,x)(k)\right\}, \\ v(T,k,x) = x'Dx. \end{cases} \tag{7}$$

The corresponding control problem is

$$\begin{cases} \text{Min } J(s,k,x,U(\cdot)) = E\Big\{\displaystyle\int_s^T \Big(x'(t)\overline{M}(\overline{\alpha}(t))x(t) + \widetilde{N}(\overline{\alpha}(t),U(t))\Big)dt \\ \qquad\qquad\qquad\qquad\qquad\qquad + x'(T)Dx(T)\Big\} \\ \text{s.t. } dx(t) = f(t,\overline{\alpha}(t),x(t),U(t))dt + \sigma dw(t),\ x(s)=x,\ s \le t \le T, \end{cases}$$

where $\overline{\alpha}(\cdot) \in \{1,2,\ldots,l\}$ is a Markov chain generated by $\overline{Q}$ with $\overline{\alpha}(s) = k$.

The optimal control for this limit problem is:

$$U^*(s,k,x) = (U^{1*}(s,x),\ldots,U^{l*}(s,x)), \quad \text{with}$$

$$U^{k*}(s,x) = (u^{k1*}(s,x),\ldots,u^{km_k*}(s,x)), \quad \text{and}$$
$$u^{kj*}(s,x) = -N^{-1}(s_{kj})B'(s_{kj})\overline{K}(s,k)x.$$

Using such controls, we construct

$$\overline{u}^*(s,\alpha,x) = \sum_{k=1}^{l}\sum_{j=1}^{m_k} I_{\{\alpha=s_{kj}\}}u^{kj*}(s,x) \tag{8}$$

for the original problem. Note that this control can also be written as

$$\overline{u}^*(s,\alpha,x) = -N^{-1}(\alpha)B'(\alpha)\overline{K}(s,k)x, \text{ if } \alpha \in \mathcal{M}_k.$$

Apparently, this control is identical to the optimal control in (5) except K^ε is replaced by $\overline{K}$.

Theorem 4.2. *The control $u^\varepsilon(t) = \overline{u}^*(t,\alpha^\varepsilon(t),x(t))$ is nearly optimal, i.e.,*

$$\lim_{\varepsilon\to 0} |J^\varepsilon(s,\alpha,x,u^\varepsilon(\cdot)) - v^\varepsilon(s,\alpha,x)| = 0.$$

Remark 4.3. The most attractive feature of such an approximation scheme is that it requires much less computation effort. For instance, if the dimension of the system is $n_1 = 3$ and the Markov chain has $m = 40$ states divided into 5 groups with each group consisting 8 states, then the optimal scheme requires compute the Riccati equations of dimension $(n_1(n_1+1)/2) \times m = 6 \times 40 = 240$. The dimension of the limit Riccati equation is only $(n_1(n_1+1)/2) \times 5 = 30$.

5 A NUMERICAL EXAMPLE

Let $\mathcal{M} = \{1,2\}$ and $Q^\varepsilon = \varepsilon^{-1}\begin{pmatrix} -0.5 & 0.5 \\ 0.5 & -0.5 \end{pmatrix}$. Consider the one-dimensional system with the following specifications: $A(1) = 0.5, A(2) = -0.1, B(1) = 1, B(2) = 2, \sigma = 1, M(1) = M(2) = N(1) = N(2) = D = 1$. To solve the problem numerically, we discretize the equations with step size h. The time horizons in the continuous-time model is $T = 5$ and in the corresponding discrete-time setting is $T_h = 10/h$ with $h = 0.0001$. The results below are based on computations using 100 sample paths.

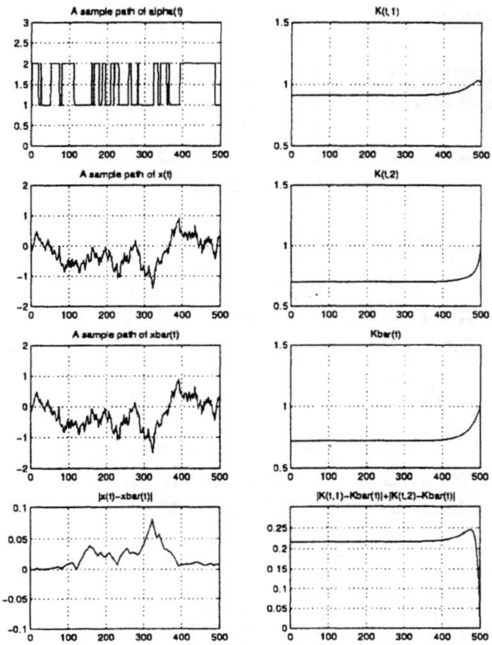

Fig. 1. Various sample paths with $\varepsilon = 0.1$

Take $s = 0$, $\alpha(0) = 1$, and $x(0) = x = 0$. Let $v^\varepsilon = v^\varepsilon(0,1,0)$, $J^\varepsilon = J^\varepsilon(0,1,0,u^\varepsilon(\cdot))$, and $v = v(0,0)$. For various ε we have the error bounds given in the following table.

ε	$\lvert K^\varepsilon - \overline{K}\rvert$	$\lvert x^\varepsilon - \overline{x}^\varepsilon\rvert$	$\lvert v^\varepsilon - v\rvert$	$\lvert J^\varepsilon - v^\varepsilon\rvert$
0.1	2.17ε	0.10ε	4.23ε	$2.21\sqrt{\varepsilon}$
0.01	2.47ε	0.19ε	5.01ε	$0.36\sqrt{\varepsilon}$
0.001	2.50ε	0.13ε	5.09ε	$0.46\sqrt{\varepsilon}$
0.0001	2.50ε	0.12ε	5.10ε	$4.17\sqrt{\varepsilon}$

Table 1. Error Bounds

Sample paths of various trajectories of $\alpha^\varepsilon(\cdot)$, $x^\varepsilon(\cdot)$, $\overline{x}^\varepsilon(\cdot)$, $K^\varepsilon(\cdot,1)$, $K^\varepsilon(\cdot,2)$, $\overline{K}(\cdot)$ are given in Fig. 1 for $\varepsilon = 0.1$ and in Fig. 2 for $\varepsilon = 0.01$. It can be seen from these graphs the smaller the ε the more rapidly $\alpha^\varepsilon(\cdot)$ jumps, which results better approximations. To summarize, the numerical simulations indicate that our algorithm gives a very good approximation to exact optimal solutions with only one half of the computation effort.

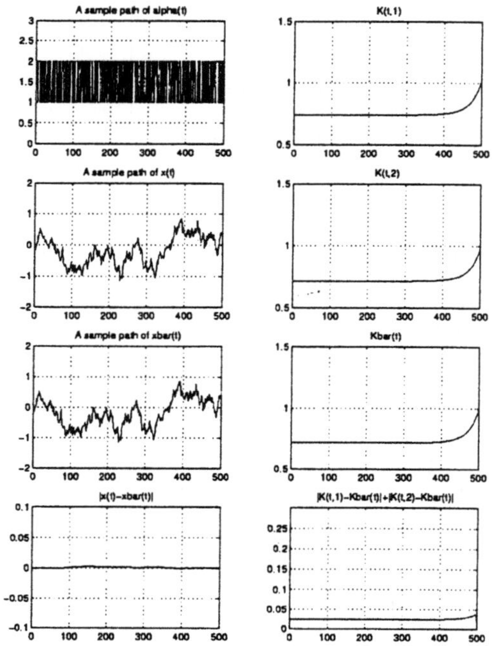

Fig. 2. Various sample paths with $\varepsilon = 0.01$

References

[1] G. Yin and Q. Zhang, *Continuous-Time Markov Chains and Applications: A Singular Perturbation Approach*, Springer-Verlag, 1998.

[2] Q. Zhang and G. Yin, On nearly optimal controls of hybrid LQG problems, preprint.

EXPLICIT EFFICIENT FRONTIER OF A CONTINUOUS-TIME MEAN-VARIANCE PORTFOLIO SELECTION PROBLEM

Xun Yu Zhou* and Duan Li†

Department of Systems Engineering and Engineering Management
The Chinese University of Hong Kong
Shatin, NT
Hong Kong

xyzhou@se.cuhk.edu.hk and dli@se.cuhk.edu.hk

Abstract: In this work we consider a continuous-time mean-variance portfolio selection problem that is formulated as a bi-criteria optimization problem. The objective is to maximize the expected return and minimize the variance of the terminal wealth. By putting weights on the two criteria one obtains a single objective stochastic control problem which is however not in the standard form. We show that this non-standard problem can be "embedded" into a class of auxiliary stochastic linear-quadratic (LQ) problems. By solving the latter based on the recent development on stochastic LQ problems with indefinite control weighting matrices, we derive the efficient frontier in a closed form for the original mean-variance problem.

1 INTRODUCTION

Portfolio selection is to seek a best allocation of wealth among a basket of securities. The mean-variance approach by Markowitz [9] provides a fundamental basis for portfolio construction in a single period. The most important contribution of this model is that it quantifies the risk by using the variance which enables investors to seek highest return after specifying their acceptable risk level. This approach becomes the foundation of modern finance theory and inspires literally hundreds of extensions and applications. In particular, in the case where the variance matrix is positive definite and short-selling is allowed, an analytic solution was obtained by Merton [10]. Perold [12] developed a more

*The research of this author was supported by RGC Earmarked Grant CUHK 4125/97E.
†The research of this author was supported by RGC Earmarked Grant CUHK 4130/97E.

general technique to locate the efficient frontier when the covariance matrix is nonnegative definite.

After Markowitz's pioneering work, the mean-variance model was soon extended to multi-period portfolio selection, see for example Mossin [11], Samuelson [13], Elton and Gruber [5], and Grauer and Hakansson [7]. The mean-variance model in a continuous-time setting has been developed a bit later, see Föllmer-Sondermann [6], Duffie and Jackson [3], and Duffie and Richardson [4]. The basic approach in these works is the dynamic programming.

The purpose of this paper is to seek optimal portfolio policy for dynamic investment problems with a mean-variance formulation in continuous-time. Different from the approach used in Duffie and Richardson [4], we employ the embedding technique, introduced by Li and Ng [8] for the multi-period, discrete-time portfolio selection problem, which leads to a stochastic linear-quadratic control problem (LQ problem, for short), and then apply the stochastic LQ theory recently developed by Chen, Li and Zhou [1] and Chen and Zhou [2] to solve the problem.

This paper is an announcement of the major results in Zhou and Li [15], where the detailed proofs are presented.

2 PROBLEM FORMULATION

Suppose there is a market in which $m+1$ assets (or securities) are traded continuously. One of the assets is the bond whose price process $P_0(t)$ is subject to the following (deterministic) ordinary differential equation:

$$\begin{cases} dP_0(t) = r(t)P_0(t)dt, & t \in [0,T], \\ P_0(0) = p_0 > 0, \end{cases} \tag{1}$$

where $r(t) > 0$ is called the *interest rate* (of the bond). The other m assets are called *stocks* whose price processes $P_1(t), \cdots, P_m(t)$ satisfy the following stochastic differential equation:

$$\begin{cases} dP_i(t) = P_i(t)\{b_i(t)dt + \sum_{j=1}^{m} \sigma_{ij}(t)dW_j(t)\}, & t \in [0,T], \\ P_i(0) = p_i > 0, \end{cases} \tag{2}$$

where $b_i(t) > 0$ is called the *appreciation rate*, and $\sigma_i(t) \equiv (\sigma_{i1}(t), \cdots, \sigma_{im}(t)) : [0,T] \times \Omega \to R^m$ is called the *volatility* or the *dispersion* of the stocks. Here, $W(t) \equiv (W^1(t), \cdots, W^m(t))'$ is a standard $\{\mathcal{F}_t\}_{t\geq 0}$-adapted m-dimensional Brownian motion defined on some fixed filtered complete probability space $(\Omega, \mathcal{F}, P, \{\mathcal{F}_t\}_{t\geq 0})$. Define the *covariance matrix*

$$\sigma(t) = \begin{pmatrix} \sigma_1(t) \\ \vdots \\ \sigma_m(t) \end{pmatrix} \equiv (\sigma_{ij}(t))_{m\times m}. \tag{3}$$

The basic assumption throughout this paper is

$$\sigma(t)\sigma(t)' \geq \delta I, \quad \forall t \in [0,T], \tag{4}$$

for some $\delta > 0$. This is the so-called *non-degeneracy* assumption. We also assume that all the functions are measurable and uniformly bounded in t.

Consider an investor whose total wealth at time $t \geq 0$ is denoted by $x(t)$. Suppose he/she decides to hold $N_i(t)$ shares of i-th asset ($i = 0, 1, \cdots, m$) at time t. Then

$$x(t) = \sum_{i=0}^{m} N_i(t) P_i(t), \quad t \geq 0. \tag{5}$$

Assume that the trading of shares takes place continuously and there is no transaction cost and consumptions during the whole time period $[0, T]$. Then one has

$$\begin{cases} dx(t) &= \sum_{i=0}^{m} N_i(t) dP_i(t) \\ &= \Big\{ r(t) N_0(t) P_0(t) + \sum_{i=1}^{m} b_i(t) N_i(t) P_i(t) \Big\} dt \\ &\quad + \sum_{i=1}^{m} N_i(t) P_i(t) \sum_{j=1}^{m} \sigma_{ij}(t) dW_j(t) \\ &= \Big\{ r(t) x(t) + \sum_{i=1}^{m} \big[b_i(t) - r(t) \big] u_i(t) \Big\} dt \\ &\quad + \sum_{j=1}^{m} \sum_{i=1}^{m} \sigma_{ij}(t) u_i(t) dW_j(t), \\ x(0) &= x_0 > 0, \end{cases} \tag{6}$$

where

$$u_i(t) \equiv N_i(t) P_i(t), \quad i = 0, 1, 2 \cdots, m, \tag{7}$$

denotes the total market value of the investor's wealth in the i-th bond/stock. If $u_i(t) < 0$ ($i = 1, 2, \cdots, m$), then the investor is *short-selling* i-th stock. If $u_0(t) < 0$, then the investor is borrowing the amount $|u_0(t)|$ at rate $r(t)$. It is clear that by changing $u_i(t)$, the investor changes the "allocation" of his wealth in these $m + 1$ assets. We call $u(t) = (u_1(t), \cdots, u_m(t))'$ a *portfolio* of the investor. Notice that we exclude the allocation to the bond, $u_0(t)$, from the portfolio as it will be completely determined by the allocations to the stocks. The objective of the investor is to maximize the mean terminal wealth, $Ex(T)$, and at the same time to minimize the variance of the terminal wealth

$$\text{Var } x(T) \equiv E\big[x(T) - Ex(T)\big]^2 = Ex(T)^2 - \big[Ex(T)\big]^2. \tag{8}$$

This is a *multi-objective optimization* problem with two criteria in conflict.

Let us denote by $L^2_{\mathcal{F}}(0, T; R^m)$ the set of all R^m-valued, measurable stochastic processes $f(t)$ adapted to $\{\mathcal{F}_t\}_{t \geq 0}$, such that $E \int_0^T |f(t)|^2 dt < \infty$.

Definition 1. A portfolio $u(\cdot)$ is said to be *admissible* if $u(\cdot) \in L^2_{\mathcal{F}}(0, T; R^m)$.

Definition 2. The mean-variance portfolio optimization problem is denoted as

$$\begin{array}{ll} \text{Minimize} & \Big(J_1(u(\cdot)), J_2(u(\cdot))\Big) \equiv \Big(- Ex(T), \text{Var } x(T) \Big), \\ \text{Subject to} & \begin{cases} u(\cdot) \in L^2_{\mathcal{F}}(0, T; R^m), \\ (x(\cdot), u(\cdot)) \text{ satisfy equation (6).} \end{cases} \end{array} \tag{9}$$

Moreover, an admissible portfolio $\bar{u}(\cdot)$ is called an *efficient portfolio* of the problem if there exists no admissible portfolio $u(\cdot)$ such that

$$J_1(u(\cdot)) \leq J_1(\bar{u}(\cdot)), \quad J_2(u(\cdot)) \leq J_2(\bar{u}(\cdot)), \tag{10}$$

and at least one of the inequalities holds strictly. In this case, $(J_1(\bar{u}(\cdot)), J_2(\bar{u}(\cdot))) \in R^2$ is called an *efficient point*. The set of all efficient points is called the *efficient frontier*.

In other words, an efficient portfolio is one that there exists no other portfolio better than it with respect to both the mean and variance criteria. The problem then is to identify the efficient portfolios along with the efficient frontier. By standard multi-objective optimization theory (see Zeleny [14]), an efficient portfolio can be found under certain convexity condition by solving a single-objective optimization problem where the objective is a weighted average of the two original criteria. The efficient frontier can then be generated by varying the weights. Therefore, the original problem can be solved via the following optimal control problem

$$\begin{array}{ll} \text{Minimize} & J_1(u(\cdot)) + \mu J_2(u(\cdot)) \equiv -Ex(T) + \mu \text{Var } x(T), \\ \text{Subject to} & \left\{ \begin{array}{l} u(\cdot) \in L^2_{\mathcal{F}}(0,T;R^m), \\ (x(\cdot),u(\cdot)) \text{ satisfy equation (6)}, \end{array} \right. \end{array} \tag{11}$$

where the parameter (representing the weight) $\mu > 0$. Denote the above problem by $P(\mu)$. Define

$$\Pi_{P(\mu)} = \{u(\cdot) | u(\cdot) \text{ is an optimal control of } P(\mu)\}. \tag{12}$$

3 CONSTRUCTION OF AUXILIARY PROBLEM

Note that Problem $P(\mu)$ is *not* a standard stochastic optimal control problem and is hard to solve directly due to the term $[Ex(T)]^2$ in its cost function which is non-separable in the sense of dynamic programming. We now propose to embed the problem into a tractable auxiliary problem that turns out to be a stochastic linear-quadratic (LQ) problem. To do this, set

$$\begin{array}{ll} \text{Minimize} & J(u(\cdot);\mu,\lambda) \equiv E\{\mu x(T)^2 - \lambda x(T)\}, \\ \text{Subject to} & \left\{ \begin{array}{l} u(\cdot) \in L^2_{\mathcal{F}}(0,T;R^m), \\ (x(\cdot),u(\cdot)) \text{ satisfy equation (6)}, \end{array} \right. \end{array} \tag{13}$$

where the parameters $\mu > 0$ and $-\infty < \lambda < +\infty$. Let us call the above Problem $A(\mu,\lambda)$. Define

$$\Pi_{A(\mu,\lambda)} = \{u(\cdot) | u(\cdot) \text{ is an optimal control of } A(\mu,\lambda)\}. \tag{14}$$

The following result tells the relationship between the problems $P(\mu)$ and $A(\mu,\lambda)$.

Theorem 1. *For any $\mu > 0$, one has*

$$\Pi_{P(\mu)} \subseteq \bigcup_{-\infty<\lambda<+\infty} \Pi_{A(\mu,\lambda)}. \tag{15}$$

Moreover, if $\bar{u}(\cdot) \in \Pi_{P(\mu)}$, *then* $\bar{u}(\cdot) \in \Pi_{A(\mu,\bar{\lambda})}$ *with* $\bar{\lambda} = 1 + 2\mu E\bar{x}(T)$.

The implication of Theorem 1 is that any optimal solution of the problem $P(\mu)$ (as long as it exists) can be found via solving Problem $A(\mu, \lambda)$. Notice that the auxiliary problem $A(\mu, \lambda)$ is a standard stochastic optimal control problem (parameterized by (μ, λ)), separable in the sense of dynamic programming and of a linear-quadratic structure.

4 SOLUTION TO AUXILIARY PROBLEM

Now let us solve Problem $A(\mu, \lambda)$. Putting

$$\gamma = \frac{\lambda}{2\mu} \text{ and } y(t) = x(t) - \gamma, \tag{16}$$

Problem $A(\mu, \lambda)$ is equivalent to minimizing

$$E[\frac{1}{2}\mu y(T)^2], \tag{17}$$

subject to

$$\begin{cases} dy(t) & = \{A(t)y(t) + B(t)u(t) + b(t)\}dt \\ & \quad + \sum_{j=1}^{m} D_j(t)u(t)dW^j(t), \\ y(0) & = x_0 - \gamma, \end{cases} \tag{18}$$

where

$$\begin{cases} A(t) = r(t), & B(t) = (b_1(t) - r(t), \cdots, b_m(t) - r(t)), \\ b(t) = \gamma r(t), & D_j(t) = (\sigma_{1j}(t), \cdots, \sigma_{mj}(t)). \end{cases} \tag{19}$$

This is a typical stochastic LQ control problem as the dynamics (18) of this problem is linear (with a nonhomogeneous term in the drift) and the cost (17) is quadratic. Moreover, the running cost (over the period $[0, T]$) in (17) is absent, so it is a problem with *indefinite* control weighting running cost, one that was recently extensively investigated by Chen, Li and Zhou [1], and Chen and Zhou [2]. Applying the general results in [1, 2] and making use of the fact that the state variable $x(t)$ in the present case is scalar-valued, we get the following result.

Theorem 2. *The optimal state feedback control for the problem* $A(\mu, \lambda)$ *is given by*

$$\begin{aligned} \bar{u}(t, x) \equiv & \ (\bar{u}_1(t, x), \cdots, \bar{u}_m(t, x)) \\ = & \ (\sigma(t)\sigma(t)')^{-1}B(t)'(\gamma e^{-\int_t^T r(s)ds} - x). \end{aligned} \tag{20}$$

5 EFFICIENT FRONTIER

Under the optimal feedback control (20), the wealth equation (6) evolves as

$$\begin{cases} dx(t) &= \left\{(r(t)-\rho(t))x(t)+\gamma e^{-\int_t^T r(s)ds}\rho(t)\right\}dt \\ &\quad +B(t)(\sigma(t)\sigma(t)')^{-1}\sigma(t)(\gamma e^{-\int_t^T r(s)ds}-x(t))dW(t) \\ x(0) &= x_0, \end{cases} \tag{21}$$

where

$$\rho(t)=B(t)\left(\sigma(t)\sigma(t)'\right)^{-1}B(t)'.$$

Moreover, applying Ito's formula to $x(t)^2$, we obtain

$$\begin{cases} dx(t)^2=\left\{(2r(t)-\rho(t))x(t)^2+\gamma^2e^{-2\int_t^T r(s)ds}\rho(t)\right\}dt \\ \quad +2x(t)B(t)(\sigma(t)\sigma(t)')^{-1}\sigma(t)(\gamma e^{-\int_t^T r(s)ds}-x(t))dW(t), \\ x(0)^2=x_0^2. \end{cases} \tag{22}$$

Taking expectations on both sides of (21) and (22), we conclude that $Ex(t)$ and $Ex(t)^2$ satisfy the following two ordinary differential equation:

$$\begin{cases} dEx(t) &= \{(r(t)-\rho(t))Ex(t)+\gamma e^{-\int_t^T r(s)ds}\rho(t)\}dt, \\ Ex(0) &= x_0, \end{cases} \tag{23}$$

and

$$\begin{cases} dEx(t)^2 &= \{(2r(t)-\rho(t))Ex(t)^2+\gamma^2e^{-2\int_t^T r(s)ds}\rho(t)\}dt, \\ Ex(0)^2 &= x_0^2. \end{cases} \tag{24}$$

Solving (23) and (24), we can express $Ex(T)$ and $Ex(T)^2$ as explicit functions of γ,

$$Ex(T)=\alpha x_0+\beta\gamma,\ \ Ex(T)^2=\delta x_0^2+\beta\gamma^2, \tag{25}$$

where

$$\alpha=e^{\int_0^T(r(t)-\rho(t))dt},\ \ \beta=1-e^{-\int_0^T\rho(t)dt},\ \ \delta=e^{\int_0^T(2r(t)-\rho(t))dt}. \tag{26}$$

By Theorem 1, an optimal solution of the problem $P(\mu)$, if it exists, can be found by selecting $\bar\lambda$ so that

$$\bar\lambda=1+2\mu E\bar x(T)=1+2\mu(\alpha x_0+\beta\frac{\bar\lambda}{2\mu}).$$

This yields

$$\bar\lambda=\frac{1+2\mu\alpha x_0}{1-\beta}=e^{\int_0^T\rho(t)dt}+2\mu x_0e^{\int_0^T r(t)dt}. \tag{27}$$

Hence the optimal control for the problem $P(\mu)$ is given by (20) with $\gamma = \bar{\gamma} = \frac{\bar{\lambda}}{2\mu}$ and $\bar{\lambda}$ given by (27). Therefore one can calculate the corresponding variance of the terminal wealth in terms of the expectation. This leads to the efficient frontier we are seeking.

Theorem 3. *The efficient frontier of the bi-criteria optimal portfolio selection problem (9), if it ever exists, must be given by the following*

$$\text{Var } \bar{x}(T) = \frac{e^{-\int_0^T \rho(t)dt}}{1 - e^{-\int_0^T \rho(t)dt}} \left(E\bar{x}(T) - x_0 e^{\int_0^T r(t)dt} \right)^2. \tag{28}$$

The relationship (28) reveals explicitly the trade-off between the mean (return) and variance (risk). For example, if one has set an expected return level, then the above can tell the risk he/she has to take; and *vice versa*. In particular, if one cannot take any risk, namely, Var $(\bar{x}(T)) = 0$, then $E\bar{x}(T)$ has to be $x_0 e^{\int_0^T r(t)dt}$ meaning that he/she can only put his/her money in the bond. Another interesting phenomenon is that the efficient frontier (28) involves a perfect square. This is due to the possible inclusion of the bond in a portfolio. In the case when the riskless bond is excluded from consideration, then the efficient frontier may no longer be a perfect square, which means one cannot have a risk-free portfolio.

References

[1] Chen, S., Li, X., and Zhou, X. Y., Stochastic linear quadratic regulators with indefinite control weight costs, *SIAM J. Control & Optim.*, 36 (1998), 1685–1702.

[2] Chen, S. and Zhou, X. Y., Stochastic linear quadratic regulators with indefinite control weight costs, II, preprint.

[3] Duffie, D. and Jackson, M., Optimal hedging and equilibrium in a dynamic futures market, *J. Econ. Dynam. Control.*, 14 (1990), 21–33.

[4] Duffie, D. and Richardson, H., Mean-variance hedging in continuous time, *Ann. Appl. Probab.*, 1 (1991), 1–15.

[5] Elton, E. J. and Gruber M. J., *Finance as a Dynamic Process*, Prentice Hall, Englewood Cliffs, 1975.

[6] Föllmer, H. and Sondermann, D., Hedging of non-redundant contingent claims, in *Contributions to Mathematical Economics*, A. Mas-Colell and W. Hildenbrand ed., North-Holland, Amsterdam, 1986, 205–223.

[7] Grauer, R. R. and Hakansson, N. H., On the use of mean-variance and quadratic approximations in implementing dynamic investment strategies: a comparison of returns and investment policies, *Manag. Sci.*, 39 (1993), 856–871.

[8] Li, D. and Ng, W. L., Optimal dynamic portfolio selection : Multi-period mean-variance formulation, submitted for publication, 1997.

[9] Markowitz, H., Portfolio selection, *J. of Fin.*, 7 (1952), 77–91.

[10] Merton, R. C., An analytic derivation of the efficient portfolio frontier, *J. Fin. Quant. Anal.*, 7 (1972), 1851–1872.

[11] Mossin J., Optimal multiperiod portfolio policies, *J. of Business*, 41 (1968), 215–229.

[12] Perold, A.d F., Large-scale portfolio optimization, *Manag. Sci.*, 30 (1984), 1143–1160.

[13] Samuelson, P. A., Lifetime portfolio selection by dynamic stochastic programming, *Rev. of Econ. and Stat.*, 51 (1969), 239–246.

[14] Zeleny, M., *Multiple Criteria Decision Making,* McGraw-Hill, New York, 1981.

[15] Zhou, X. Y. and Li, D., Continuous-time mean-variance formulation in dynamic portfolio selection, preprint.

MIX
Papier aus verantwortungsvollen Quellen
Paper from responsible sources
FSC® C105338

If you have any concerns about our products,
you can contact us on
ProductSafety@springernature.com

In case Publisher is established outside the EU,
the EU authorized representative is:
Springer Nature Customer Service Center GmbH
Europaplatz 3, 69115 Heidelberg, Germany

Printed by Libri Plureos GmbH
in Hamburg, Germany